STRUCTURAL ANALYSIS FOR ENGINEERING TECHNOLOGY

Structural Analysis for Engineering Technology

Jack D. Bakos, Jr.

Youngstown State University

CHARLES E. MERRILL PUBLISHING COMPANY

A Bell & Howell Company

Columbus, Ohio

Published by
Charles E. Merrill Publishing Co.
A Bell & Howell Company
Columbus, Ohio 43216

International Standard Book Number: 0-675-09063-6

Library of Congress Catalog Card Number: 72-81673

1 2 3 4 5 6—78 77 76 75 74 73
Printed in the United States of America

PREFACE

This book is intended for use in an elementary course in structural theory. In preparing the presentation, the author has assumed that the reader is not familiar with the subject. Ideally, however, the reader should have had the first courses in Engineering Mechanics (Statics) and Strength of Materials. The elementary theory of structures is not a particularly difficult subject and, thus, requires only a limited amount of higher mathematics. The only preparation needed is a knowledge of arithmetic, algebra, and trigonometry.

The book has been designed for use as a classroom textbook in two-year colleges and technical institutes, and in survey courses in four-year colleges and four-year technical programs. Because of the general nature of structural analysis itself and due to the general structural theory presented herein, this book can also be of help to non-civil engineering students such as architects, mining engineers, and mechanical engineers.

Chapter 1 contains a review of the basic principles of statics with an emphasis on those principles and techniques most often encountered when analyzing structures. The next six chapters contain the basic concepts of structural theory and its applications to statically determinate structures. Chapter 8 deals with statically indeterminate structures including the Theorem of Three Moments and the Slope Deflection Method. The scope of the chapter is broad enough to provide the reader with a sufficient background to analyze most commonly occurring statically indeterminate beams and rigid frames. In addition, the chapter can provide a stepping-stone for advanced study if desired.

Although numerous example problems are presented throughout the chapters, emphasis is not placed on the sheer number presented to convey the theory. Emphasis is placed on a detailed step-by-step analysis of the basic theory and of each example problem. Because of this approach, this book lends itself well to the self-teaching needs of many technicians and draftsmen. In addition, the answers to a vast majority of the problems can be found at the end of the text.

The author wishes to express his appreciation to those who have assisted in the preparation of this text. He is indebted to his wife, Patsy, who typed all of the manuscript, and who provided the necessary inspiration and encouragement. Thanks also are due to Guy Returra for his assistance in solving the problems. Finally, he is indebted to all his students, who have taught him more than he would like to admit.

January, 1973 *Jack D. Bakos, Jr.*

CONTENTS

STRUCTURAL ANALYSIS FOR ENGINEERING TECHNOLOGY

CHAPTER ONE

Introduction

1.1 Theory of Structures

Engineers design structures such as bridges, buildings, power poles, transmission towers, machine parts, ships, aircraft, and many others. Structural engineering has always been an integral part of civil engineering, but one can readily see from the list of structures mentioned that structural engineering is much too broad to lie within civil engineering alone. The dividing lines between the interests of the various types of engineers are not well defined; thus, considerable overlap exists. In order to cope with these overlaps of interests, and with various types of structures, structural engineering must not be confined to any specific branch of engineering. The **theory of structures,** therefore, deals with the principles, methods, and fundamentals which can apply to all structures regardless of use, shape, or size.

Structural theory is subdivided into **analysis** and **design.** Structural analysis is the determination of all the loads and forces acting on the structure as well as the axial force, shear, bending moment, and deflection at any section for each member in the structure. Analysis of a structure is often accomplished analytically by application of the principles of engineering mechanics. Analysis may be supplemented by experimental procedures and, in rare cases, accomplished solely by experimental means.

Structural design is the process of determining the **form** and **proportions** of a structure, subject to a load-carrying performance requirement, i.e., design code requirements, and based on an established measure of efficiency. The proportions are the quantities which define the size for each member and for the structure on the whole.

In order to avoid limiting the scope of structural theory, we may generally define a **structure** as **a material body whose primary purpose is to internally**

contain and/or externally support other bodies while sustaining applied loads.

1.2 Structural Form

The form, or configuration, of a structure describes the shape and relative arrangements of its component members, if any. The selection of the most suitable structural form is one of the most difficult and important decisions made by the structural engineer. The most feasible structural form is that one which satisfies economic, aesthetic, and functional requirements.

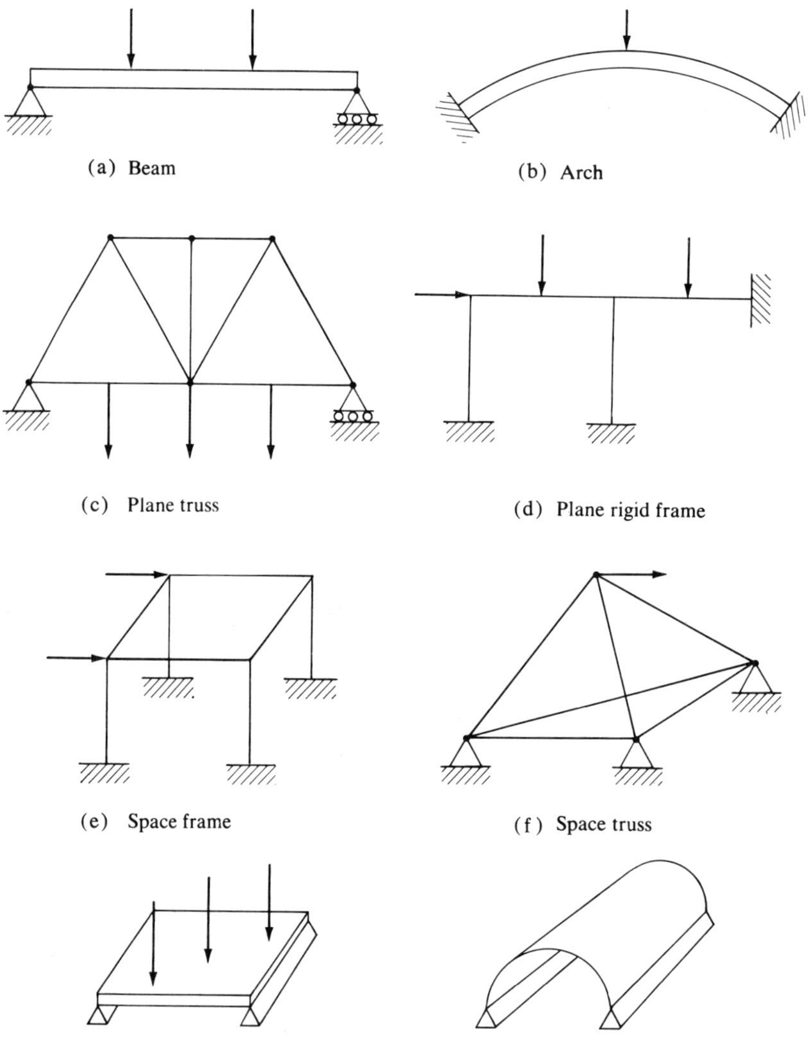

(a) Beam (b) Arch

(c) Plane truss (d) Plane rigid frame

(e) Space frame (f) Space truss

(g) Plate (h) Shell

Figure 1.1

Some of the most common structural forms are shown in Figure 1.1. Most structures, however, fall into one of the following classifications: **beams, trusses,** or **rigid frames.** A beam is a structural member subjected only to transverse loads. A truss is a structure in which all the members are joined by frictionless hinges. A rigid frame is a structure composed of members which are rigidly connected at the joints. The analysis of such structures is generally referred to as **elementary structural analysis** since only the principles of mechanics and a basic background in mathematics are required of the analyst. The analysis of plate and shell structures, however, requires a highly sophisticated background in mathematics and engineering mechanics; thus, such an analysis is referred to as **advanced structural analysis.**

1.3 Loads on Structures

The forces applied to a structure, or a structural member, are referred to as **loads.** Loads are classified as either **dead loads** or **live loads.** Dead loads are those which are always on or attached to the structure. The weight of the structure itself and of any items that are permanently attached to the structure are dead loads. Other examples of dead loads are the weights of floor or roof slabs, walls, fixtures, and heating equipment. Since a structure is analyzed before it is designed, the dead load must be assumed initially and then revised if necessary after the design is completed.

Live loads are loads **not** permanently applied to the structure, i.e., loads which may be removed, reapplied, or in motion during application. Live loads usually encountered are:

(1) Weight of vehicular traffic on a bridge.
(2) Weight of people, furniture, and storage goods in a building.
(3) Weight of snow or ice accumulations on the exterior of a structure.
(4) Pressures from action of the wind.
(5) Dynamic effects accompanying moving loads, i.e., **impact.**
(6) Hydrastatic pressures, i.e., pressures from fluids in storage tanks.

1.4 Ideal Structures

Actual structures are very often idealized by **line diagrams** where the lines represent the centerlines of the members. Such a representation is shown in Figure 1.2. The actual beam shown in Figure 1.2a is represented by its centerline in Figure 1.2b. The support offered by the brick walls is also idealized by the equivalent concentrated loads. This convenient representation does deviate from the actual conditions, but the errors introduced are generally within tolerable limits. Since the beam and possibly the walls are not designed before the analysis is accomplished, it is necessary to idealize the structure in this way in order to carry out a practical analysis. Experience is the prime factor in determining the correct ideal structure. If doubt exists in the analyst's mind regarding the choice of the correct ideal structure, several choices and corresponding designs can be made and compared.

The loads, to some degree, are also represented idealistically. The wood flooring and beam weight, for example, are estimated and represented by the uniformly distributed load w_0 acting along the entire length of the beam as shown in Figure 1.2b. After the design, the estimated weights can be checked and an adjustment made if necessary. Similarly, the stored goods give rise to a live uniform load w_1 acting on that length of beam directly beneath the goods.

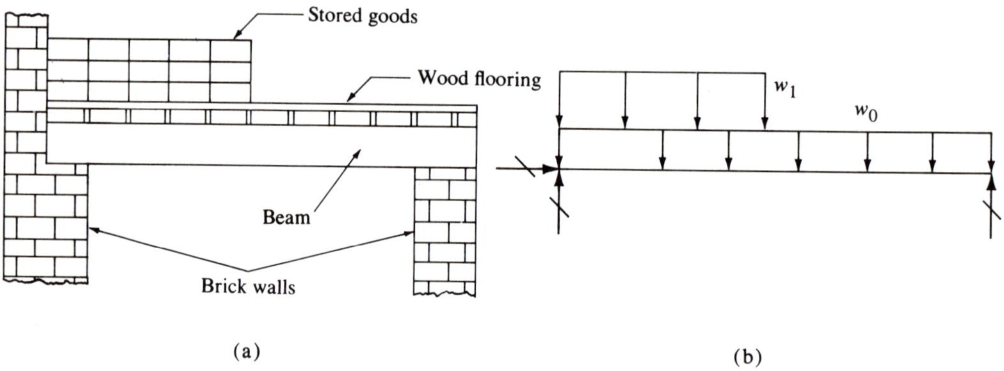

Figure 1.2

CHAPTER TWO

Forces, Equilibrium, and Reactions

2.1 General

The basic tools utilized in the solution of structural analysis problems are the fundamentals, definitions, and techniques presented in the basic statics course. Before proceeding with the solution of such problems, a review of the techniques presented in statics will be presented here. The major portion of this text will be devoted to the analysis of **planar structures,** i.e., structures that may be considered to lie in a plane that contains all the forces acting on the structure. Such an emphasis will not limit the scope of this text since planar structures are by far the most commonly occurring structures treated or represented in structural analysis. The principles presented here for planar structures may be easily extended for the case of three-dimensional structures, or **space structures.** A limited treatment of space structures will be presented for comparison in later portions of the text.

2.2 Coplanar Force Systems

The forces acting on planar structures are classified as **coplanar forces,** i.e., forces whose lines of action are contained in one plane. Coplanar force systems may be further broken down as shown in Figure 2.1.

Figure 2.1a shows a **concurrent coplanar force system.** Such a force system consists of two or more forces the lines of action of which all pass through the same point. A **general coplanar force system** is shown in Figure 2.1b. In this system, which is probably the most commonly encountered in structural analysis, all the forces lie in the same plane but do not necessarily pass through a

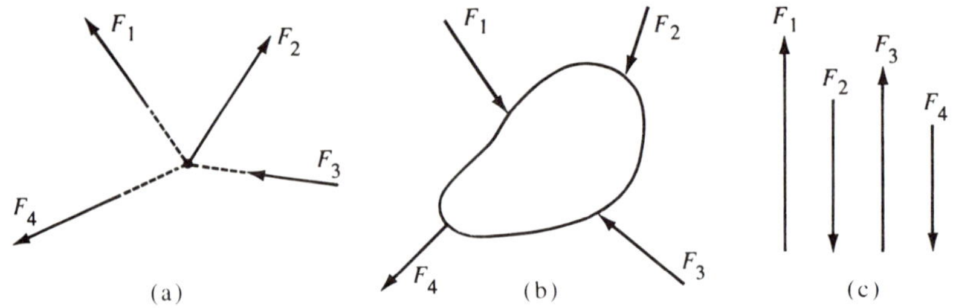

Figure 2.1

common point. Figure 2.1c shows a rather special system known as a **parallel coplanar force system,** i.e., all the forces have lines of action which are parallel and contained in the same plane.

There is another certain system of forces which is extremely important in structural analysis. This coplanar system consists of two forces having equal magnitudes, opposite but parallel lines of action, and noncollinear, i.e., not lying along the same line of action. Such a system is called a **couple** and is shown in Figure 2.2. The moment of the couple is equal to the product of the magnitude of the force and the perpendicular distance between the two forces, or

$$M = Fd$$

It is important to note that the resulting moment is independent of the choice of the origin O of the selected coordinate axes. In this text, the words **couple** and **moment** will be used interchangeably.

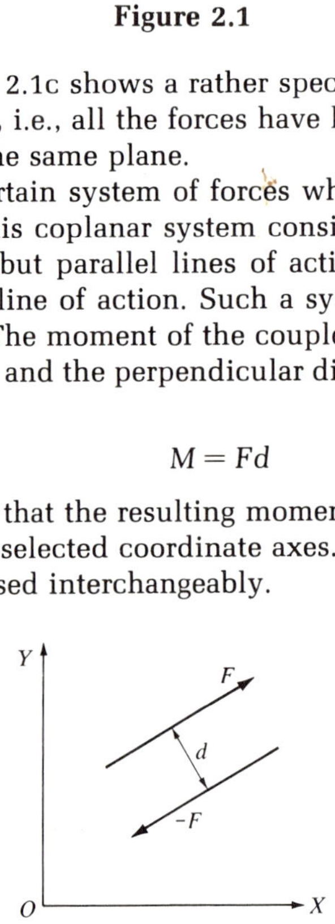

Figure 2.2

2.3 Equivalent Systems of Forces

Consider a force F acting on a rigid body at a point A as shown in Figure 2.3a. Before proceeding any further, a **rigid body,** as used in this text, is defined as a body which does not deform under the action of the applied loads. In all actuality, structures do deform and a considerable portion of this text will be devoted to determining the magnitude and direction of such deformations. These deformations, however, are usually small and, hence, do not appreciably affect the external conditions of the problem or equilibrium.

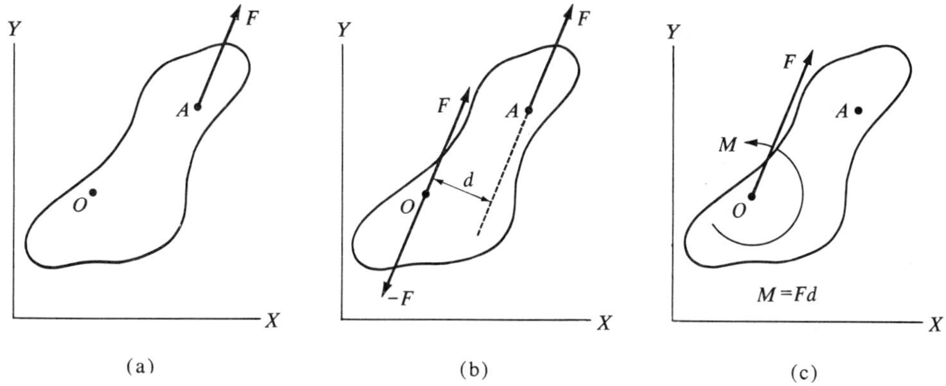

Figure 2.3

For some reason, it may be desired to have the force F act at point O rather than at point A. It will be shown later that it may be convenient to change the point of application of a force as a computational aid. It should be evident that we cannot simply pick up the force F and translate it to point O without changing the conditions of the problem. At point O, however, we may apply two forces, F and $-F$, as shown in Figure 2.3b. The conditions of the problem are not changed since the application of the two opposite, collinear forces at O are equivalent to zero. Notice that there now exists in addition to the force at O, a couple formed by the two equal and opposite parallel forces acting at O and A. The force system shown in Figure 2.3b may be replaced by the force-couple system shown in 2.3c. The force system shown in Figure 2.3a is said to be replaced by the **equivalent force-couple system** shown in Figure 2.3c. What has been accomplished by this manipulation of forces is that the force F was moved to point O on the rigid body, but the move could not have been accomplished without the addition of a moment. It must be emphasized that the effect on the body is the same for the action of F at point A or the action of F at point O with the addition of the moment M. Thus, any force F acting on a rigid body may be arbitrarily moved to any point O provided that a couple is added of moment equal to the moment of F about O. Although only a single force was treated, an entire system of forces may be treated in a similar fashion, i.e., each force of the system may be moved to the same point provided that each force movement is accompanied by the addition of a couple. The forces then become concurrent (passing through the same point) and may be vectorially added to yield a single resultant force. The moments, or the couples, are also in the same plane and, hence, may also be added vectorially to produce a single resultant moment. This technique for a force system is shown in Figure 2.4.

2.4 Principle of Transmissibility

The **principle of transmissibility** also allows for a relaxation of the requirements that a force (vector) must have a well defined point of application. This

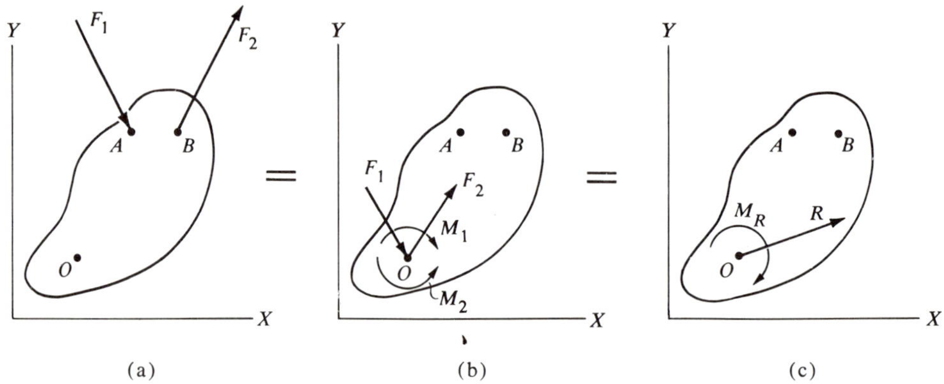

Figure 2.4

principle simply states that the external effect of a force on a rigid body is the same if the force is applied anywhere along its line of action. The principle is demonstrated in Figure 2.5. The external effect on the beam is not altered for either loading condition, since all that is changed is that the force P has been moved along its line of action. Consider, for example, men towing a disabled truck. The conditions of motion and all the other external forces acting on the truck are not changed if the men push on the rear bumper instead of pulling on the front bumper.

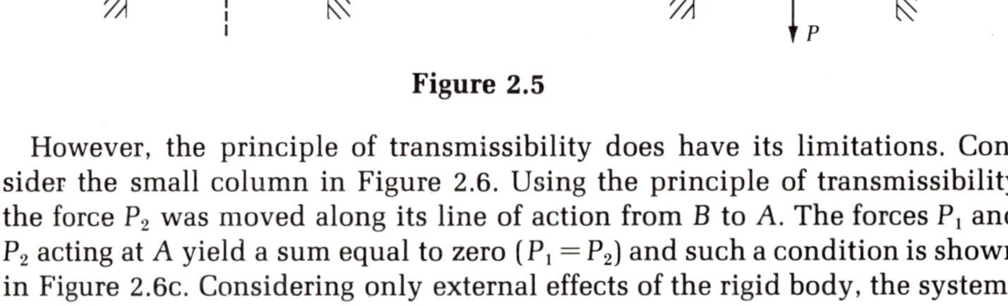

Figure 2.5

However, the principle of transmissibility does have its limitations. Consider the small column in Figure 2.6. Using the principle of transmissibility the force P_2 was moved along its line of action from B to A. The forces P_1 and P_2 acting at A yield a sum equal to zero ($P_1 = P_2$) and such a condition is shown in Figure 2.6c. Considering only external effects of the rigid body, the systems shown in Figures 2.6a and 2.6c are equivalent. The body shown in Figure 2.6a is in compression and will shorten slightly and at the same time experience compressive stress. Although this may be evident from Figure 2.6a, it is impossible to detect the internal forces and deformations from Figure 2.6c. Hence, the principle of transmissibility may be used freely when considering the external effects of forces acting on bodies, but should be avoided when considering internal forces, or stresses, and deformations.

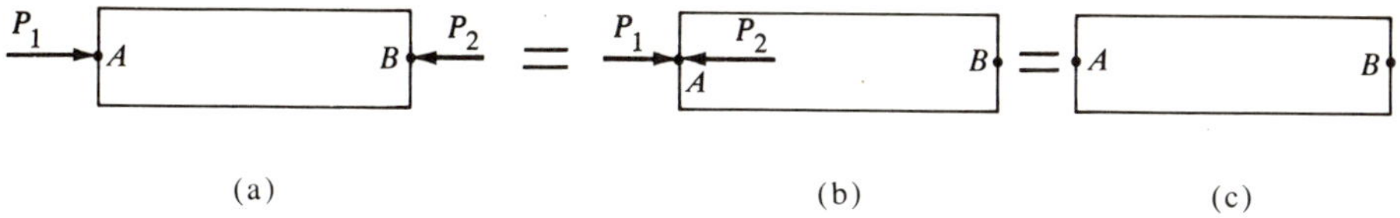

Figure 2.6

Example 2.1

For the bracket shown in Figure 2.7, determine the equivalent force-couple system at point A.

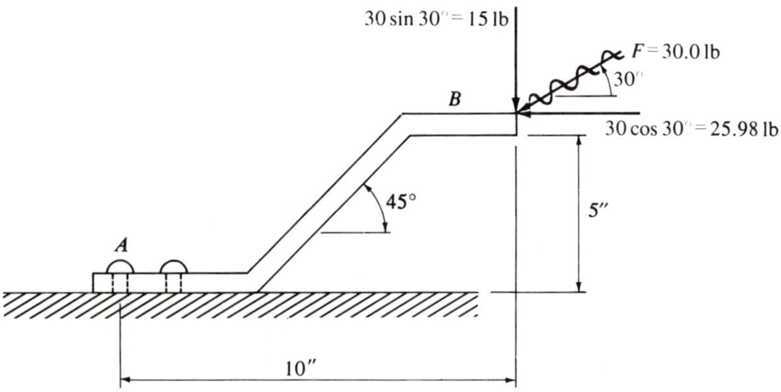

Figure 2.7

Solution:

1. To replace the 30-lb force at B with an equivalent force-couple system at A simply means moving the force from B to A, maintaining the same sense and direction, and adding a couple of moment equal to the moment of F acting at B about the point A.
2. To calculate the moment of F about point A, it is necessary to resolve F into vertical and horizontal components as shown in Figure 2.7.
3. The moment about point A becomes 15×10 lb-in. $- 25.98 \times 5$ lb-in. $= 150 - 129.9 = 20.1$ lb-in. where the negative sign indicates a counterclockwise (ccw) rotation about A, and the positive sign associated with a clockwise (cw) rotation about A. It should be noted that these signs on the moments can be selected arbitrarily, but some sign convention must be selected to distinguish between the two different directions of possible rotation about A. The net result yields a clockwise moment of 20.1 lb-in. about A.
4. The equivalent force-couple system is shown in Figure 2.8. The two systems as shown by Figures 2.7 and 2.8 are equivalent, i.e., the external

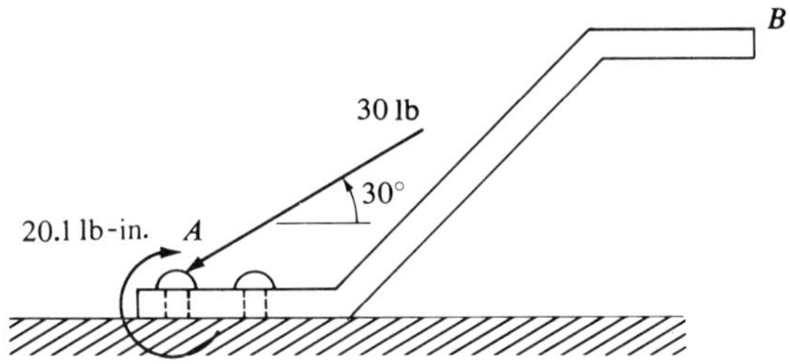

Figure 2.8

conditions of the problem are identical considering the single force system or its equivalent force-couple system.

Example 2.2

For the bracket shown in Figure 2.9, determine the equivalent force-couple system at point A.

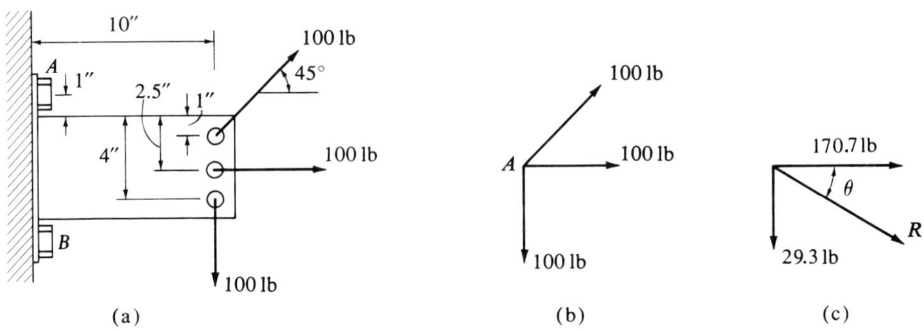

Figure 2.9

Solution:

1. The first step is to move each of the three forces to point A, maintaining the same sense and line of action. This gives rise to the concurrent force system shown in Figure 2.9b. The polygon law may be used to find the resultant of the three forces, or it may be calculated by resolving all the forces into vertical and horizontal components.

$$\begin{aligned}\Sigma F_Y &\equiv \text{Sum of Forces in Vertical Direction (upward positive)}\\ &= +(\sin 45^\circ)100 - 100 = 70.7 - 100 = 29.3 \text{ lb down}\end{aligned}$$

$$\begin{aligned}\Sigma F_X &\equiv \text{Sum of Forces in Horizontal Direction (to right positive)}\\ &= (\cos 45^\circ)(100) + 100 = 170.7 \text{ lb to right}\end{aligned}$$

$$\text{Resultant,}\quad R = \sqrt{(29.3)^2 + (170.7)^2}$$

$$R = 173.2 \text{ lb}$$

$$\Theta = \tan^{-1}\frac{29.3}{170.7} \quad \text{(See Figure 2.9c)} = 9^\circ - 43' \text{ down and to the right}$$

2. A moment must accompany each move, and since each is in the same plane, they may be added algebraically to obtain the resultant moment about A. Moments which would tend to impart a clockwise rotation about A will be called positive; counterclockwise will be negative.

$$\Sigma M_A \equiv \text{Sum Moments about } A = [-(\sin 45°)(100)(10) - (\cos 45°)(100)(2) - 100(3.5) + 100(10)] \text{ lb-in.}$$
$$= -707 - 141 - 350 + 1000 = -198 \text{ lb-in.}$$

$$\Sigma M_A = 198 \text{ lb-in.} \circlearrowleft$$

3. The system of forces shown in Figure 2.9 may be replaced with a force of

173.2 lb ↘ 9°43′ at A and a moment of 198 lb-in. ccw.

Example 2.3

For the footing shown in Figure 2.10, find the equivalent force-couple system at point A.

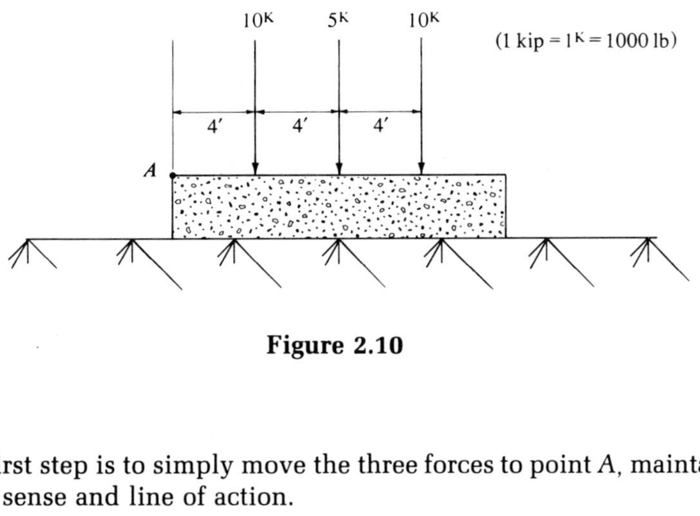

Figure 2.10

Solution:

1. The first step is to simply move the three forces to point A, maintaining the same sense and line of action.

$$\Sigma F_Y \;@\; A = 10 + 5 + 10 = 25 \text{ kips down}$$

2. Again, each move must be accompanied by a couple of moments equal to the moment of each force about the point to which the move is to be made, point A. Moments which tend to impart a clockwise rotation will be arbitrarily called positive.

$$\Sigma M_A = +10(4) + 5(8) + 10(12) = 40 + 40 + 120 = 200 \text{ kip-ft} \circlearrowright +$$

3. The equivalent force-couple system is shown in Figure 2.11.

Example 2.4

For the force-couple system shown in Figure 2.11, find a single equivalent force to replace the force-couple system.

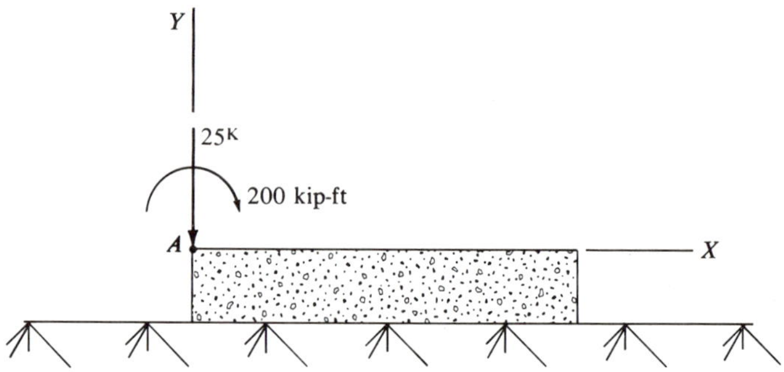

Figure 2.11

Solution:

1. Up to this point, we have been finding a force-couple system equivalent to an entire system of forces. The problem now asks to do the reverse of this procedure. Irregardless of the loading system used, the total downward force on the footing must be 25 kips down as can be seen from both Figures 2.10 and 2.11. The only choice the analyst has is how and where to distribute the applied loads and still maintain an equivalent system. Both equivalent loading systems, Figures 2.10 and 2.11, produce a downward load of 25 kips and tend to impart a rotation of 200 kip-ft about point *A*. To fulfill the requirements of this problem, we must determine the point to which we may move the 25-kip resultant load and yet still tend to impart a 200 kip-ft rotation about point *A*.

2. The moment about *A* must be conserved at 200 kip-ft. Hence, the moment of the 25-kip force must be equal to this amount. Choosing *A* as an origin,

$$\Sigma M_A = 200^{k-'} = 25x$$

$$x = \frac{200}{25} = 8 \text{ ft}$$

3. Figure 2.12 shows the equivalent system. All three systems, Figures 2.10, 2.11, and 2.12, are equivalent.

4. The technique utilized in this problem is commonly encountered in foundation engineering.

Example 2.5

For the block shown in Figure 2.13, find at least three other equivalent couples.

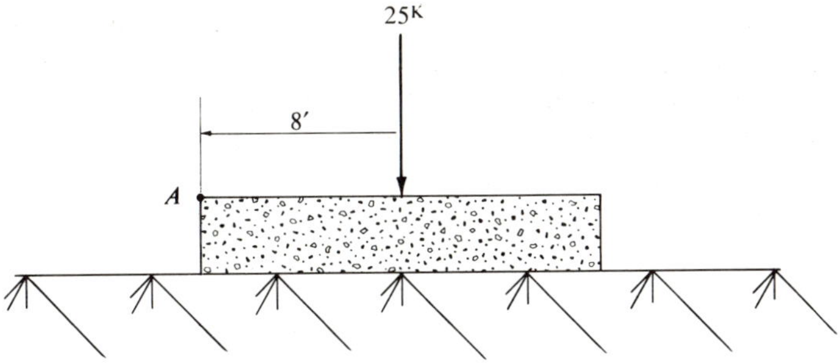

Figure 2.12

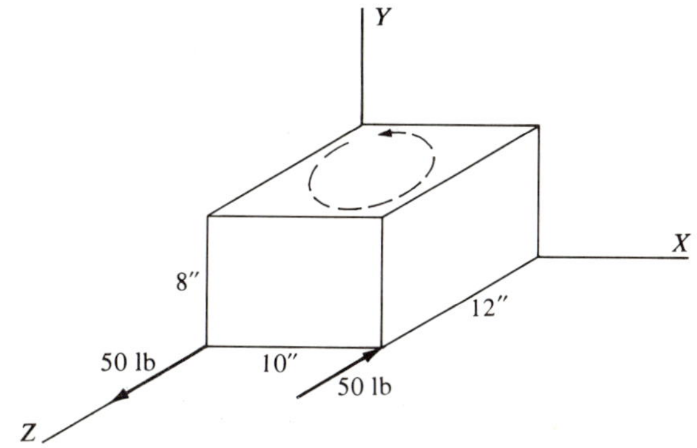

Figure 2.13

Solution:

1. In section 2.2, it was shown that the moment of a couple is independent of the choice of axes and that the moment is equal to the magnitude of the couple force times the perpendicular distance between the two couple forces. For the problem at hand, the magnitude of the moment of the couple is $50 \times 10 = 500$ lb-in. It would tend to impart a rotation in the direction shown in Figure 2.13.
2. Considering external effects only, the couple forces would not tend to translate the block since they are equal and oppositely sensed forces. Hence, the only external effect on the block by the couple is the tendency for the couple to cause a rotation in the direction shown of magnitude equal to 500 lb-in. Any couple which would tend to impart a rotation of same magnitude and direction is an equivalent couple. We have in equation form

$$Fd = 500 \text{ lb-in.}$$

where F is the magnitude of the equivalent couple force and d is the perpendicular distance between the couple forces.

3. All three couples shown in Figure 2.14 tend to rotate the block in the same direction and with the same magnitude and, hence, are said to be equivalent couples. It should be retained for later reference that when considering external effects, the points of application of the couple forces are immaterial as long as the moment of the equivalent couple tends to impart a rotation of the same direction and magnitude.

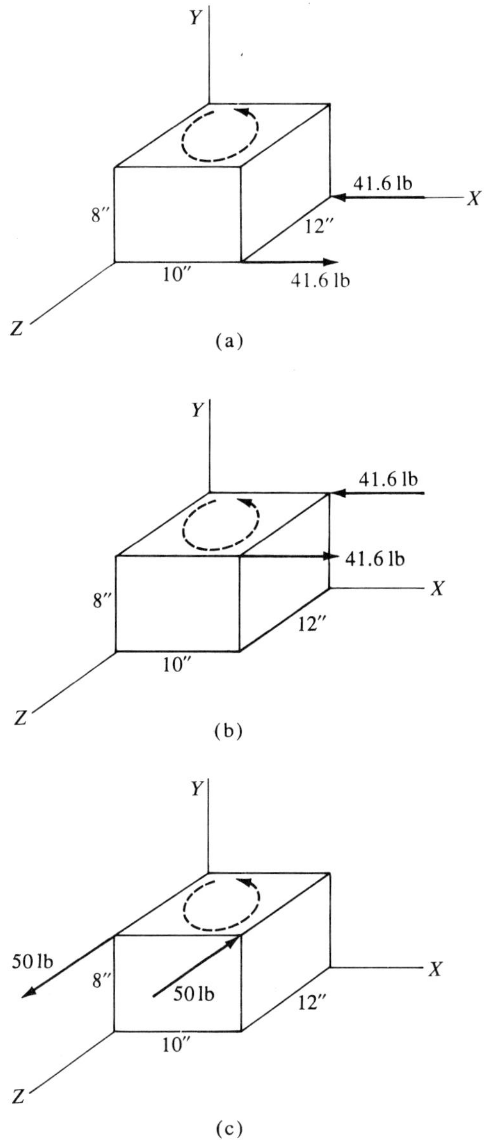

$F(12) = 500$ lb-in.

$F = \frac{500}{12} = 41.6$ lb

$M = 41.6(12) = 500$ lb-in.

$M = 50(10) = 500$ lb-in.

Figure 2.14

2.5 Equilibrium

Consider the rigid body shown in Figure 2.15 and the forces and moments acting upon it. If the body was at rest before the application of the forces and moments, one of the following four conditions must occur:

1. The body will translate, i.e., move in a straight line
2. the body will rotate, i.e., tumble about an axis
3. the body will translate and rotate simultaneously
4. the body will remain at rest

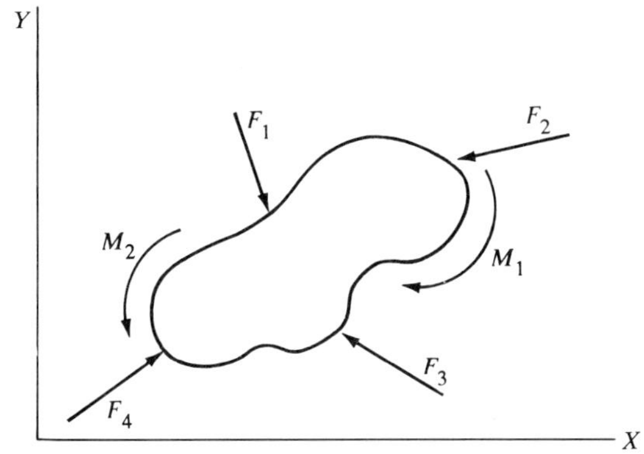

Figure 2.15

For most practical situations occurring in structural engineering, condition four is the desirable end result of a component subjected to the action of forces and moments. A body initially at rest will remain at rest if and only if the sum of the forces acting on the body is zero and if the sum of the moments about any point is zero. The body is said to be in a state of **static equilibrium.** The condition that the sum of the forces acting upon the body be zero is just a way of stating that the resultant force at the system of forces must be zero. With the resultant force being zero, it is impossible to translate the body. The condition that the sum of the moments about any point be zero prevents any rotation of the body by ensuring that the resultant effect of the system of forces is not a couple which would be unresisted and, hence, cause rotational motion. For the case of coplanar forces, the forces are generally resolved into components in the vertical and horizontal direction. The forces could be resolved into components in any direction, but it is generally accepted practice to use the vertical and horizontal direction as shown by the coordinate system in Figure 2.16a. The conditions of static equilibrium represented in equation form and using rectangular Cartesian coordinates are as follows for a plane structure lying in the X-Y plane:

$$\Sigma F_X = 0 \qquad \Sigma F_Y = 0 \qquad \Sigma M_Z = 0$$

Observe that the Z-axis becomes a point when projected on the X-Y plane (see Figure 2.16a); hence, when considering planar structures, the moment about any point is identical with the moment about an axis perpendicular to the X-Y plane passing through that point. In addition, when dealing with planar structures, the sum of the moments about any point will be carried out omitting the Z subscript. Expressing the conditions of equilibrium for a space structure and using the rectangular Cartesian coordinates shown in Figure 2.16b, we obtain the following **equations of static equilibrium:**

$$\Sigma F_X = 0 \qquad \Sigma F_Y = 0 \qquad \Sigma F_Z = 0$$
$$\Sigma M_X = 0 \qquad \Sigma M_Y = 0 \qquad \Sigma M_Z = 0$$

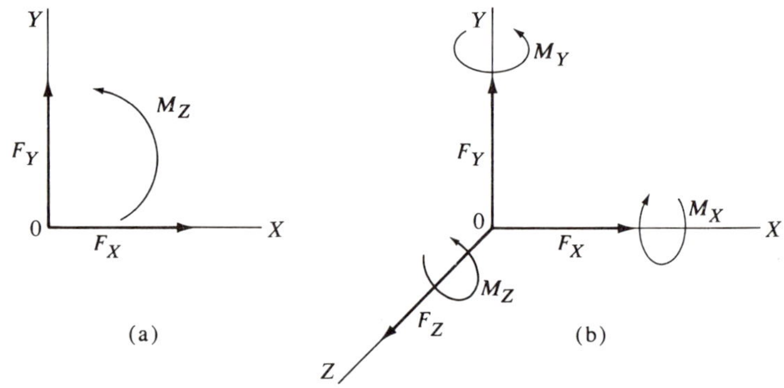

Figure 2.16

The forces and moments acting on a rigid body consist of known and unknown quantities. In general, the applied loads, i.e., wind loads, snow loads, floor loadings, and other live and dead loads, are known or estimated by the engineer. The structural engineer must determine the unknown forces acting on the body, which generally consist of the constraining forces, i.e., the forces exerted by the supports of the structure or body, upon the body, to prevent any translation or rotation. Since there exist three equations of static equilibrium for plane structures, a maximum of three unknown forces or moment quantities may be determined. Recall that a force possesses both magnitude and direction; hence, its magnitude, its direction, or both may be unknown giving rise to the possibility of two unknown quantities for each unknown force. Generally speaking, the directions of the unknown constraining forces are known, leaving only the magnitudes to be determined. For space structures, there are six equations of equilibrium allowing for the solution of a maximum of six unknowns. As stated previously, the emphasis in this text will be on planar structures unless stated otherwise. The methods and techniques presented here can be extended to space structures and in some instances will be.

When applying the three equations of static equilibrium, $\Sigma F_X = 0$, $\Sigma F_Y = 0$, and $\Sigma M = 0$, it is always permissible to replace them by the three independent moment equations, $\Sigma M_A = 0$, $\Sigma M_B = 0$, and $\Sigma M_C = 0$, when the points A, B, and C do not lie on a straight line. This may be verified as follows:

A. If the applied and constraining forces satisfy the moment equation $\Sigma M_A = 0$, then the resultant force of the system of forces must be a resultant force whose line of action passes through point A.
B. If the system of acting forces also satisfies the equation $\Sigma M_B = 0$, then the resultant force of the system of acting forces must have a line of action which passes through both points A and B. This can only be accomplished if the line of action of the resultant force and the line joining points A and B are collinear, i.e., the line of action of the resultant lies along the line joining points A and B.
C. If the system of forces also satisfies the equation $\Sigma M_C = 0$ and point C does lie on a line joining points A and B, then it is impossible to have a resultant force action along the line joining A and B.
D. It must be concluded that if the system of forces satisfies the three moment equations, the effect of the force system cannot be a resultant force nor a couple, since either would produce a moment about one or more of the points chosen, and therefore the system must be in a state of static equilibrium. It should be noted that the substitution of two additional moment equations for the conventional equilibrium equations $\Sigma F_X = 0$ and $\Sigma F_Y = 0$ does not increase either the number of independent equations or the allowable number of unknown quantities. Only three independent equations of static equilibrium exist, and the fact that we may substitute moment equations for the conventional force summation equations means that we have provided a computational aid and not a source of additional independent equations. It can also be easily shown that we may substitute one moment equation for either conventional force summation equation. The equations of equilibrium may then take the following form:

   1. $\Sigma F_X = 0$, $\Sigma F_Y = 0$, $\Sigma M = 0$
   2. $\Sigma M_A = 0$, $\Sigma M_B = 0$, $\Sigma M_C = 0$ where points A, B, and C are not collinear.
   3. $\Sigma M_A = 0$, $\Sigma M_B = 0$, $\Sigma F_X = 0$ where line AB is not perpendicular to X-axis.
   4. $\Sigma M_A = 0$, $\Sigma M_B = 0$, $\Sigma F_Y = 0$ where line AB is not perpendicular to Y-axis.

   A numerical example illustrating the use of the various available forms of the equilibrium equations will be presented in a later section.

2.6 Conventional Supports

A rigid body under the action of an applied force system must be restrained to prevent motion of any type. This restriction against motion of the body is provided by restraints or supports. It is generally assumed that the supports are affixed to a rather stationary member; hence, the supports are considered

unyielding, i.e., a negligible amount of settlement or deformation occurs at the supports. This is, in fact, a conservative assumption in most cases. If the supports must prevent any type of motion of the body under the applied loads, they must, therefore, exert resisting forces and couples on the body. These forces, as previously mentioned, are usually unknown and must be determined utilizing the equations of equilibrium. The forces exerted by the support on the body are called **reactions,** and their solutions are foremost in structural engineering problems. There are several types of supports frequently encountered in structural applications. The standard symbols used for each conventional support and the reactions exerted on a body by each are shown in Figure 2.17. The hinges represented in the hinge and link supports are assumed to be comprised of smooth frictionless pins. The rollers in the roller support are also assumed to be frictionless. The reaction forces shown in Figure 2.17 have a point of application at the location of the support. Also, the sense of the forces and moment shown in Figure 2.17 may be as shown or reversed depending upon the conditions of the problem and loading conditions.

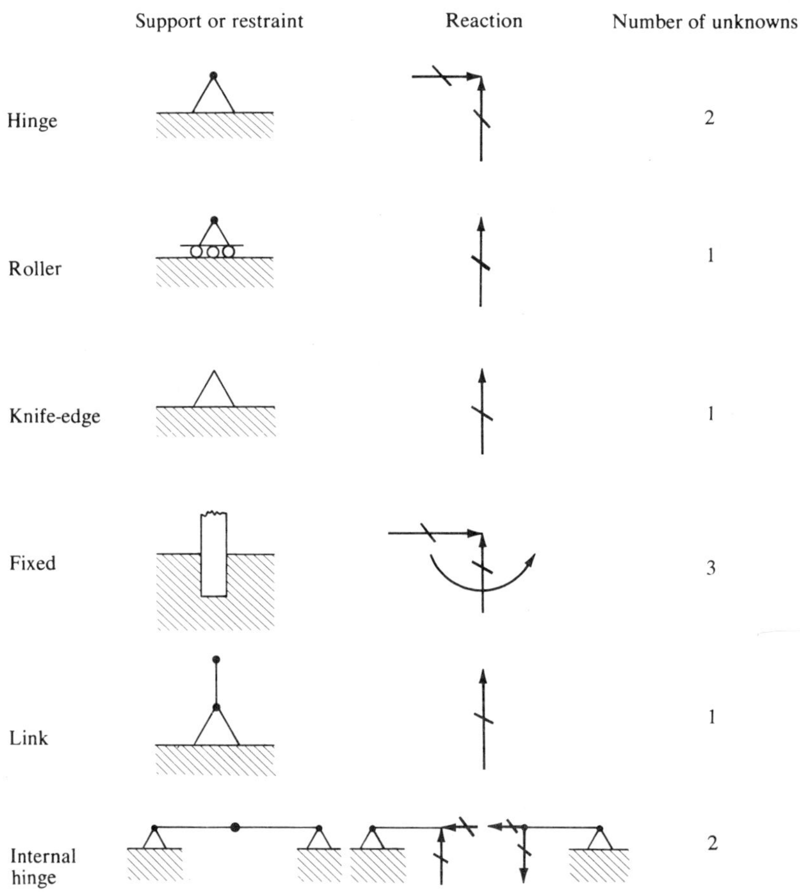

Figure 2.17

2.7 Free-Body Diagrams

The goal of the structural analyst is to determine all the forces and moments acting on a structural member. An additional or omitted force would completely negate the conditions of equilibrium. In addition, the external forces and moments determined by the structural analyst are used to determine the internal forces and, hence, internal stresses which ultimately control the design of the member. The importance of accurately determining all the forces acting on a body cannot be overemphasized. Therefore, every problem solution must be initiated by drawing a **free-body diagram.** A free-body diagram consists of the rigid body separated from the ground, its supports, or any other body and all the forces acting upon it to include its own weight, all magnitudes, points of application, and pertinent dimensions. A problem cannot be correctly solved without a proper free-body diagram, and the time consumed in drawing one is never wasted. Trying not to overemphasize its importance, it can confidently be stated that if a proper free-body diagram is drawn, the remaining solution becomes secondary in nature and effort.

As an example, consider the system shown in Figure 2.18 consisting of two spheres in a container. It will be assumed that the frictional effects are small enough to be neglected; hence, the only forces involved are the weights and contact forces. Each sphere has been isolated and all the acting forces have been shown. The contact forces, F_1, F_2, F_3, and F_4, are taken normal to the surface of contact. Notice that the contact forces have the same magnitude but opposite senses on the adjacent contacting surfaces. This is obvious since the action and reaction of one body on another must be numerically equal but opposite in sense. In other words, as sphere B "pushes" down on sphere A with a force of F_3, sphere A "pushes" back with the same force F_3. Furthermore, as the two spheres are brought back into contact, the two external contact forces must cancel, as they do, since upon contact they become internal forces. This is shown in Figure 2.18d. Observe that Figure 2.18b, 2.18c, and 2.18d are proper free-body diagrams, and that Figure 2.18e is not, the reason being that the container has not been separated from its support and all the acting forces are not shown. Before leaving this example it should be noted that four unknown forces, F_1, F_2, F_3, and F_4, have been introduced and we have only three equations of equilibrium available for their solution. Although it appears that we may be short one equation, such is not the case. Each free-body diagram is a separate problem in itself, and no one diagram has more than three unknown forces present. The four unknown forces, therefore, can be obtained.

The sphere problem has been represented by many authors and educators to demonstrate the techniques of drawing a proper free-body diagram. Although several more examples will appear later in this chapter, the student should linger on this example until it is completely understood.

2.8 Statically Determinate and Indeterminate Structures, Structural Stability

Consider now the beam shown in Figure 2.19a which is held by a hinge at A and by a smooth roller at B. It can be noted from the free-body diagram, also

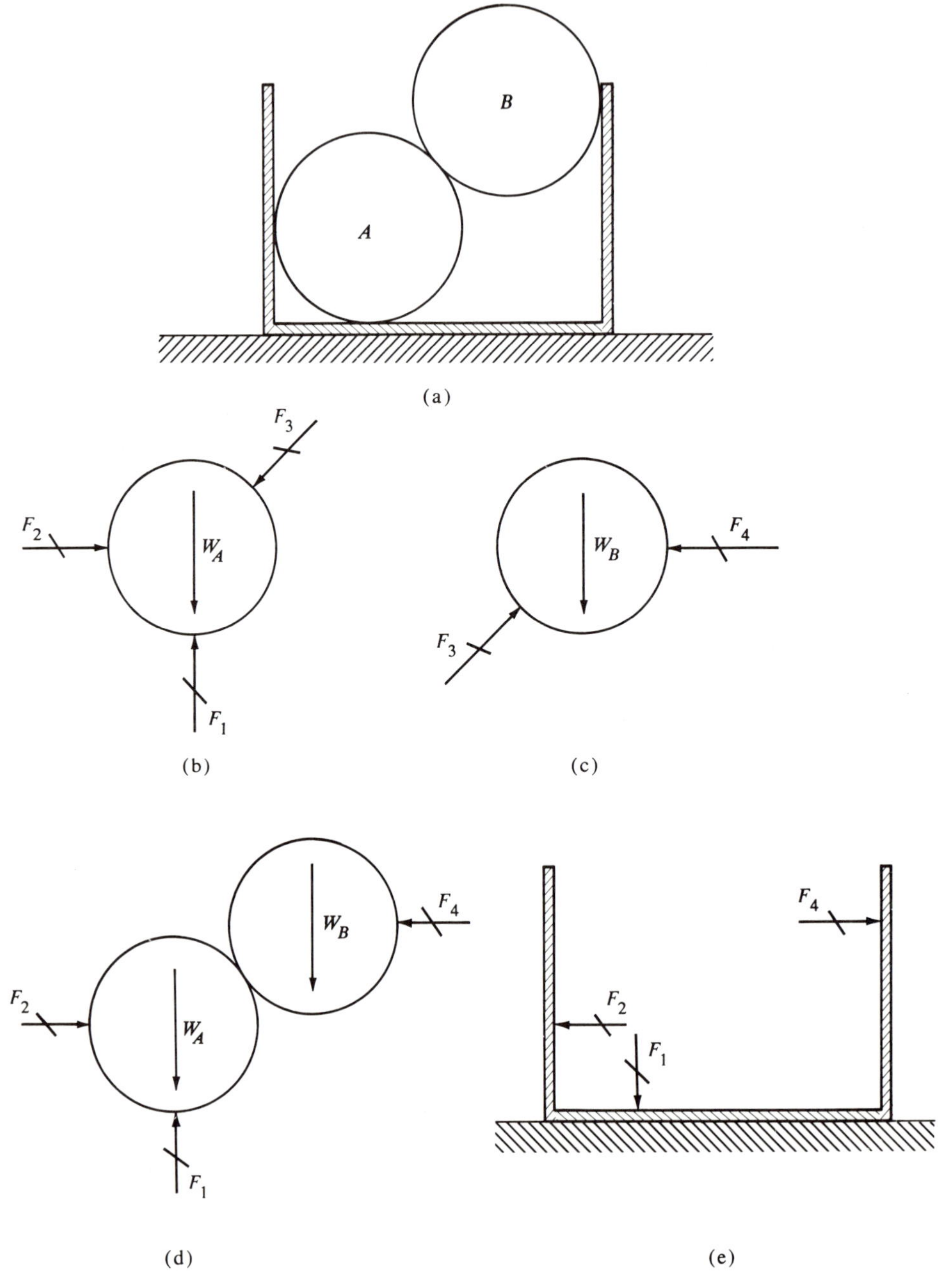

Figure 2.18

shown in Figure 2.19a, that the reactions yield three unknowns which may be obtained by simultaneously solving the three equations of static equilibrium. At the same time, it should also be noted that it is impossible for the beam to

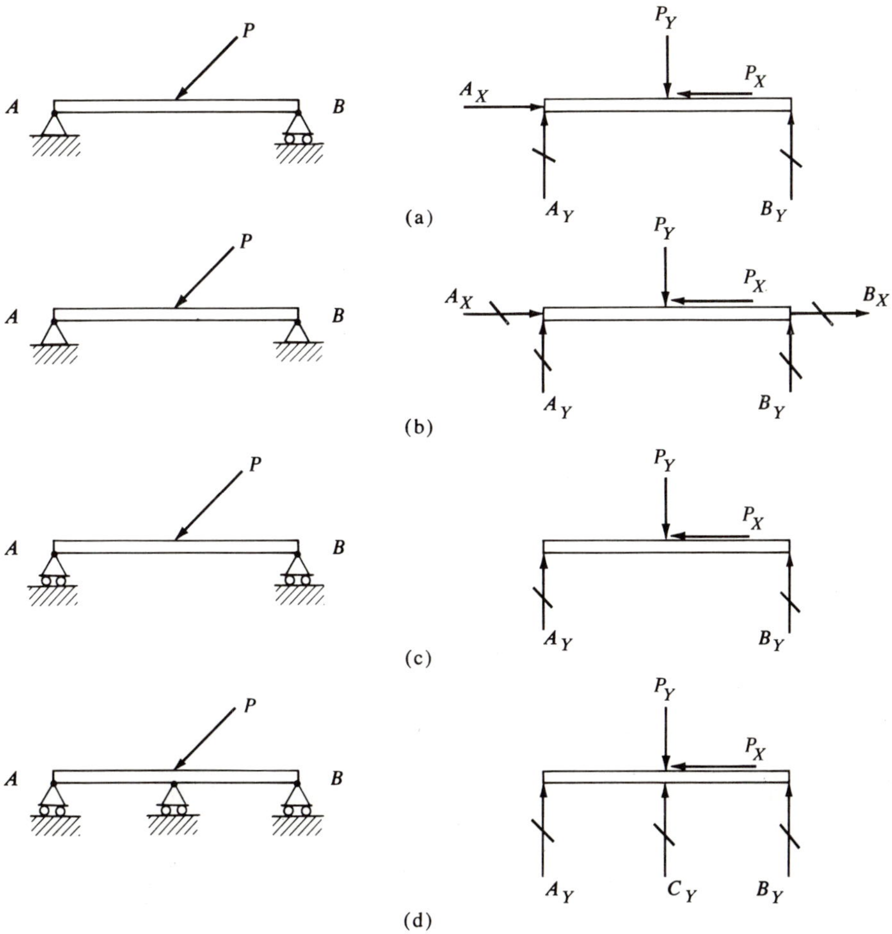

Figure 2.19

move with the supports provided. This beam is said to be a **statically determinate structure,** i.e., the reactions may be obtained by using only the equations of static equilibrium. In addition, the beam is said to be completely restrained. or **stable.** Let us now replace the roller support at *B* with a hinge support and then draw a proper free-body diagram as shown in Figure 2.19b. The addition of the hinge support at *B* has introduced an additional unknown, thus giving a total of four unknown reaction forces. Having four unknowns and only three available equilibrium equations makes the beam shown in Figure 2.19b a **statically indeterminate structure,** i.e., the equations of static equilibrium alone are not sufficient for a complete solution to the problem. A complete solution may be obtained by considering the deformations, but any explanation of this method will be postponed until a later chapter. Any motion of the beam is prevented and the structure is said to be completely restrained or stable. Suppose now we replace the hinge supports at *A* and *B* with smooth rollers and then draw the corresponding free-body diagram as shown in Figure 2.19c. There are now only two unknown reactions and three available equations of static equilibrium. Since there are fewer unknowns than available equilibrium equations, one of the equilibrium conditions will not be satisfied. In this case, the equation $\Sigma F_X = 0$ will not be satisfied; therefore, the beam will move in the X direction under the application of the load. This can be seen from the free-body diagram (Figure 2.19c), which shows no horizontal reaction forces to resist, or balance, the horizontal component of the applied load. The beam is said to be improperly restrained or **unstable.** It appears that in order for a structure to be statically determinate and properly restrained, there must be as many unknown reactions as there are equations of static equilibrium. This condition is necessary but not necessarily sufficient as can be seen from Figure 2.19d. Examination of the free-body diagram (Figure 2.19d) shows that there are three unknown reactions to be solved for with the three equations of equilibrium. The equation $\Sigma F_X = 0$ cannot be satisfied since there are no reactions to balance the horizontal component of the applied load. The structure will move in the horizontal direction and is, therefore, said to be improperly restrained, or unstable. In addition, there are three unknown reactions and only two available equations of equilibrium, $\Sigma M = 0$ and $\Sigma F_Y = 0$, since $\Sigma F_X = 0$ is not satisfied, thus making the reactions statically indeterminate. It is, therefore, possible for improper restraints to create statical indeterminacy.

2.9 Illustrative Examples

The following examples are intended for prolonged student study. The student should not memorize each individual example and then try to search for similarities when solving other problems, but rather seek an understanding of the techniques and principles utilized in the solution of the examples.

Example 2.6

Determine the reactions of the beam shown in Figure 2.20. Neglect weight of the beam.

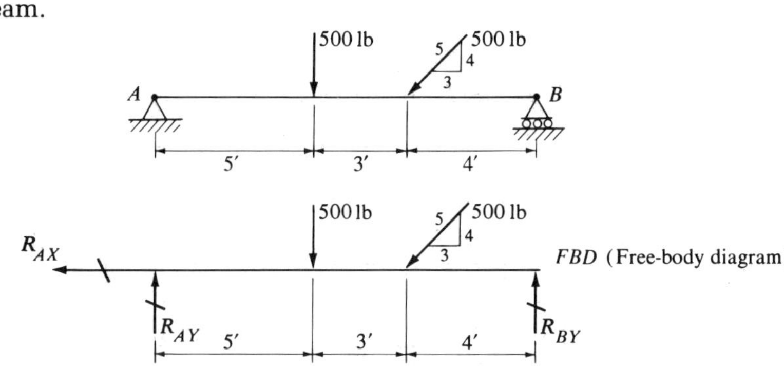

Figure 2.20

Solution:

1. The first step is to draw a proper free-body diagram (FBD) as shown in Figure 2.20. It will be emphasized at this time that the number, type, and line of action of the reactions available for each type of support are known, but the sense of the reaction is not necessarily known. For example, it is known that at the hinge support at A there exists the possibility of a vertical restraining force and a horizontal restraining force. Whether the vertical force is directed up or down and whether the horizontal force is directed to the left or right are not known because the sense of the forces is a function of the particular loading condition. Therefore, when drawing the FBD, the sense of the reactions must be assumed and treated as such in the solution. If a positive answer is obtained for the magnitude of the unknown reaction, then the analyst has assumed the sense correctly. If a negative value is obtained for the reaction magnitude, then the sense must be reversed. Thus, the analyst should not spend any excess time trying to "guess" the sense of the reactions but assume a sense and proceed with the solution.
2. The equations of equilibrium are applied as follows:

$$\Sigma F_X = 0, \text{ forces to the right are assumed positive } \overset{+}{\rightarrow}$$
$$-R_{AX} - (\tfrac{3}{5})500 = 0$$
$$R_{AX} = -300 \text{ lb}$$

The negative answer means that we have treated R_{AX} as if it were directed to the left, and yet we obtained a negative magnitude, meaning that R_{AX} is really directed to the right. Hence:

$$R_{AX} = -300 \text{ lb} \leftarrow = \underline{300 \text{ lb} \rightarrow}$$

the arrows indicating the direction of the force. When the forces in the X-direction were summed, forces directed to the right were taken as positive. Some sign convention must be assumed to distinguish between forces to the left and right. Forces directed to the left could also have been taken as positive and the same result would have been obtained as follows:

$$\Sigma F_X = 0 \leftarrow + \text{ (arrow symbol indicating direction of positive forces)}$$
$$+R_{AX} + \tfrac{3}{5}(500) = 0$$
$$R_{AX} = -300 \text{ lb} \leftarrow = \underline{300 \text{ lb} \rightarrow}$$

Continuing,

$$\Sigma F_Y = 0 \uparrow + \text{ (upward forces arbitrarily taken as positive)}$$
$$R_{AY} + R_{BY} - 500 - (\tfrac{4}{5})500 = 0$$
$$R_{AY} + R_{BY} = 900 \qquad \textbf{(2–1)}$$

$\Sigma M_B = 0$ (+ ↻) (clockwise rotations arbitrarily taken as positive)

$R_{AY}(12) - 500(7) - (\frac{4}{5})(500)(4) = 0$

$$R_{AY} = \frac{(3500 + 1600)}{12} = \frac{5100}{12}$$

$R_{AY} = 425$ lb. The answer is positive, hence
$\underline{R_{AY} = 425 \text{ lb} \uparrow}$ as assumed.

Substitution for R_{AY} in equation (2–1) yields:

$425 + R_{BY} = 900$
$\underline{R_{BY} = 475 \text{ lb} \uparrow}$ since positive magnitude obtained.

3. The reactions are as follow:

$R_{AX} = 300 \text{ lb} \rightarrow \qquad R_{AY} = 425 \text{ lb} \uparrow \qquad R_{BY} = 465 \text{ lb} \uparrow$

As a check, the sum of moments about any point must be zero for equilibrium. Choosing point A, we have:

$\Sigma M_A = 0$ (+ ↻)

$500(5) + (\frac{4}{5})(500)(8) - 475(12) = 0$
$2500 + 3200 - 5700 = 0$
$0 = 0$ Check

4. The solution was obtained by solving the equations $\Sigma F_X = 0$, $\Sigma F_Y = 0$, and $\Sigma M_B = 0$ for the three unknown reactions. We could have just as easily used $\Sigma M_A = 0$ as a replacement for $\Sigma F_Y = 0$. We would then have:

$\Sigma M_A = 0$ (+ ↻)

$500(5) + (\frac{4}{5})(500)(8) - R_{BY}(12) = 0$

$$R_{BY} = \frac{2500 + 3200}{12}$$

$R_{BY} = 475$ lb. The answer is positive, hence
$R_{BY} = 475 \text{ lb} \uparrow$

5. As a check, the $\Sigma F_Y = 0$ may now be used.

$\Sigma F_Y = 0 \uparrow +$ (This check only checks vertical reactions.)
$425 - 500 - \frac{4}{5}(500) + 475 = 0$
$0 = 0$ Check

Example 2.7

Determine the reactions of the truss shown in Figure 2.21. Neglect weight of the truss.

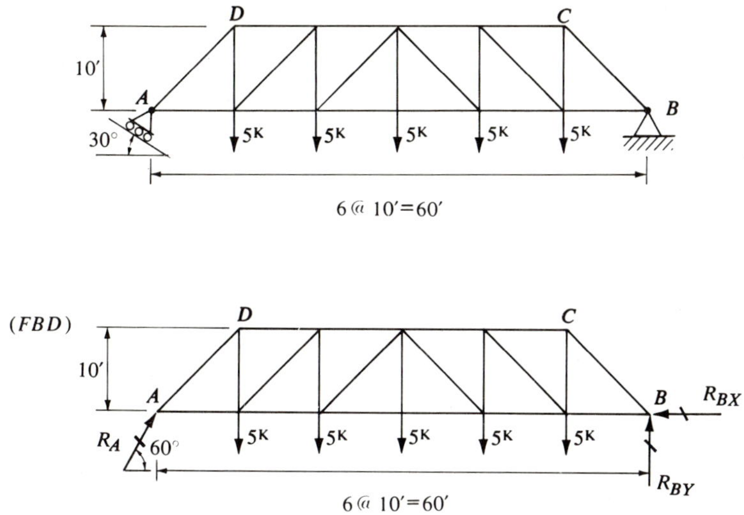

Figure 2.21

Solution:

1. As for all problems, the first step is to draw the FBD as shown in Figure 2.21. The roller support at A is on a 30° incline giving rise to a reaction whose line of action is perpendicular to the surface upon which the rollers roll. In this case, the line of action of the reaction forms a 60° angle with the horizontal. The sense of each reaction was assumed as shown in the FBD.

2. The equations of equilibrium may be applied as follows:

$$\Sigma M_A = 0 \quad (+\curvearrowright)$$

$$5(50 + 40 + 30 + 20 + 10) - 60R_{BY} = 0$$

$$R_{BY} = \frac{750}{60}$$

$$R_{BY} = +12.50 \text{ kips}$$

$$\underline{R_{BY} = 12.50^{K} \uparrow}$$

$$\Sigma F_Y = 0 \uparrow +$$

$$-5 - 5 - 5 - 5 - 5 + 12.50 + R_A(\sin 60°) = 0$$

$$R_A = \frac{12.50}{0.866}$$

$$R_A = +14.44 \text{ kips}$$

$$\underline{R_A = 14.44^{K} \uparrow}$$

$$\Sigma F_X = 0 \rightarrow +$$
$$(14.44)\cos 60° - R_{BX} = 0$$
$$R_{BX} = +7.22 \text{ kips}$$
$$\underline{R_{BX} = 7.22^K \leftarrow}$$

3. As a check, the sum of the moments about any point, say about point C, must be zero.

$$\Sigma M_C = 0 \ (+\curvearrowright)$$

$$14.44(\sin 60°)(50) - 14.44(\cos 60°)(10) - 5(40 + 30 + 20 + 10)$$
$$-12.50(10) + 7.22(10) = 0$$
$$625 - 72.2 - 500 - 125 + 72.2 = 0$$
$$0 = 0 \quad \text{Check}$$

This check is a check on all the reactions since the sum of the moments was taken about a point which does not lie on a line through A and B.

4. An alternate solution is as follows:

use $\Sigma M_A = 0$ to obtain R_{BY}
use $\Sigma M_B = 0$ to obtain R_A
use $\Sigma F_X = 0$ to obtain R_{BX}

Example 2.8

Determine the reactions of the beam shown in Figure 2.22. Neglect weight of the beam.

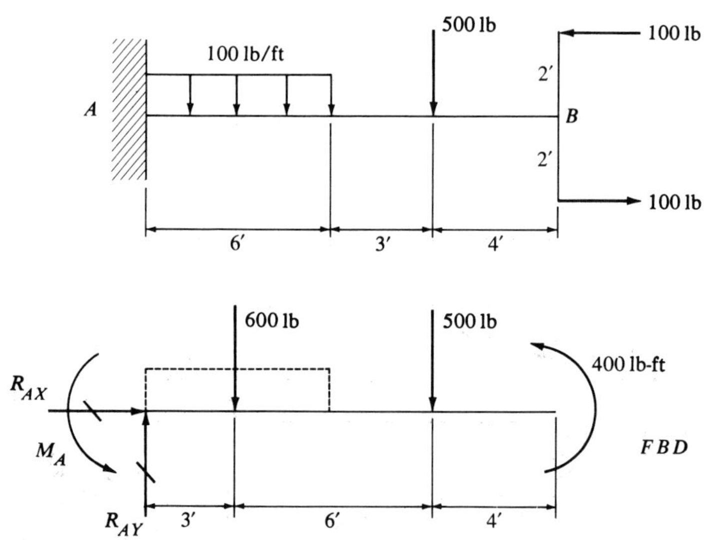

Figure 2.22

Solution:

1. As usual, the first step is to draw a proper FBD as shown in Figure 2.22. The 100-lb-per-foot load is known as a uniformly distributed load. A distributed load is one which is spread over a considerable area such as that resulting from sand in a bin, water on the face of a dam, or snow on the roof of a building. A distributed load may vary in any manner along the length of the member, i.e., uniformly, linearly, parabolically, etc., depending upon the cause of the load. In any case, when considering the external equilibrium of a rigid body, a distributed load may always be replaced by its equivalent resultant force whose magnitude is equal to the area under the loading curve and whose line of action passes through the centroid of the loading curve and is parallel to the loading ordinates. For the problem at hand, the resultant force of the uniformly distributed load is

$$F = \text{Area under loading curve}$$
$$F = 100 \text{ lb/ft} \times 6 \text{ ft} = 600 \text{ lb}$$

   The line of action of the 600-lb resultant is parallel to the loading ordinates and is 3 ft from point A as shown in Figure 2.22.

2. At point B there are two equal, parallel but oppositely sensed forces which we have previously defined as a couple. The moment of the couple is

$$M = 100(4) = 400 \text{ lb-ft}$$

   The couple tends to rotate the beam in a counterclockwise manner with a magnitude of 400 lb-ft as shown in the FBD. Recall that the position of the couple on the beam does not affect the external equilibrium.

3. Applying the equations of equilibrium we have:

$$\Sigma F_X = 0 \rightarrow +$$
$$\underline{R_{AX} = 0}$$

$$\Sigma M_A = 0 \quad (\circlearrowright +)$$

$$-M_A + 600(3) + 500(9) - 400 = 0$$
$$M_A = +5900 \text{ lb-ft}$$
$$\underline{M_A = 5900 \text{ lb-ft} \circlearrowleft}$$

$$\Sigma F_Y = 0 \uparrow +$$
$$R_{AY} - 600 - 500 = 0$$
$$R_{AY} = +1100 \text{ lb}$$
$$\underline{R_{AY} = 1100 \text{ lb} \uparrow}$$

Check:

$$\Sigma M_B = 0 \quad \circlearrowright +$$

$$-5900 + 1100(13) - 600(10) - 500(4) - 400 = 0$$
$$-5900 + 14300 - 6000 - 2000 - 400 = 0$$
$$14300 - 14300 = 0$$
$$0 = 0 \quad \text{Check}$$

4. Note that the horizontal reaction at A, R_{AX}, is zero. Any reaction may be zero, meaning that no restraining force is required for the given loading condition. The support at A has the potential to exert a horizontal restraining force if the loading condition should change. As an example, if the 500-lb concentrated load in Figure 2.22 were not plumb on the beam, a horizontal component would be present. The horizontal reaction at A would, therefore, be utilized to prevent motion.

Example 2.9

The frame in Figure 2.23 is held by a hinge at A and by a cable BDC which passes over a frictionless pulley at D. Determine the reaction at A and the tension in the cable.

Solution:

1. A cable, or rope, is only capable of transmitting a force in tension and, therefore, the direction of a force introduced by a cable, or rope, is always known. In addition, the frictionless pulley at D simply changes the direction of the cable and does not alter the force transmitted by the cable. The magnitude of the cable force at C and the magnitude of the force at B are, therefore, equal but their lines of action are different due to the presence of the frictionless pulley at D. Figure 2.23 shows the proper FBD.
2. The equations of static equilibrium may be applied as follows:

$$\Sigma M_A = 0 \quad \circlearrowright +$$

$$200(4) + T(\sin 30°)(2) + T(\cos 30°)(4) - 8T = 0$$
$$800 + T + 3.464T - 8T = 0$$
$$T = \frac{800}{3.54}$$
$$\underline{T = 226 \text{ lb in directions assumed}}$$

$$\Sigma FY = 0 \uparrow +$$
$$226(\sin 30°) + 226 - R_{AY} = 0$$
$$R_{AY} = +339 \text{ lb}$$
$$\underline{R_{AY} = 339 \text{ lb} \downarrow}$$

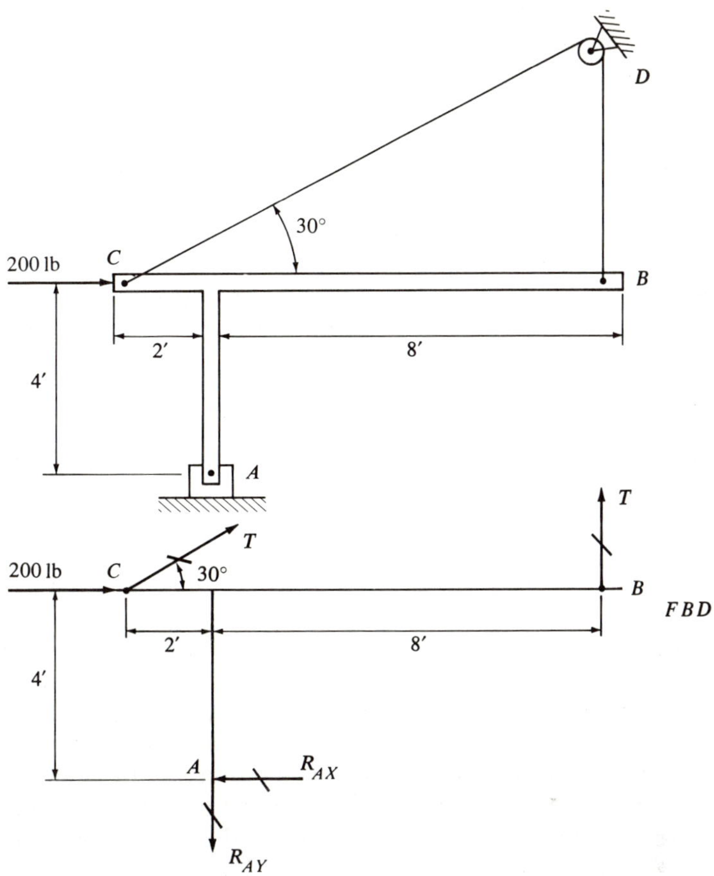

Figure 2.23

$$\Sigma F_X = 0 \rightarrow +$$

$$200 + 226(\cos 30°) - R_{AX} = 0$$
$$R_{AX} = +395.5 \text{ lb}$$
$$\underline{R_{AX} = 395.5 \text{ lb} \leftarrow}$$

Check:

$$\Sigma M_E = 0 \quad (+\curvearrowright)$$

$$226(\sin 30°)2 - 8(266) + 395.5(4) = 0$$
$$226 - 1808 + 1582.0 = 0$$
$$0 = 0 \quad \text{Check}$$

3. Before leaving this example, a little time should be spent on the sequence of solution, i.e., which equation of equilibrium should be used first, or about which point should the moments be summed to yield the solution

with the least effort. For example, in this problem the moments were summed about point A first, thereby eliminating R_{AX} and R_{AY} from the resulting equation and producing an equation with only T as an unknown. Solving for T and then using $\Sigma F_Y = 0$ and $\Sigma F_X = 0$ yielded two equations, each of which contained only one unknown. If point A had not been used as a moment center, the resulting solution could have only been accomplished by solving the equilibrium equations simultaneously since summing the moments about any other point would have yielded an equation with at least two unknowns. The analyst should, therefore, spend some time deciding upon a sequence of solution since it can save time as well as eliminate a source of error introduced by using a more complex solution.

Example 2.10

Determine the reactions of the beam shown in Figure 2.24. Neglect the weight of the beam.

Solution:

1. Examination of the FBD shown in Figure 2.24 reveals that the structure has four unknown reactions, which is one more in number than available equations of equilibrium. It would appear that the structure is statically indeterminate, but upon further inspection it is noted that the structure is actually two rigid bodies, AD and DC, joined at D by an internal hinge. We do, therefore, know something about the internal forces at D, i.e., a hinge may transmit only a force acting in any direction through the center of the hinge, since a frictionless hinge cannot transmit a couple. Figure 2.24c shows a FBD of both rigid bodies, AD and DC, with the force at the hinge represented by its possible vertical and horizontal components. Both rigid bodies, shown in Figure 2.24c, are statically determinate, i.e., the three equations of static equilibrium alone are sufficient to solve for the reactions. The insertion of an internal hinge gives the analyst an additional condition that the action of one member on the other can only be a force, acting in any direction through the center of the hinge. Note that this condition allows us to solve for the additional unknown reaction appearing in the FBD shown in Figure 2.24b.
2. Observe that the hinge force components, D_Y and D_X, on member AD are shown in the opposite direction on member DC since the action and reaction of one body on another must be equal in magnitude but opposite in direction, or sense (see Section 2.7). The resultant force of the distributed load acting on span BC is equal to the area under the loading curve, which was found as follows:

$$\text{Area} = R = \tfrac{1}{2}(7\text{ ft})\,(1\text{ kip/ft})$$
$$R = 3.5\text{ kips}\ \downarrow$$

This 3.5-kip force acts through the centroid of the area which is 7/3 ft from support C.

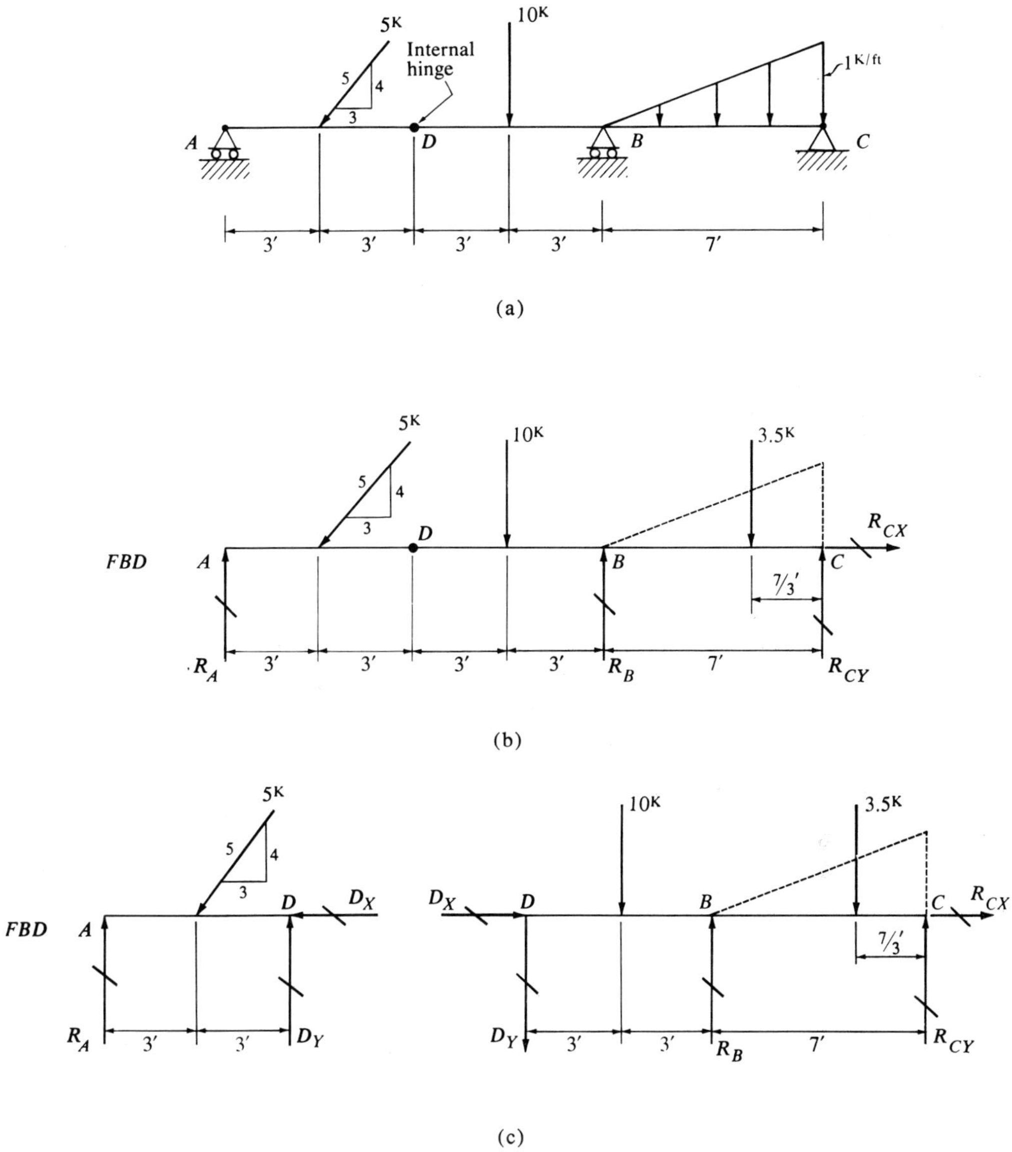

Figure 2.24

3. The remaining solution is straightforward as shown in the following:

Member AD

$$\Sigma F_X = 0 \rightarrow +$$

$$(-\tfrac{3}{5})(5) - D_X = 0$$

$$D_X = -3 \text{ kips}$$

$$\therefore \underline{D_X = 3 \text{ kips} \rightarrow \text{on } AD}$$

$$\Sigma M_A = 0 \quad \circlearrowright +$$

$$\tfrac{4}{5}(5)(3) - 6D_Y = 0$$
$$D_Y = +2 \text{ kips}$$
$$\underline{D_Y = 2 \text{ kips} \uparrow \text{ on } AD}$$

$$\Sigma F_Y = 0 \uparrow +$$
$$R_A - (\tfrac{4}{5})(5) + 2 = 0$$
$$R_A = +2 \text{ kips}$$
$$\underline{R_A = 2 \text{ kips} \uparrow}$$

Member DC

$$\Sigma F_X = 0 \rightarrow +$$
$$-D_X + R_{CX} = 0,$$

D_X is taken as negative since it acts to the left on member *DC*. Its direction was incorrectly assumed in Figure 2.24c since a negative answer was obtained for its magnitude when member *AD* was analyzed.

$$-3 + R_{CX} = 0$$
$$R_{CX} = +3 \text{ kips}$$
$$\underline{R_{CX} = 3 \text{ kips} \rightarrow}$$

$$\Sigma M_C = 0 \quad \circlearrowright +$$

$$-2(13) - 10(10) + R_B(7) - 3.5(\tfrac{7}{3}) = 0$$
$$R_B = +\frac{134.16}{7} \text{ kips}$$
$$\underline{R_B = 19.16 \text{ kips} \uparrow}$$

$$\Sigma F_Y = 0 \uparrow +$$
$$-2 - 10 + 19.16 - 3.5 + R_{CY} = 0$$
$$R_{CY} = -3.66 \text{ kips}$$
$$\underline{R_{CY} = 3.66 \text{ kips} \downarrow}$$

Check:

$$\Sigma F_Y = 0 \uparrow + \text{entire structure}$$
$$2 - \tfrac{4}{5}(5) - 10 + 19.16 - 3.5 - 3.66 = 0$$
$$21.16 - 21.16 = 0$$
$$0 = 0 \quad \text{Check}$$

Problems

2–1. Draw the Free-Body Diagrams for all the bodies shown in the figure.

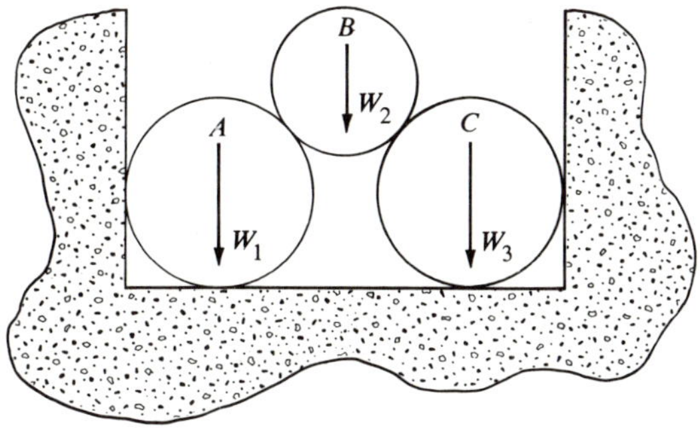

Figure P 2–1

2–2. Draw the Free-Body Diagram for member AB in the figure.

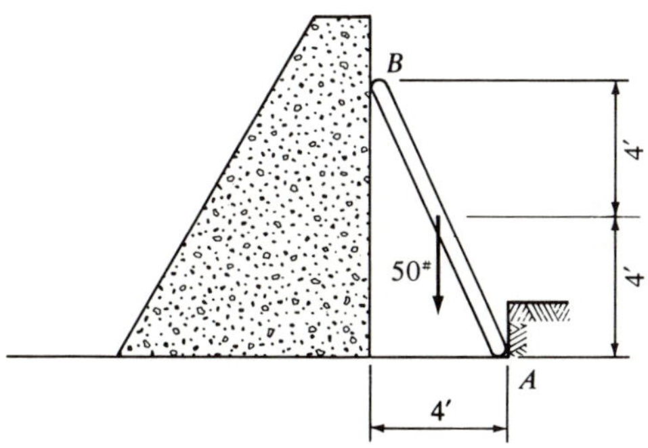

Figure P 2–2

2–3. Draw a Free-Body Diagram for each member of the structure shown in the figure.

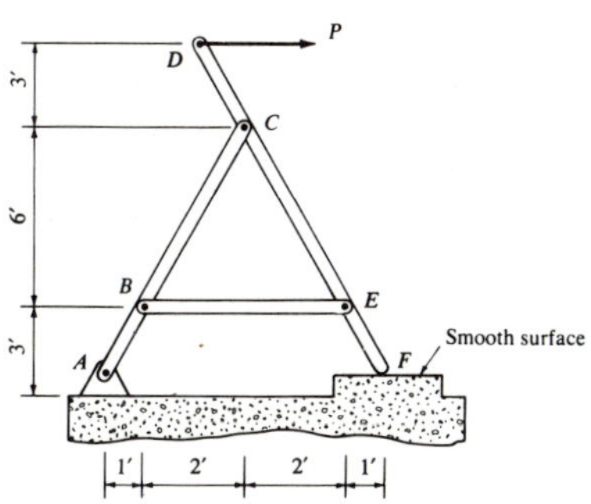

Figure P 2–3

2–4. Draw a Free-Body Diagram for each of the structures in the figure.

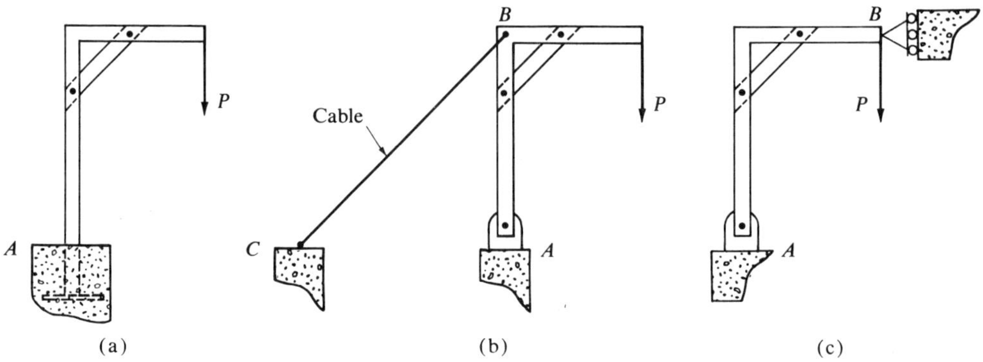

Figure P 2–4

2–5. Draw a Free-Body Diagram for both members *AB* and *CD* shown in the figure.

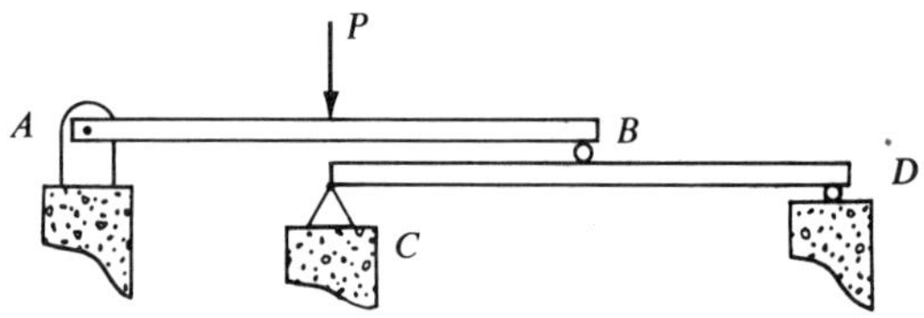

Figure P 2–5

2–6. Draw a Free-Body Diagram for the beam shown in the figure.

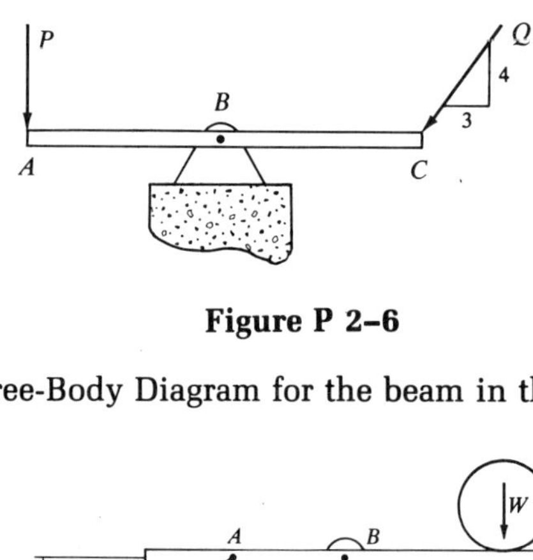

Figure P 2–6

2–7. Draw the Free-Body Diagram for the beam in the figure.

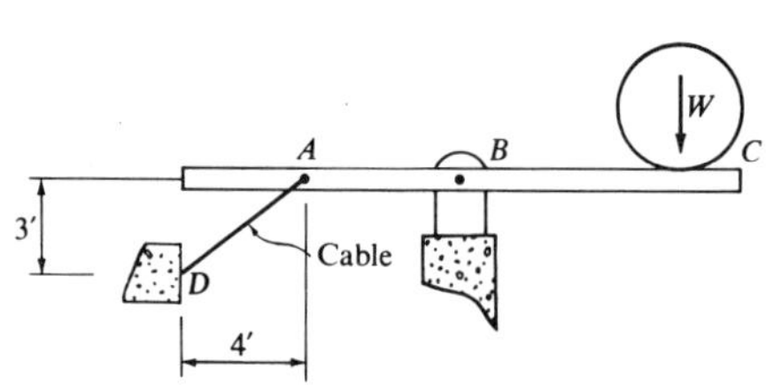

Figure P 2–7

2–8. Draw the Free-Body Diagram for the cantilever beam shown in the figure.

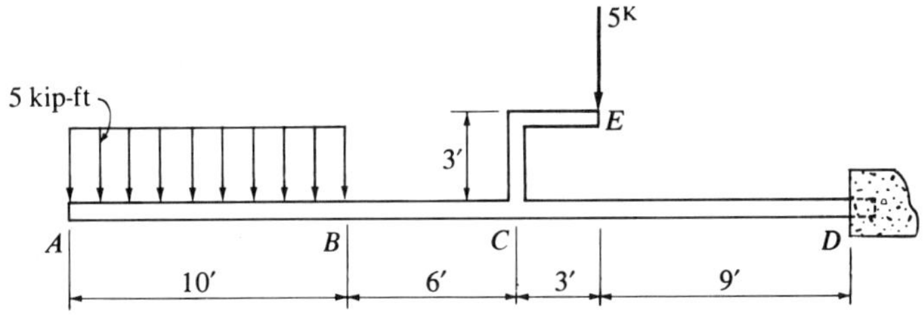

Figure P 2–8

2–9. Draw the Free-Body Diagrams for members *AC* and *CD* of the compound beam shown in the figure.

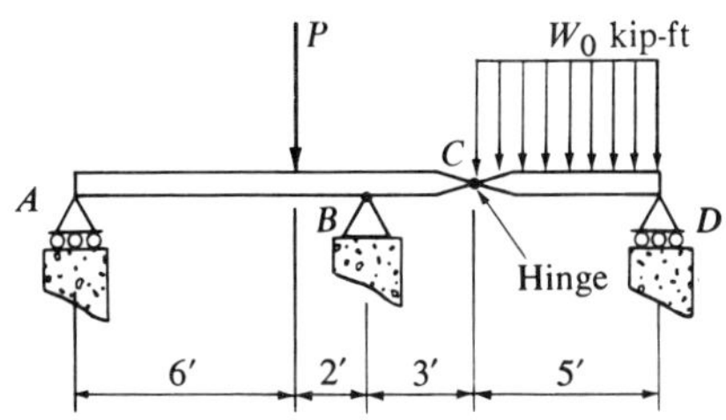

Figure P 2–9

2–10. Determine the resultant of the coplanar, parallel force system.

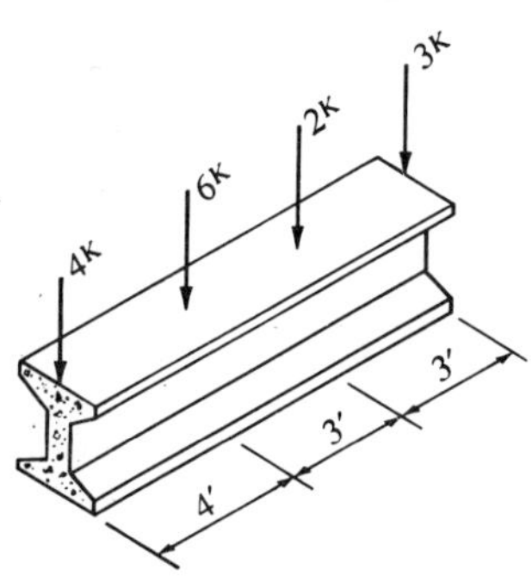

Figure P 2–10

2–11. What is the resultant effect of the two couples on the 8-ft diameter wheel?

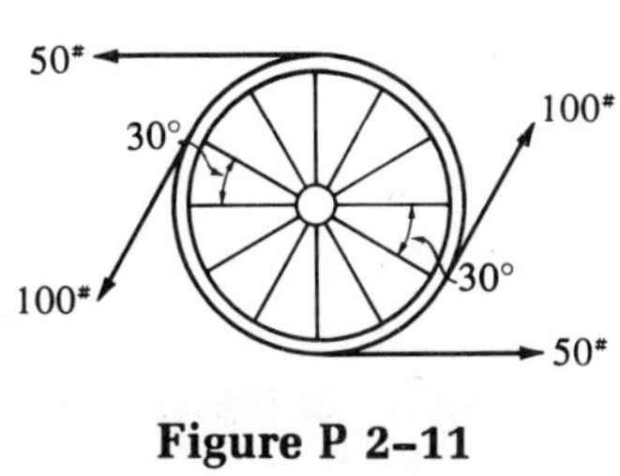

Figure P 2–11

2–12. Replace the 200-lb force by a vertical force at the origin A and a couple whose forces are horizontal and pass through A and B.

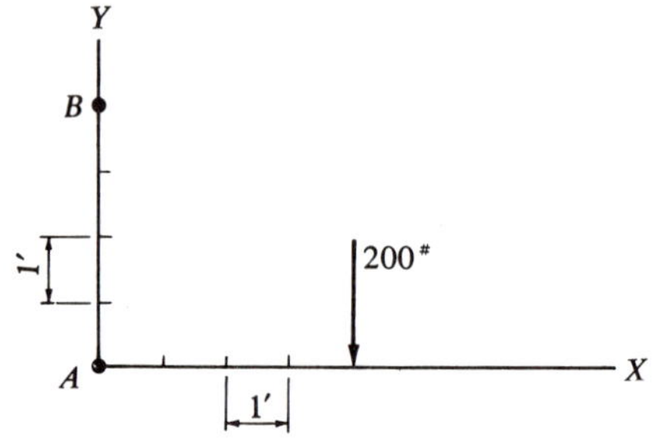

Figure P 2–12

2–13. Determine the third force that will produce a zero resultant.

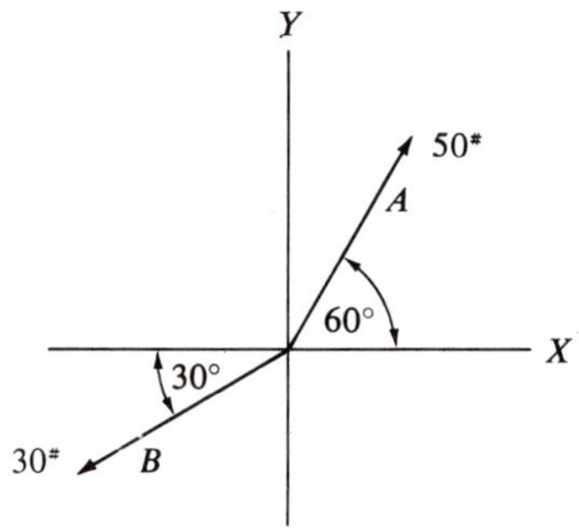

Figure P 2–13

2–14. Find the resultant of the parallel force system shown acting on the slab.

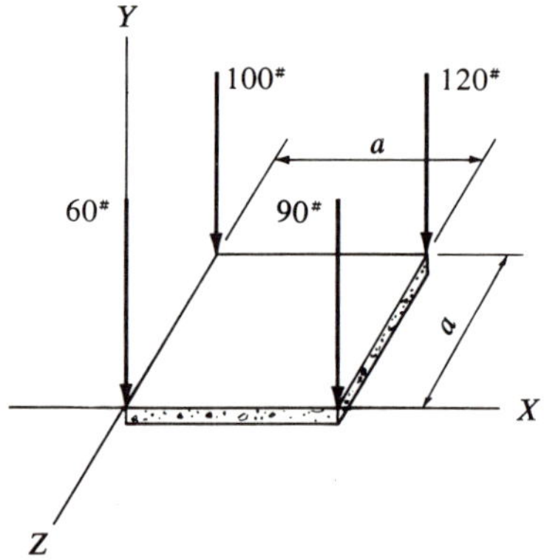

Figure P 2–14

2–15. Find the equivalent force-couple system acting at A for the coplanar, parallel force system shown.

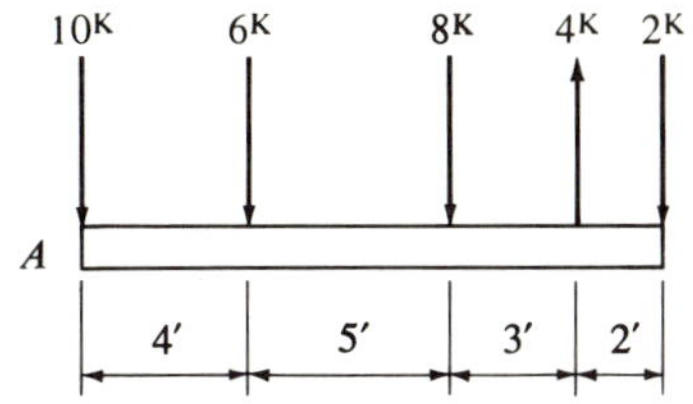

Figure P 2–15

2–16. Determine the resultant of the noncoplanar, parallel force system shown.

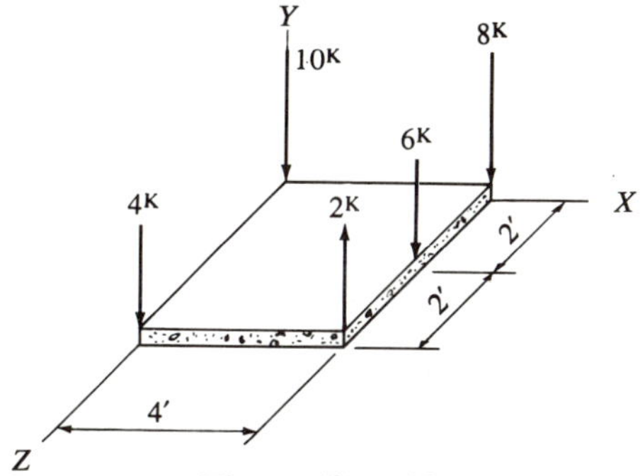

Figure P 2–16

2–17. Replace the forces and couple shown with a single equivalent force-couple system at *A*.

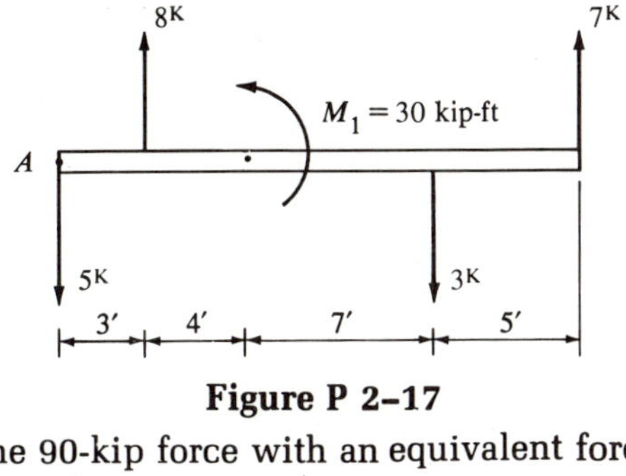

Figure P 2–17

2–18. Replace the 90-kip force with an equivalent force-couple system at the column centerline.

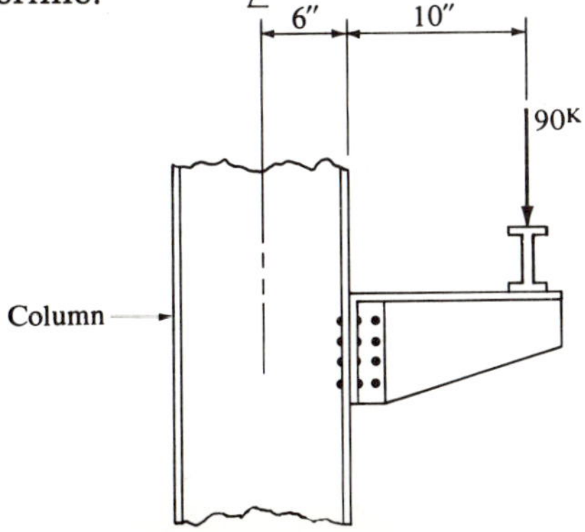

Figure P 2–18

2–19. Replace the 100-kip force with an equivalent force-couple system at the column centerline.

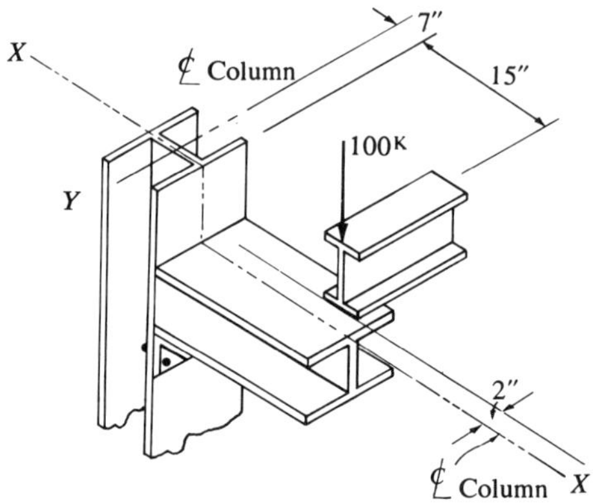

Figure P 2–19

2–20. Replace the force and couple acting on the angle bracket with an equivalent force-couple system at the column centerline.

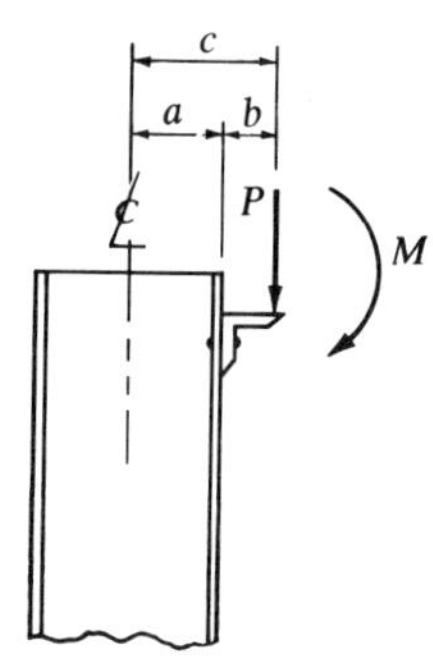

Figure P 2–20

2–21. The loading shown varies linearly from zero at the left end to 600-lb-per-ft at the right end. What is the resultant load and where does it act?

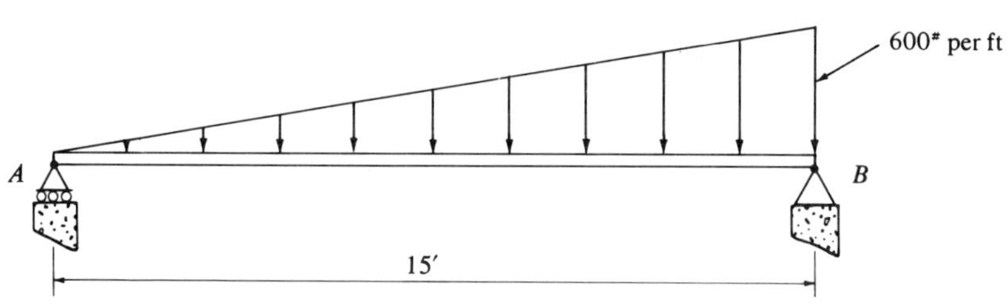

Figure P 2–21

2–22. For the loading shown, determine the resultant and the location of its line of action.

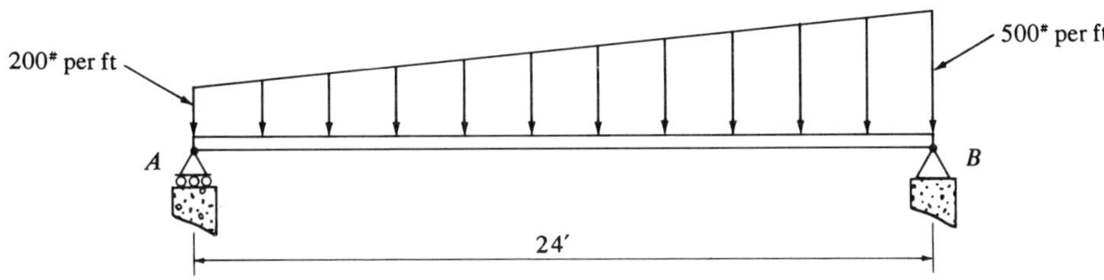

Figure P 2–22

2–23. Replace the loading system shown with an equivalent force-couple system at *A*.

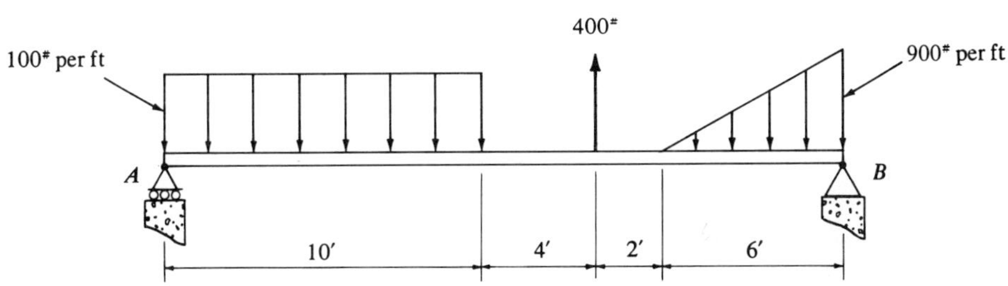

Figure P 2–23

2–24. The traffic light shown in Figure P 2–24 weighs 150 lb. Find the force in each cable.

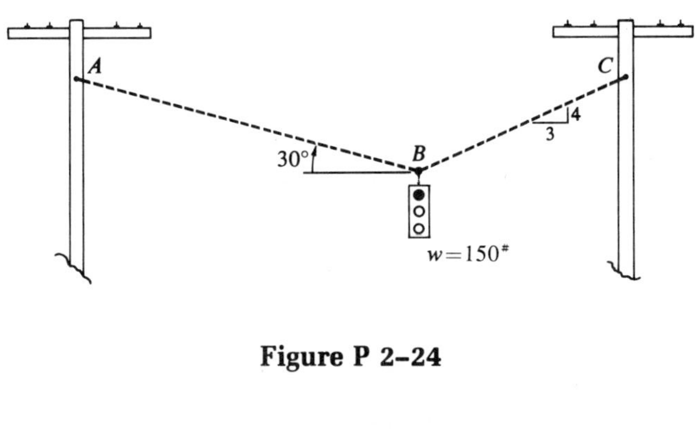

Figure P 2–24

2–25. If the maximum permissible force in the cables is 650 lb, determine the maximum load W that can be safely supported.

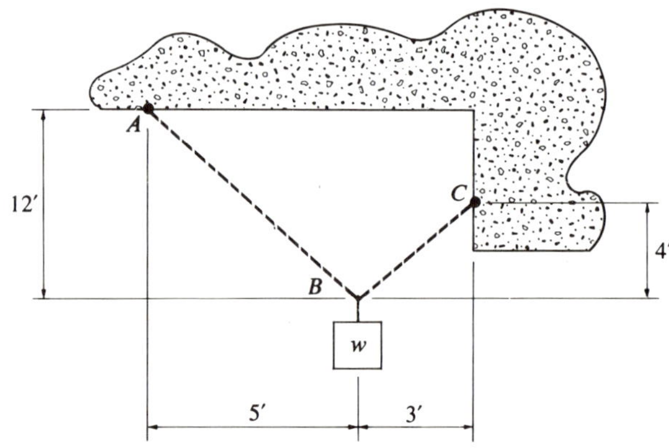

Figure P 2–25

2–26. Classify the structures shown as to statical determinacy and stability.

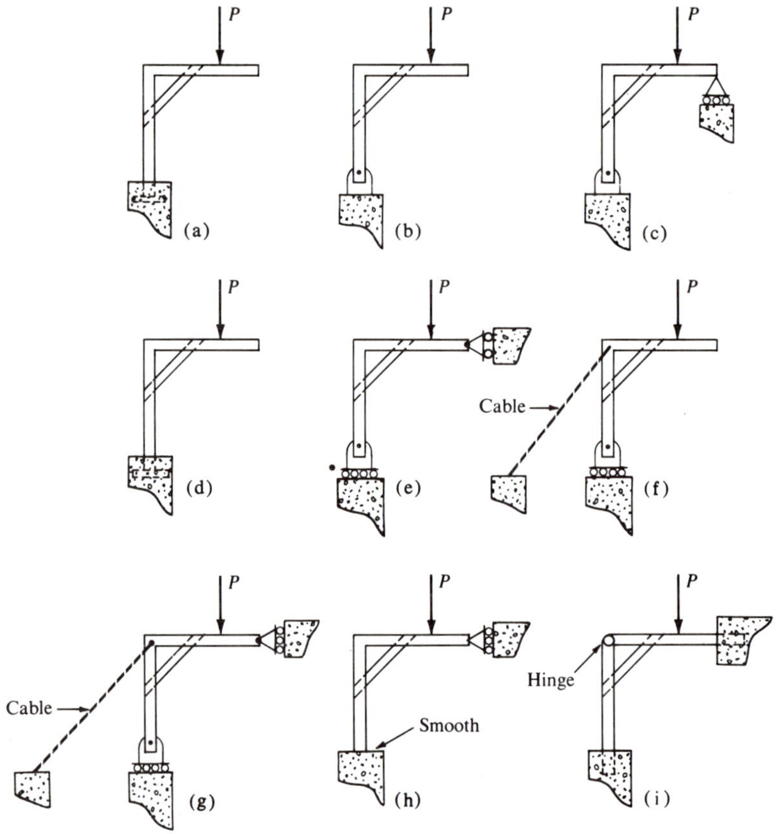

Figure P 2–26

2–27 to 2–42. Find the reactions and cable forces, when applicable, for the structures shown.

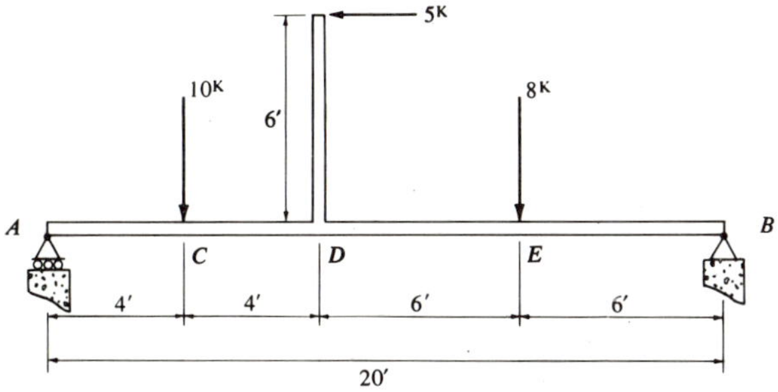

Figure P 2–27

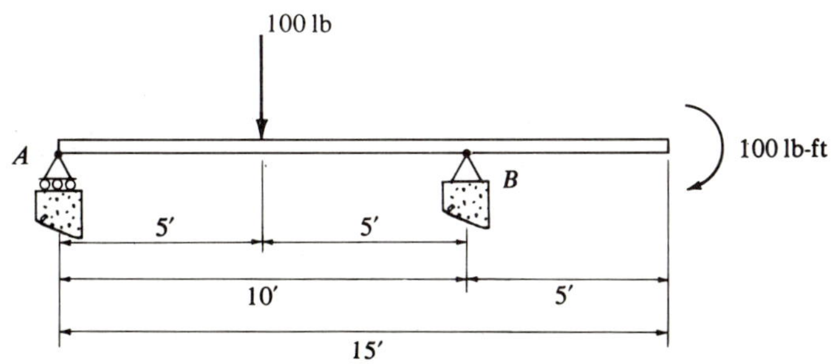

Figure P 2–28

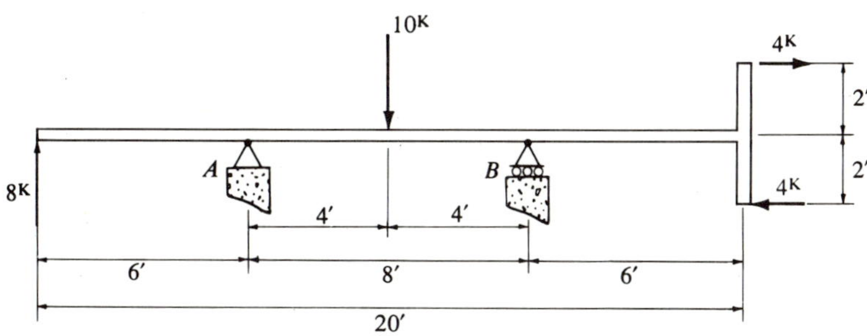

Figure P 2–29

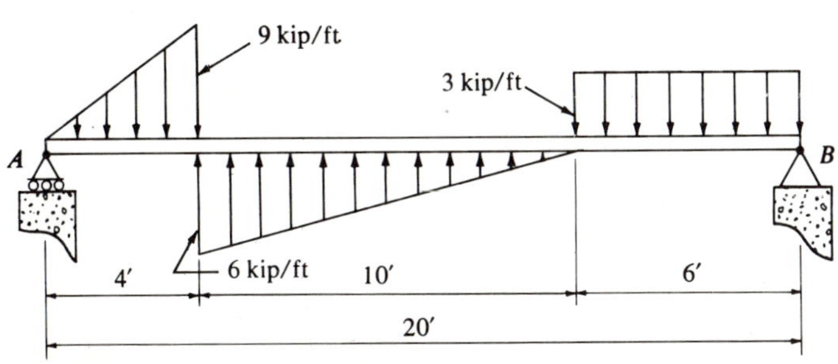

Figure P 2–30

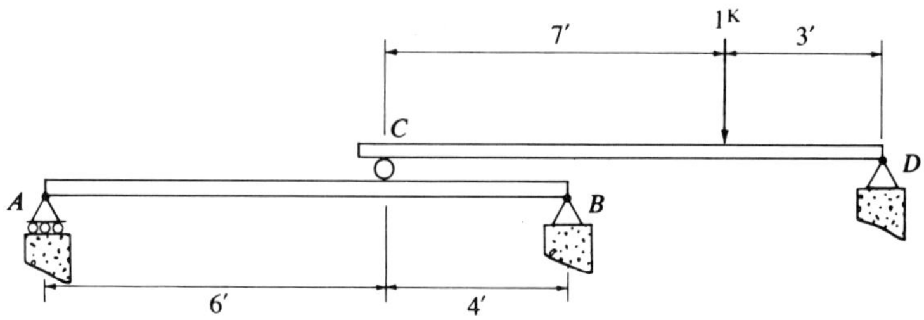

Figure P 2–31

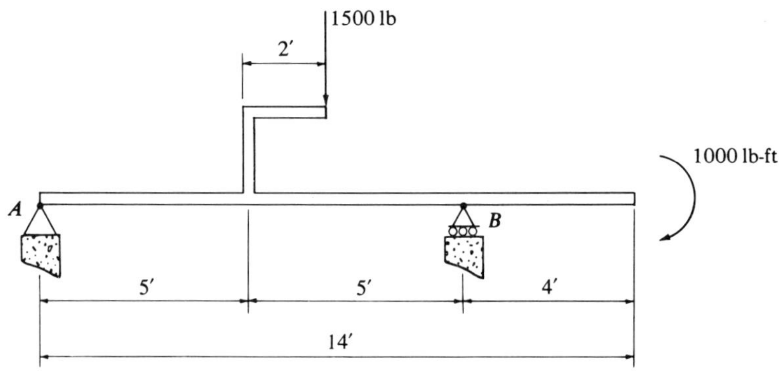

Figure P 2–32

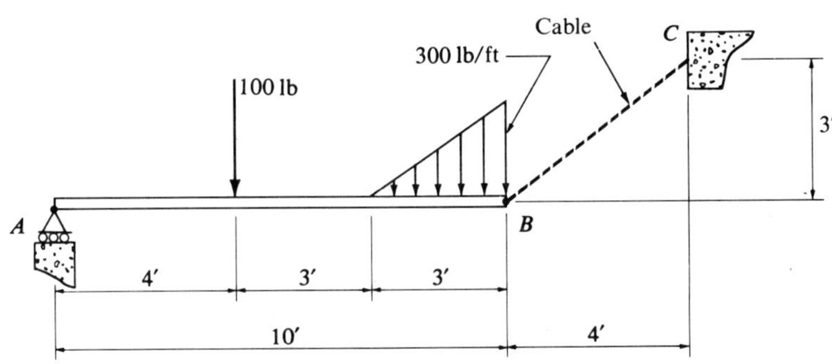

Figure P 2–33

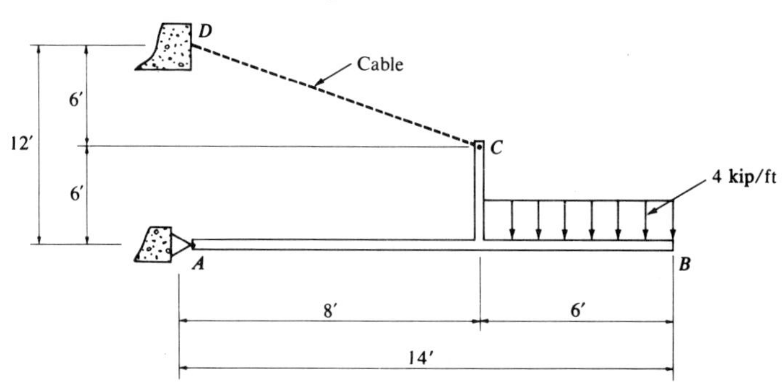

Figure P 2–34

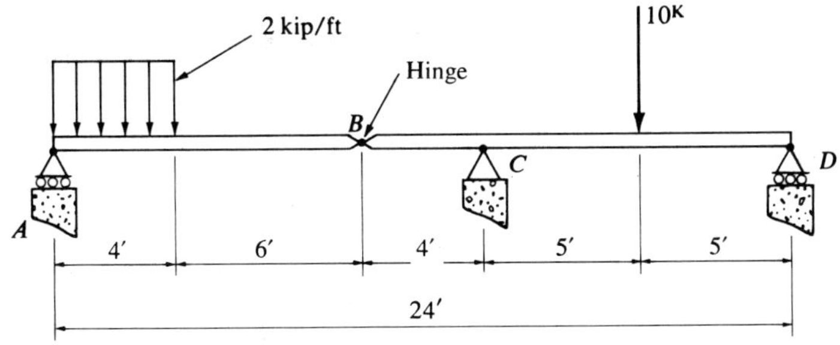

Figure P 2–35

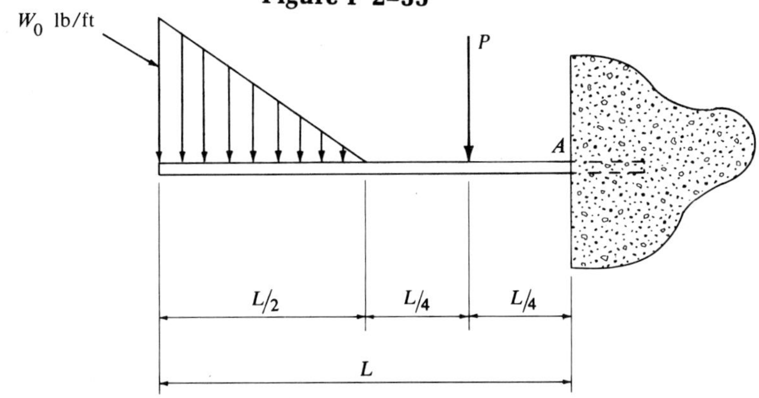

Figure P 2–36

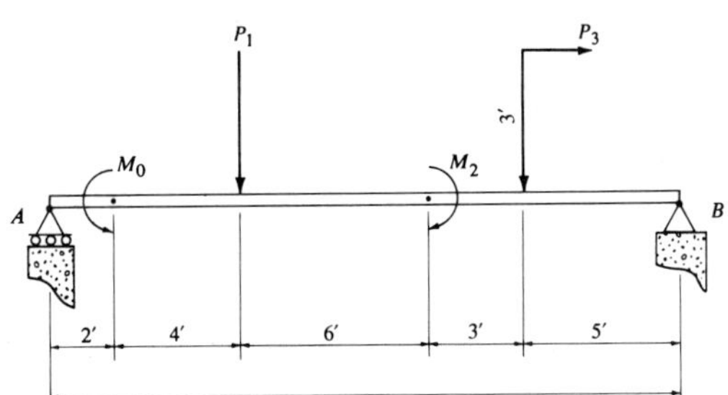

Figure P 2–37

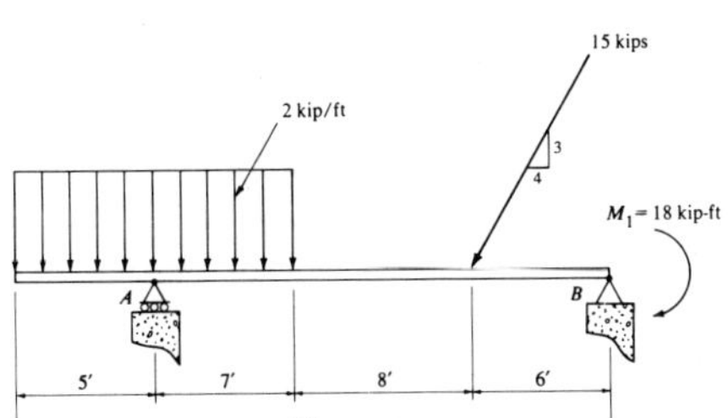

Figure P 2–38

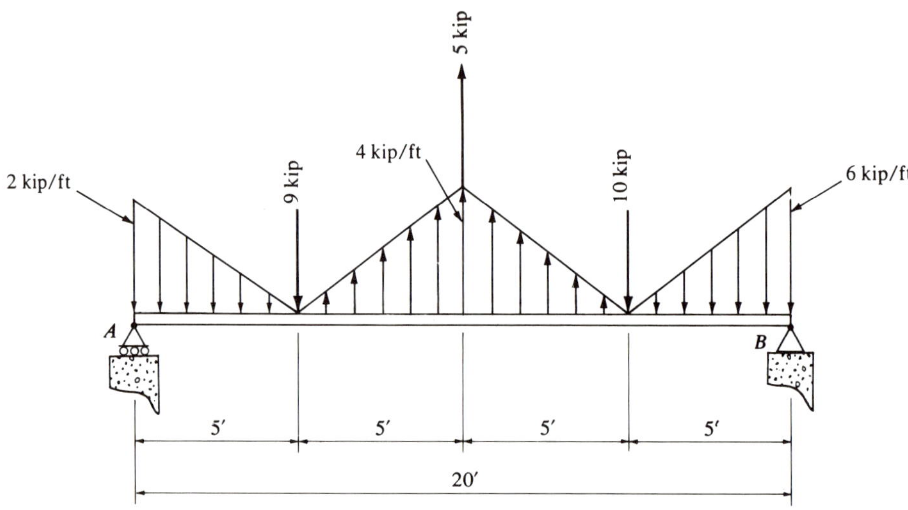

Figure P 2–39

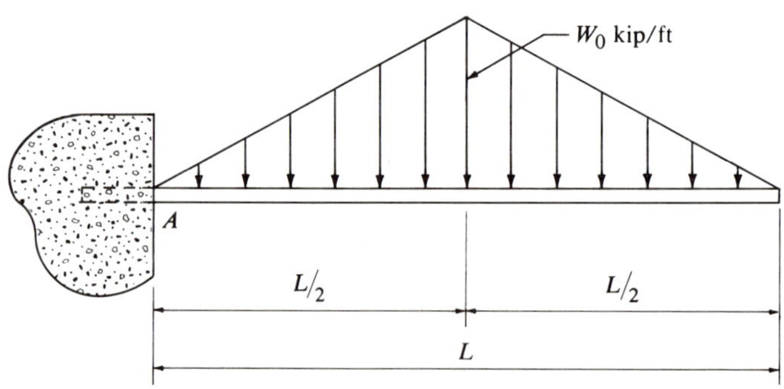

Figure P 2–40

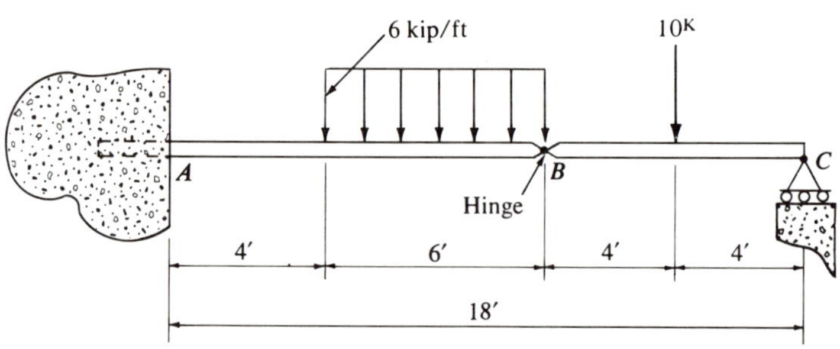

Figure P 2–41

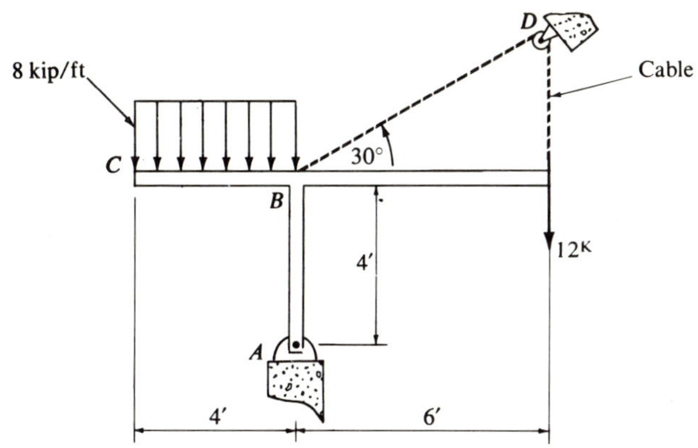

Figure P 2–42

CHAPTER THREE

Shears and Moments in Beams

3.1 General

After the structural analyst has determined the loads acting on the structure and solved for the reactions of the structure, his attention must be focused upon the internal forces and moments, particularly at points within the structure where these quantities are maximum. Using the maximum values of the internal forces and moments, the stresses are computed and the member can be adequately designed. Structural analysis, therefore, involves determining all the forces and moments acting on the structure, to include the reactions, as well as determining the internal forces and moments along with their variation in and along various members of the structure and the location and magnitude of their maximum values. The most frequently encountered planar structures are beams, rigid frames, arches, and trusses. Plane trusses will be treated separately in Chapter Four since trusses are a widely used structural system and require some special treatment and assumptions. **Beams,** which are structural members subjected to loads applied perpendicular to their longitudinal axis, are by far the most frequently occurring structural member and, therefore, a majority of this chapter will be devoted to them. In some cases beams may be subjected to loads acting both transversely and parallel to the longitudinal axis. Such beams may be referred to as **beam columns.**

Many different types of beams are used in practice and each is classified according to the type of supports utilized. Figure 3.1 shows the beam classification generally accepted in structural engineering.

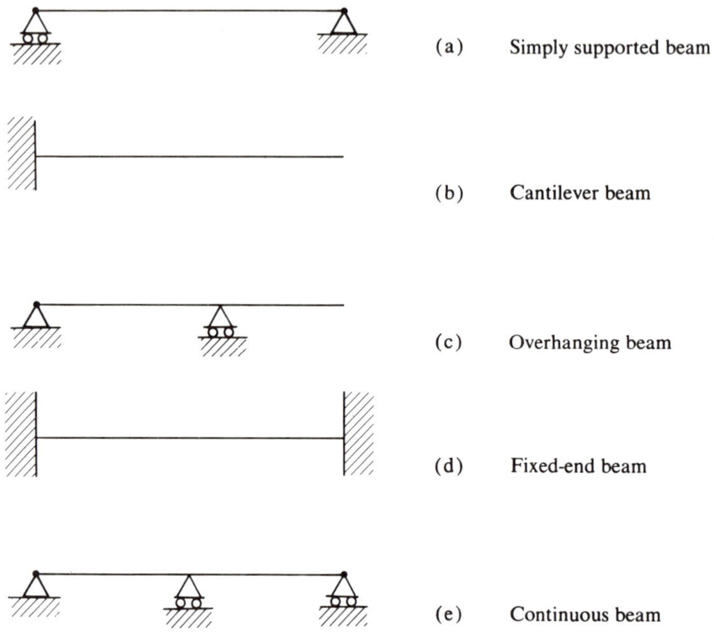

Figure 3.1

3.2 Internal Forces and Moments

In order to determine the internal forces and moments acting at a specific point along a beam's span, it is necessary to pass an imaginary cutting plane through the beam at that point. This cutting plane isolates a portion of the beam and at the same time exposes the cross section of the beam at the cut section. The internal forces and moments acting at the cut section are, therefore, those forces and moments, which are the resultant effects of the stresses acting on the cross section being investigated, necessary to keep the isolated beam portion in equilibrium. Suppose that it is required to know the internal forces and moment acting at point *C* for the beam shown in Figure 3.2a. Figure 3.2b shows a FBD of the beam along with the correct magnitudes and directions for the reactions. Next, the cutting plane *AA* was passed through point *C* and all of the beam to the left of the cut was isolated as shown in Figure 3.2c. Note that the beam portion *AC* is not in equilibrium, and, therefore, forces and a moment are required at the cut section, *C*, to maintain equilibrium. These forces and the moment required at *C* are the resultant effects of the stresses acting on the cross section of the beam at *C* and are called **stress resultants.** For simple planar structures, there exists the possibility of three stress resultants acting at a cut section. Figure 3.2d shows the three possible stress resultants acting at section *C* necessary to maintain equilibrium of the beam portion *AC*. The horizontal force *N* (Figure 3.2d) may be required to satisfy the equation $\Sigma F_x = 0$; the verti-

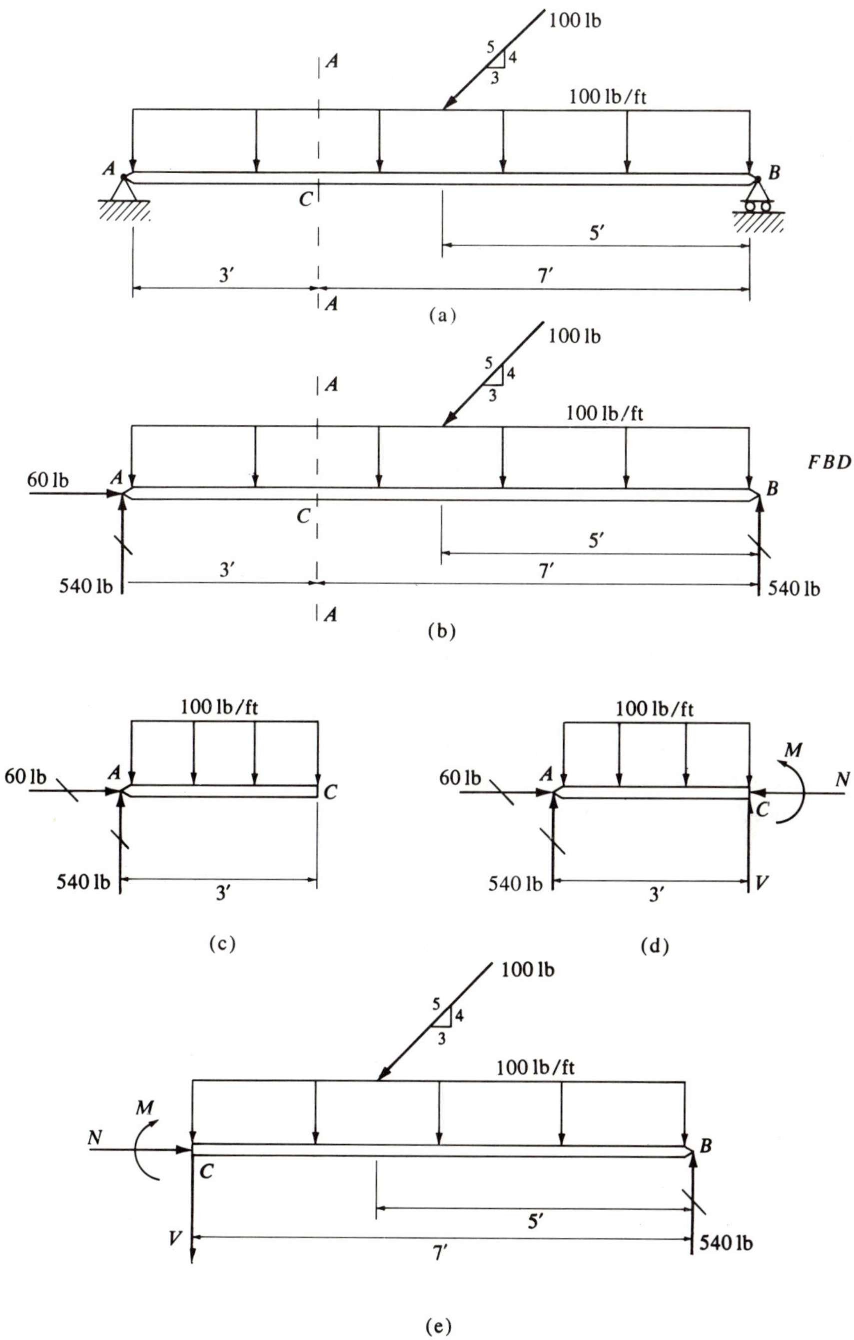
100 lb
5
4
3
A
100 lb/ft
A
B
C
5′
3′
7′
A
(a)
100 lb
5
4
3
A
100 lb/ft
FBD
60 lb
A
B
C
5′
3′
7′
540 lb
540 lb
A
(b)
100 lb/ft
60 lb
A
C
540 lb
3′
(c)
100 lb/ft
M
60 lb
A
N
C
540 lb
3′
V
(d)
100 lb
5
4
3
100 lb/ft
M
N
B
C
5′
V
7′
540 lb
(e)

Figure 3.2

cal force V may be required to satisfy the equation $\Sigma F_Y = 0$; and the moment M may be required to satisfy the equation $\Sigma M = 0$. Any one of these (N, V, or M) may be zero depending upon the cut section selected and/or the loading condition. For example, if the 100-lb **concentrated load,** i.e., a load applied at a specific point (shown in Figure 3.2a) was vertical, then the horizontal reaction at A would not be required since there would be no horizontally applied loads on the beam. The 60-lb horizontal reaction would then not appear on the FBD shown in Figure 3.2d; therefore, the horizontal stress resultant N would not be utilized, i.e., its magnitude would be zero. These stress resultants shown in Figure 3.2d are also referred to as an **axial force** N, a **transverse shear force** V, and a **bending moment** M. Some textbooks refer to these stress resultants as resisting forces and resisting moments since they are the result of the member's attempt to resist the effects of the applied loads. It is common practice to refer to the transverse shear force and the bending moment simply as **"shear"** and **"moment"** when dealing with plane structures. Such an abbreviation will be used in this text. Before solving for the shear, moment, and axial force acting at point C, it should be noted that the isolated portion of the beam shown in Figure 3.2d is, in fact, a FBD and, as before, must be properly drawn before a solution can be obtained.

After a proper FBD is drawn for the isolated portion of the structure (as shown in Figure 3.2d), the next step in the solution is to solve for the shear, moment, and axial force using the three equations of static equilibrium. As before, the sense, or direction, of the stress resultants is assumed when shown on the FBD and treated as such in the corresponding solution. If a positive answer is obtained for the magnitude, then the analyst has correctly assumed the direction. If a negative answer is obtained, then the analyst has incorrectly assumed the direction and the direction must be reversed. For the beam shown in Figure 3.2a, the solution is as follows:

Member AC (Figure 3.2d)

$$\Sigma F_X = 0 \rightarrow +$$
$$60 - N = 0$$
$$N = +60 \text{ lb}$$
$$\underline{N = 60 \text{ lb} \rightarrow}$$

$$\Sigma F_Y = 0 \uparrow +$$
$$540 - 100(3) + V = 0$$
$$V = -240 \text{ lb}$$
$$\underline{V = -240 \text{ lb} \uparrow = 240 \text{ lb} \downarrow}$$

Note that the direction of the shear force, V, was incorrectly assumed on the FBD to be directed upward. By summing the vertical forces, however, its direction was found to be downward.

In order to eliminate the unknown force V in the following moment equation, it is suggested that the moments be summed about a point located in the plane of the cut section such as point C.

$$\Sigma M_C = 0 \quad (\circlearrowright +)$$

$$540(3) - 100(3)\tfrac{3}{2} - M = 0$$
$$M = +1170 \text{ lb-ft}$$
$$\underline{M = 1170 \text{ lb-ft} \circlearrowleft}$$

After passing the imaginary cutting plane through point C, only that portion of the beam to the left of the cut was considered. We could have just as easily considered that portion of the beam to the right of the cut. The FBD of the beam segment CB, which is that portion of the beam to the right of the cut, is shown in Figure 3.2e. The solution for the stress resultants is as follows:

Member CB (Figure 3.2e)

$$\Sigma F_X = 0 \rightarrow +$$
$$N - (\tfrac{3}{5})100 = 0$$
$$N = +60 \text{ lb}$$
$$\underline{N = 60 \text{ lb} \rightarrow}$$

$$\Sigma F_Y = 0 \uparrow +$$
$$-V - \tfrac{4}{5}(100) - 100(7) + 540 = 0$$
$$V = -240 \text{ lb}$$
$$\underline{V = -240 \text{ lb} \downarrow \; = 240 \text{ lb} \uparrow}$$

$$\Sigma M_C = 0 \quad (\circlearrowright +)$$

$$-540(7) + 100(7)(\tfrac{7}{2}) + \tfrac{4}{5}(100)(2) + M = 0$$
$$M = +1170 \text{ lb-ft}$$
$$\underline{M = 1170 \text{ lb-ft} \; (\circlearrowright +)}$$

The same numerical values were obtained for the stress resultants acting at section C on members AC and CB but their directions were opposite. This is evident, recall, since the action and reaction of one body on another must be numerically equal but opposite in sense. It can be stated, therefore, that when the stress resultants at a specific point are required, the analyst may pass an imaginary cutting plane through that point and consider that portion, or segment, of the cut member to the left or to the right of the cut as a means of accomplishing the solution. The decision whether to consider everything to the left or to the right of the cut is based upon which treatment yields a solution with the least effort. In general, the segment subjected to the least number of applied forces and moments should be the segment selected.

3.3 General Shear and Moment Equations

It would appear from the problem treated in Section 3.2 (Figure 3.2) that each time the shear and moment at a section are required, the analyst must cut the member, draw the corresponding FBD, and apply the equations of equilibrium. In order to obtain a vivid picture of how the shear and moment vary along the length of the beam, a large number of cut sections would have to be analyzed. Fortunately, better methods are available to determine the variation of the shear and moment in a structural member. One such method is to write general shear and moment equations for portions of the structural member. Consider the cantilever beam shown in Figure 3.3a. In order to write general equations for the shear and moment, a coordinate system must be selected. The FBD for the beam is shown in Figure 3.3b along with a selected coordinate system which places the origin at the left of the beam. Considering shear first, we observe that if we mentally pass a cutting plane through the beam anywhere along the span between the origin and the 1000-lb concentrated load the only vertical forces encountered are the shear force itself and the uniformly distributed load. If, however, we mentally pass a cutting plane through the beam anywhere between the 1000-lb concentrated load and the fixed support at A we have now encountered the 1000-lb load in addition to the uniformly distributed load and the shear force itself. It should be noted that these mental pictures were made considering all the vertical forces to the left of our imaginary cut since the origin is located at the left end of the beam; hence, all pertinent dimensions and references must be taken from the selected origin. Since the shear force at any section is a function of sum of the external vertical forces acting on the isolated portion, the same general expression for the shear force from the origin to the 1000-lb concentrated load would not be applicable for the span between the 1000-lb load and the fixed support at A. To accurately represent the variation in the shear for the span from O to A (Figure 3.3a), one general equation would be required for the interval from O to B and another general equation for the interval from B to A. As a general rule it can be stated that each time the vertical loading along the span changes, the general expression for the shear must also change. Figures 3.3c and 3.3d show the cuts and FBDs required to obtain the general equations for the shear force in the problem at hand. Note that the cuts were arbitrarily selected within each interval, i.e., the location of the cut section is not specific but rather representative of any point within the interval. The general equations are obtained simply by applying the equations of equilibrium to the FBDs shown in Figures 3.3c and 3.3d as follows:

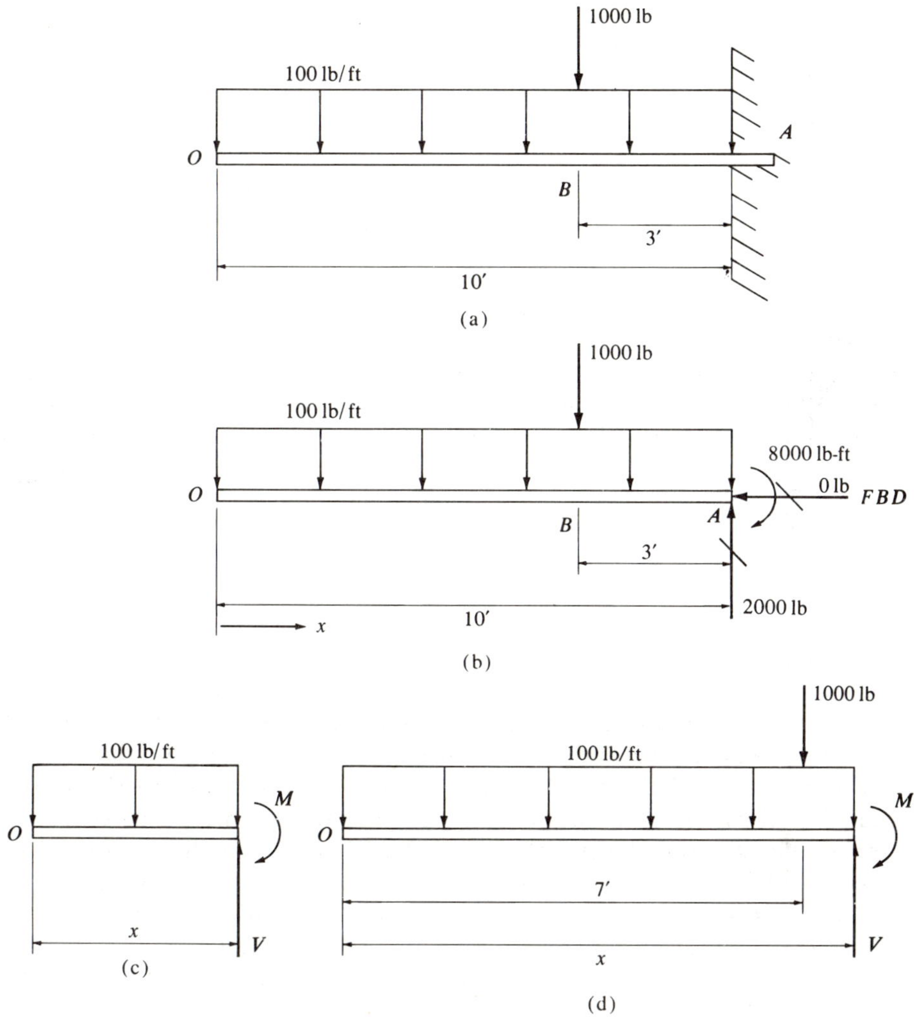

Figure 3.3

$$\Sigma F_Y = 0 \uparrow + \text{(Figure 3.3c)}$$
$$-(100 \text{ lb/ft})(x \text{ ft}) + V = 0$$
$$\underline{V = 100x \text{ lb}}$$

This equation is valid for $0 \leq x \leq 7'$ (this symbol is read **for the interval from** x **equal to and greater than zero to** x **less than and equal to 7 feet**). Continuing with the solution we have:

$$\Sigma F_Y = 0 \uparrow + \text{(Figure 3.3d)}$$
$$-(100 \text{ lb/ft})(x \text{ ft}) - 1000 \text{ lb} + V = 0$$
$$\underline{V = (100x + 1000) \text{ lb}}$$

This equation is valid for the interval $7' \leqslant x \leqslant 10'$, and the general equations for shear along the entire span may be expressed as follows:

$$V = 100x \qquad 0 \leqslant x \leqslant 7'$$
$$V = 100x + 1000 \qquad 7' \leqslant x \leqslant 10'$$

It should be noted that the shear force, V, was assumed to be directed upward in the FBDs shown in Figures 3.3c and 3.3d and a positive magnitude obtained in the general equations will validate this assumption. A negative value, on the other hand, would indicate that the shear force is actually downward. In addition, if $x = 7$ feet is substituted into the two general equations, two different values for the shear force are obtained as follows:

$$@\ x = 7';\ V = 100x = 100(7) = +700 \text{ lb} = 700 \text{ lb} \uparrow$$
$$@\ x = 7';\ V = 100x + 1000 = 700 + 1000 = +1700 \text{ lb} = 1700 \text{ lb} \uparrow$$

A close examination of the FBDs shown in Figure 3.3c and 3.3d reveals that a shear force of 700 lb is obtained if the cut is taken an infinitesimal distance, i.e., an immeasurably or incalculably small distance, to the left of the 1000-lb concentrated load at B and a shear force of 1700 lb is obtained if the cut is taken an infinitesimal distance to the right of the 1000-lb load. The 1000-lb concentrated load causes an abrupt change to occur in the shear. This substantiates the previously stated general rule that each time that the vertical loading changes, the general expression for the shear force must also change.

General equations for the moment must now be determined. Referring to the cut section shown in Figure 3.3c, we may solve for the moment, M, acting at the arbitrary section by summing the moments about a point in the plane of the cut section. Before summing moments, however, let us determine what are the contributors to this summation. We note that the only contributors are the unknown couple, M, acting at the cut section and the uniformly distributed load acting at a moment arm of $x/2$ feet. If we now consider the cut section shown in Figure 3.4d, the moment, M, acting at the cut may be determined by summing the moments about a point at the cut section. The contributions to this moment equation are the unknown couple, M, the uniformly distributed load at a moment arm of $x/2$ feet, and the 1000-lb concentrated load at a moment arm of $(x - 7)$ feet. The same general equation for moment cannot be used for both intervals, i.e., for spans OB and BA. It may be stated that each time that the applied loading on a member changes, the general equations for the moment must also change. The general equations for moment take the following form for the beam shown in Figure 3.3:

$$\Sigma M_X = 0 \ \circlearrowright\!+ \qquad \text{(Figure 3.3c)},$$

where ΣM_X refers to the sum of the moments about a point x feet from the origin

$$-(100 \text{ lb/ft})(x \text{ ft})(x/2 \text{ ft}) + M = 0$$
$$\underline{M = 50x^2 \text{ lb-ft}}$$

$$\Sigma M_X = 0 \quad \circlearrowright + \qquad \text{(Figure 3.4c)}$$

$$-(100 \text{ lb/ft})(x \text{ ft})(x/2 \text{ ft}) - (1000 \text{ lb})(x - 7) \text{ ft} + M = 0$$
$$\underline{M = [50x^2 + 1000(x - 7)] \text{ lb-ft}}$$

The general expressions for moment for the entire span OA may be expressed as:

$$M = 50x^2 \qquad 0 \leq x \leq 7'$$
$$M = 50x^2 + 1000(x - 7), \qquad 7' \leq x \leq 10'$$

As a check, the moment at x = 7 feet (point B) will be determined using both equations since both are valid at that point.

$$M = 50(7)^2 = 2450 \text{ lb-ft}$$
$$M = 50(7)^2 + 1000(7 - 7) = 2450 \text{ lb-ft}$$

Note that the moment is the same as determined by both equations, while the two general shear equations yielded different values for shear depending upon the cut section, whether it was an infinitesimal distance to the left or right of the concentrated load. The application of a concentrated load causes an abrupt change in shear, but only initiates a change in moment since moment is a function of both load and distance. If a positive magnitude for the moment is obtained, then the direction assumed on the FBDs was correct.

Example 3.1

Determine the general expressions for both shear and moment for the beam shown in Figure 3.4.

Solution:

1. The first step in the solution is to draw a proper FBD of the entire beam. A cutting plane was passed through the vertical bracket at C and the corresponding FBD of the isolated bracket as shown in Figure 3.4b. The FBD of the entire beam is shown in Figure 3.4c. The bracket at C was replaced by showing its resultant effect on the beam. Note that the force and couple, which are the stress resultants acting at the base of the bracket, are an equivalent force-couple system for the 5-kip force acting at E. The stress resultants at C and the reactions are obtained as follows:

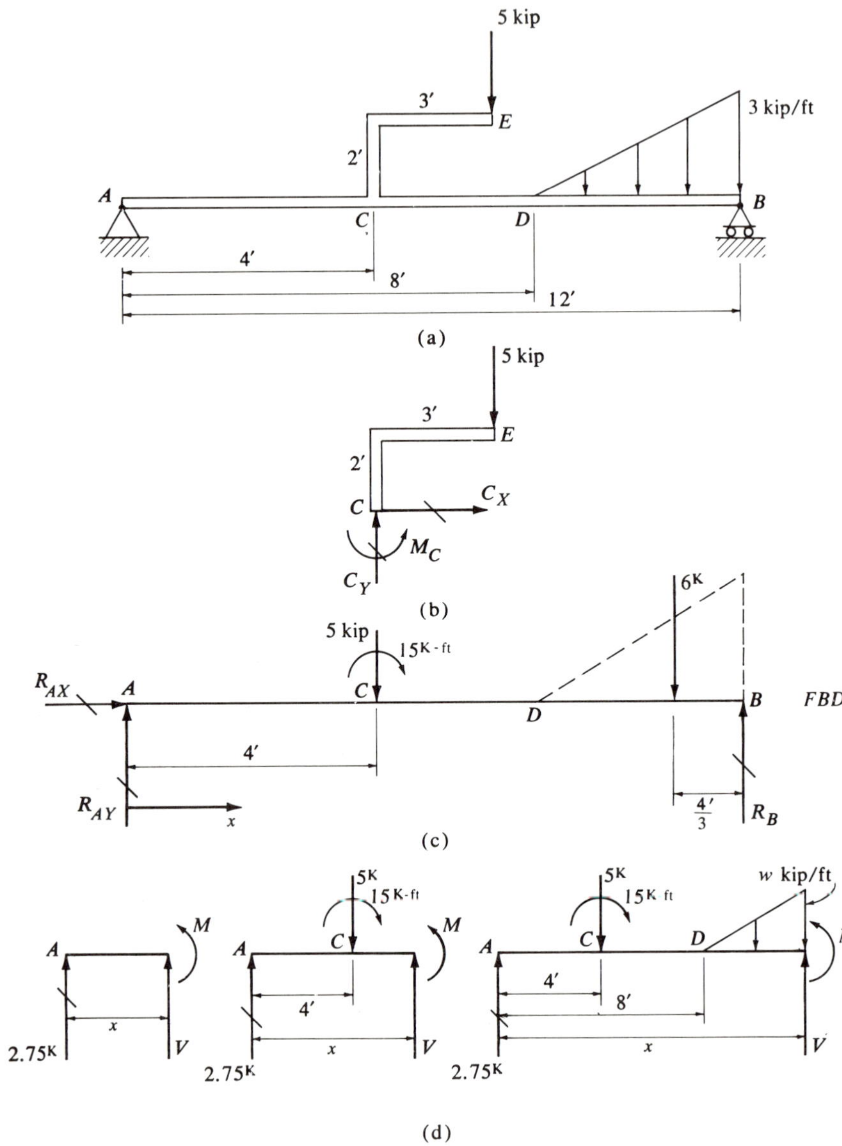

Figure 3.4

Vertical Bracket (Figure 3.4b)

$$\Sigma F_Y = 0 \uparrow +$$
$$-5 + C_Y = 0$$
$$\underline{C_Y = 5 \text{ kips} \uparrow}$$

$$\Sigma F_X = 0 \rightarrow +$$
$$\underline{C_X = 0}$$

$$\Sigma M_C = 0 \quad \circlearrowright +$$

$$5(3) - M_C = 0$$
$$\underline{M_C = 15 \text{ kip-ft} \circlearrowleft}$$

Beam AB (Figure 3.4c)

$$\Sigma M_A = 0 \quad \circlearrowright +$$

$$5(4) + 15 + (6)(12 - \tfrac{4}{3}) - 12\, R_B = 0$$
$$20 + 15 + 64 - 12\, R_B = 0$$
$$R_B = +8.25 \text{ kips}$$
$$\underline{R_B = 8.25 \text{ kips} \uparrow}$$

$$\Sigma F_Y = 0 \uparrow +$$
$$R_{AY} - 5 - 6 + 8.25 = 0$$
$$R_{AY} = +2.75 \text{ kips}$$
$$\underline{R_{AY} = 2.75 \text{ kips} \uparrow}$$

$$\Sigma F_X = 0 \rightarrow +$$
$$\underline{R_{AX} = 0}$$

2. Because of changes in the applied loading along the span, separate shear and moment equations must be written for segments *AC*, *CD*, and *DB* (see Figure 3.4a). The origin was selected at *A* and arbitrary cuts were made within each segment as shown in Figure 3.4d. The general equations for shear and moment are obtained as follows (see Figure 3.4d):

Segment AC

$$\Sigma F_Y = \uparrow +$$
$$2.75 + V = 0$$
$$V = -2.75 \text{ kips}$$
$$\therefore \underline{V = 2.75 \text{ kips} \uparrow}$$

(Note that the shear force is constant in this segment)

$$\Sigma M_X = 0 \quad \circlearrowright +$$

$$2.75x - M = 0$$
$$\underline{M = 2.75x \text{ kip-ft}}$$

Segment CD

$$\Sigma F_Y = 0 \uparrow +$$
$$2.75 - 5 + V = 0$$
$$V = +2.25 \text{ kips}$$
$$\underline{V = 2.25 \text{ kips} \uparrow}$$ (Again, the shear force is constant in this segment)

$$\Sigma M_X = 0 \; \circlearrowright +$$

$$2.75x - 5(x - 4) + 15 - M = 0$$
$$2.75x - 5x + 45 - M = 0$$
$$\underline{M = (2.25x - 45) \text{ kip-ft}}$$

Segment DB

Since the cut was taken through the distributed load, the loading ordinate at the cut must be determined (see Figure 3.4d). Using similar triangles, the following proportion is obtained:

$$\frac{3 \text{ kips/ft}}{4 \text{ ft}} = \frac{w \text{ kips/ft}}{(x - 8) \text{ ft}}$$

$$w = [(\tfrac{3}{4})x - 6] \text{ lb/ft}$$

$$\Sigma F_Y = 0 \uparrow +$$
$$2.75 - 5 - \tfrac{1}{2}(x - 8)[(\tfrac{3}{4})x - 6] + V = 0$$
$$\underline{V = (\tfrac{3}{8})x^2 - 6x + 26.25}$$

$$\Sigma M_X = 0 \; \circlearrowright +$$

$$2.75x - 5(x - 4) + 15 - \tfrac{1}{2}(x - 8)[(\tfrac{3}{4})x - 6](\tfrac{1}{3})(x - 8) - M = 0$$
$$\underline{M = 99 - 26.25x + 3x^2 - (\tfrac{1}{8})x^3}$$

The general expressions for shear and moment may be summarized as follows:

Interval	Shear	Moment
$0 \leqslant x \leqslant 4'$	$V = 2.75^K(\uparrow)$	$M = 2.75x$
$4' \leqslant x \leqslant 8'$	$V = 2.25^K(\uparrow)$	$M = (2.25x - 45)$
$8' \leqslant x \leqslant 12'$	$V = (\tfrac{3}{8})x^2 - 6x + 26.25$	$M = 99 - 26.25x + 3x^2 - (\tfrac{1}{8})x^3$

These equations may be readily checked. For example, at point *B* (see Figure 3.4a), which is a support point and the end of the 12-ft span, we know that the moment is zero (see the FBD shown in Figure 3.4c to confirm that at $x = 12$ ft, point *B*, the only acting force is R_B). We may substitute into the proper moment equation as a check as follows:

$$M = 99 - 26.25x + 3x^2 - \tfrac{1}{8}x^3$$
$$@\ x = 12';\ M = 99 - 26.25(12) + 3(12)^2 - \tfrac{1}{8}(12)^3$$
$$M = 0 \quad \text{Check}$$

3.4 Shear and Moment Diagrams

If the general equations for shear and moment have been determined for an entire span of a beam, then the magnitudes at any point may be obtained by substitution of the coordinate of the desired point into the general equations. If enough points along the span are selected, plots of shear vs. distance along span, and moment vs. distance along span may be obtained. These plots, known as **shear and moment diagrams,** would provide the analyst with maximum magnitudes and their location, a visual description of the variance along the span, and the direction, or sign, of the shear and moment at any point. In many instances, several points would have to be selected in order to provide enough points through which a smooth curve could be drawn; hence, using the general equations for shear and moment as a means of plotting shear and moment diagrams could become a laborious task. Better methods have, therefore, been devised. Consider, for example, the simply supported beam shown in Figure 3.5a with an origin, O, selected at the left support. Next, we must cut the beam at two distances from the selected origin, X and $X + \Delta X$, where, again, ΔX is an infinitesimal distance to the right of the cut at X (see Figure 3.5a). A proper FBD of the small portion of the beam of length, ΔX, is shown in Figure 3.5b. Figure 3.5b shows a shear, V_X, and a moment, M_X, acting at the

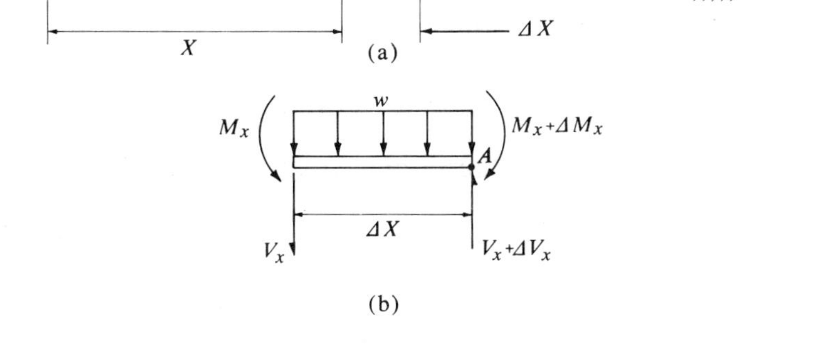

Figure 3.5

cut which is some distance X from the origin and a shear force, $V_X + \Delta V_X$, and a moment, $M_X + \Delta M_X$, acting at the cut which is at some distance $X + \Delta X$ from the origin. Recall that earlier it was shown that both the shear and moment can vary along the span and, hence, if the shear is equal to V_X in magnitude at some distance X from the origin, it is quite possible for the shear, V_X, to change by some amount ΔV_X by moving along the span a small distance ΔX. The same reasoning can be applied to the moment. (There are no axial forces acting on the beam portion shown in Figure 3.5b since there are no horizontally applied loads acting on the entire beam). It is possible for ΔM_X and ΔV_X both to be zero, both to possess some magnitude, or for either to be zero depending upon how the shear and moment vary along that portion of the span. Applying the equations of equilibrium to the small beam shown in Figure 3.5b we have:

$$\Sigma F_Y = 0 \uparrow +$$
$$-V_X - w\Delta X + (V_X + \Delta V_X) = 0$$

where $w\Delta X$ is the area under the loading curve acting on the beam portion.

$$\Delta V_X = w\Delta X \tag{3-1}$$

Equation (3–1) actually says that the change in shear between cuts at distances of X and $(X + \Delta X)$ from the origin is equal to the load between the two cuts since ΔV_X is the change in shear between the two cuts and $w\Delta X$ is the acting load between the two cuts. The reader may say, however, that the beam selected in Figure 3.5 is not a general case. Suppose that the small section of beam shown in Figure 3.5b has no load acting on it at all. Then w is equal to zero and, by Eq. (3–1), ΔV_X is equal to zero, which simply implies that the shear does not change between the two selected cuts. Equation (3–1) is, therefore, a general equation. Although we have been dealing with a beam of infinitesimal length (Figure 3.5b), we can extend to the case of beams of more practical length, by stating that **the change in shear between points A and B along the span of a beam is equal to the load between those same two points.** Equation (3–1) may be written as follows:

$$\frac{\Delta V_X}{\Delta X} = w \tag{3-2}$$

Equation (3–2) can best be explained by referring to Figure 3.6, which is a plot of shear vs distance along the span. The quantity $\Delta V_X/\Delta X$ is the slope of the shear diagram at point X; therefore, Eq. (3–1) actually states that **the slope of the shear diagram at any point along the span is equal to the intensity of the loading at that point (w).**

Consider again the FBD shown in Figure 3.5b of the small portion of beam of length ΔX. Taking the sum of the moments about point A the following is obtained:

$$\Sigma M_A = 0 \quad \circlearrowright +$$

$$-M_A - V_X\Delta X - w\Delta X\left(\frac{\Delta X}{2}\right) + (M_X + \Delta M_X) = 0$$

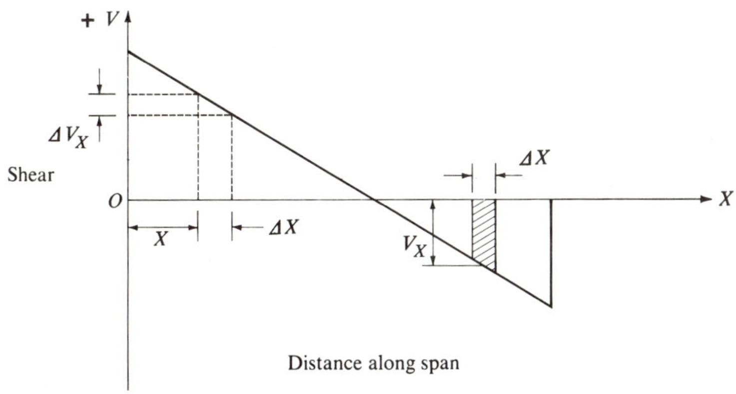

Figure 3.6

If ΔX is infinitesimal, then $(\Delta X)^2/2$ is so small that it may be disregarded. Hence, we have:

$$\Delta M_X = V_X\Delta X \tag{3–3}$$

Equation (3–3) may be stated as follows: the change in moment, ΔM_X, between cuts at distances of X and $(X + \Delta X)$ from the origin is equal to the area under shear diagram, $V_X\Delta X$, between the two cuts. Figure 3.6 shows a shaded strip whose height (av) is V_X and whose base has a width of ΔX. The area of the strip is therefore $V_X\Delta X$. Although the equation was derived for a beam of infinitesimal length, we may again extend to the case for beams of practical length by stating that **the change in moment between two points** A **and** B **along the span of a beam is equal to the area under the shear diagram between the same two points.** Equation (3–3) may also be rewritten as:

$$\frac{\Delta M}{\Delta X} = V_X \tag{3–4}$$

Equation (3–4) simply states that **the slope of the moment diagram, $\Delta M/\Delta X$, at any point is equal to the shear at that point.** The word slope at a point is defined as the slope of the tangent to the curve at that point. Before proceeding any further, a sign convention must be established for shear and moment diagrams. The establishment of a sign convention is a matter of convenience and its use gives a degree of reliability to the interpretation of the shear and moment diagrams. Figure 3.7 shows the sign convention that has been most com-

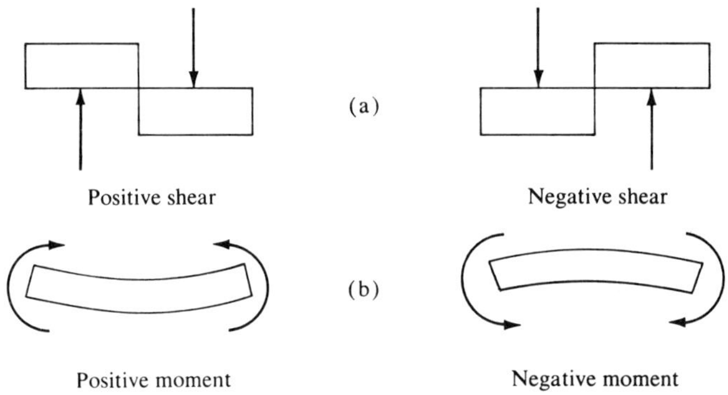

Figure 3.7

monly adopted in structural engineering. In Figure 3.7a, the pair of shear forces is classified as positive if it tends to push the left side of a member upward, and is negative if it pushes the left side of the member downward. In Figure 3.7b, the moment is said to be positive if the member tends to bend concave upward causing tension (stretching) in the bottom fibers of the member, and is negative if it causes the member to bend concave downward. It should be noted that this sign convention assumes that the member is horizontal. Figure 3.8 shows the sign convention for slope that will be used throughout this text. These sign conventions, along with the four equations derived earlier in this section, will enable the analyst to draw shear and moment diagrams with a minimum of effort.

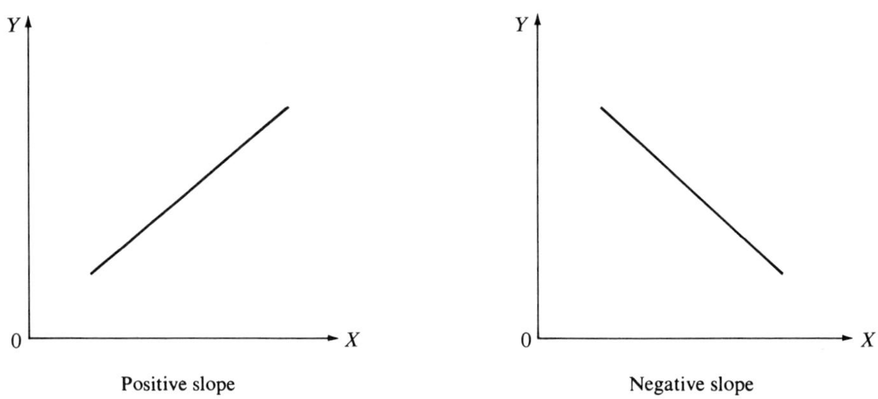

Figure 3.8

3.5 Illustrative Examples

The following examples are intended for prolonged study. The first few examples are presented with detailed explanations. The student should fully understand these initial examples before proceeding to those presented with less detailed explanation. It should be pointed out that these shear and moment diagrams (frequently referred to as V and M diagrams) are sketches with emphasis on shape and magnitudes. Very detailed plotting techniques are not necessary.

Example 3.2

Draw the shear and moment diagrams for the beam shown in Figure 3.9a.

Solution:

1. The first step is to draw a FBD as shown in Figure 3.9b and then solve for the reactions as follows:

$$\Sigma M_A = 0 \quad (+\curvearrowright)$$

$$100(5) - 10R_B = 0$$

$$\underline{R_B = 50 \text{ lb} \uparrow}$$

$$\Sigma F_Y = 0 \uparrow +$$

$$R_{AY} - 100 + 50 = 0$$

$$\underline{R_{AY} = 50 \text{ lb} \uparrow}$$

2. It is conventional to plot V and M diagrams from left to right. Starting with the V diagram (Figure 3.9c) we may proceed as follows:

   a. Starting at point A, imagine a cut a small distance to the right of the hinge support. By the previously adopted sign convention, the 50-lb reaction would tend to move that portion of the beam to the left of the cut upward and a shear force would be required at the cut to "hold" the cut section down. This, by definition, is positive shear. The magnitude of the shear at A is, therefore, +50 lb and is plotted as shown in Figure 3.9c.
   b. The change in V between any two points is equal to the load between those two points. Since there is no load between A and C, there is no change in shear between A and C. If there exists a shear of +50 lb at A and there is no change in shear up to point C, then the shear must be a constant +50 lb in that interval. A horizontal line at $V = +50$ lb is then drawn from A to C.
   c. It was also shown that the slope of the V diagram at any point is equal to the intensity of the load at that point. The intensity of the load at any point between A and C is zero; therefore, the slope of the V diagram at any point between A and C must also be zero. A curve that has a slope of zero at every point is a horizontal line as shown by the V diagram between A and C (Figure 3.9c).
   d. Imagine, next, a cut an infinitesimal distance to the right of the 100-lb force at C. The net effect of all the vertical forces acting on the member to the left of the imaginary cut is a downward force of 50 lb. A shear force of 50 lb is required, therefore, at the cut for equilibrium and its

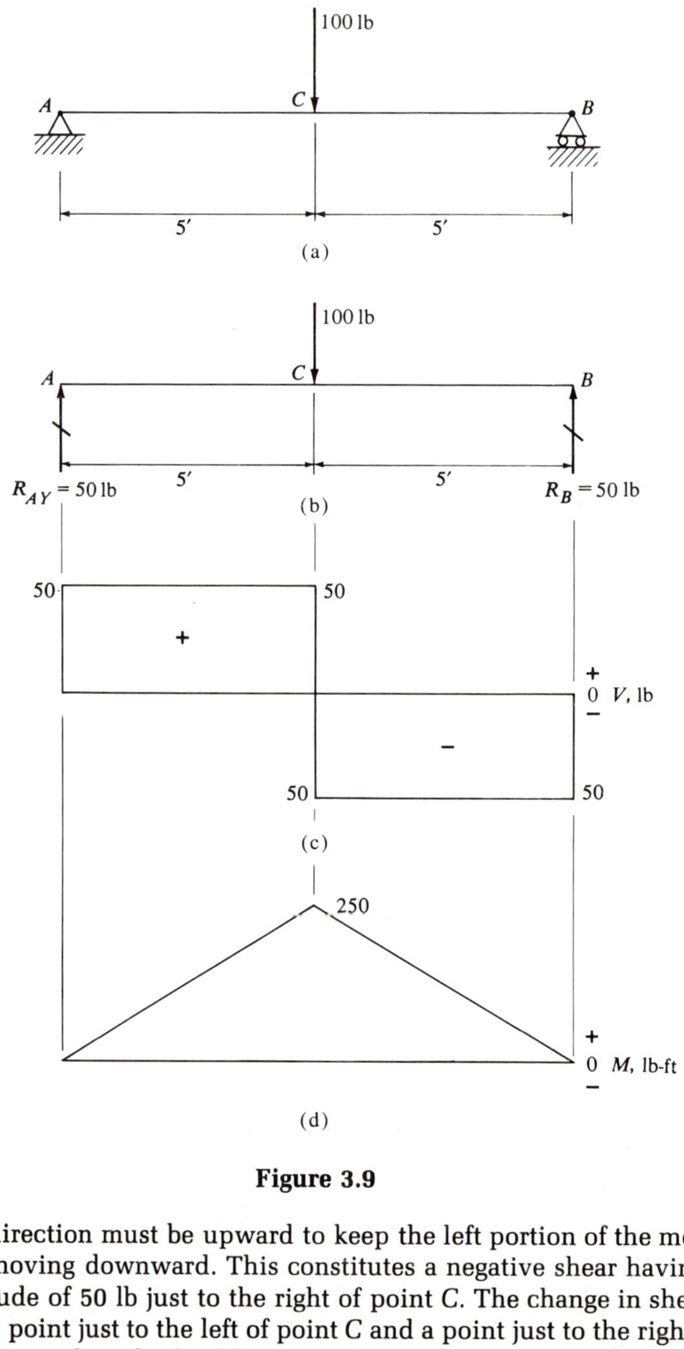

Figure 3.9

direction must be upward to keep the left portion of the member from moving downward. This constitutes a negative shear having a magnitude of 50 lb just to the right of point C. The change in shear between a point just to the left of point C and a point just to the right of point C is equal to the load between those two points, which in this case is 100 lb. Note that the shear ordinate is +50 lb just to the left of point C and is −50 lb just to the right of C, which indicates a change equal in magnitude to 100 lb. This change occurs at a point, i.e., it is an abrupt change. Therefore, at the point of application of a concentrated load the shear changes abruptly with a magnitude equal to the magnitude of the concentrated load. If the load is downward, the change is negative. Checking, we see that the change in this case is −100 lb at point *C* and the 100-lb concentrated load was, indeed, downward.

e. There is no load between *C* and *B* and, therefore, there is no change in the shear between *C* and *B*. A horizontal line is drawn through $V = -50$ lb to point *B* from point *C*. To verify this observe that the load intensity at any point between *C* and *B* is zero, giving rise to a shear curve having zero slope at any point, i.e., at horizontal line.

f. The reaction of 50 lb at point *B* is a concentrated load directed upward, which is positive by the convention mentioned in step d. An abrupt change in shear of +50 lb is, therefore, encountered at *B*, which gives a net shear force of zero as we pass point *B*. The reaction at *B* must close the shear diagram to zero since if we move a small distance to the right of *B* the beam is nonexistent and it is impossible to have a shear force.

3. Continuing with the moment diagram, we may proceed as follows:

a. Beginning at point *A*, we recall that at a hinge support there exists the possibility of having a vertical and horizontal reaction and that a hinge cannot transmit a couple. The moment, therefore, at *A* is zero. The change in moment between any two points along the beam's span is equal to the area under the shear diagram between those same two points. The change in moment between *A* and *C* is, therefore, equal to the area under the shear diagram between *A* and *C* (Figure 3.9c). The area between *A* and *C* is rectangular in shape, having a height of +50 lb and a length of 5 ft, thereby yielding an area of +250 lb-ft. Starting with a moment of zero at *A* and adding a change of +250 lb-ft from *A* to *C* yields a moment of +250 lb-ft at C. The positive sign preceding the moment indicates that the member is tending to bend concave upward at C. The sign convention presented in Figure 3.7 was followed for the shear diagram and, therefore, the sign of the moment obtained from the area under the shear diagram is consistent with the sign convention in Figure 3.7.

b. The shape of the moment diagram between *A* and *C* must now be determined. The slope of the moment diagram at any point is equal to the shear at that point. The shear between *A* and *C* (Figure 3.9c) is positive and constant, i.e., the shear is always +50 lb between *A* and *C*. The slope of the moment diagram between *A* and *C* is then positive and constant. This indicates a straight line directed up and to the right by the sign convention adopted in Figure 3.8. We are not concerned with the magnitude of the slope, but we are concerned with the sign of the slope and whether the slope is constant or variable. A straight line is drawn from zero at point *A* (Figure 3.9d) to +250 at point C.

c. The change in moment between *C* and *B* is again the area under the shear diagram between *C* and *B*. This change is −50 lb × 5 ft, or −250 lb-ft. A +250 lb-ft at C plus a change of −250 lb-ft from C to B gives a net of zero at point B.

d. The shear between C and B is negative and constant, and hence the slope of the moment diagram between C and B is constant and negative, indicating a straight line directed down and to the right. A line from +250 at C is drawn to zero at B as shown in Figure 3.9d.

4. Before leaving this example, the following characteristics of the V and M diagrams shown in Figure 3.9 should be noted:

   a. When $V = C$, where C is a constant, for a given interval, then $M = f(x)$, i.e., the moment is a linear function of x which plots as a straight line.
   b. For a concentrated load acting on a simply supported beam, the M diagram is always triangular with the maximum moment occurring under the load and the V diagram always consists of two rectangular shapes.
   c. For symmetrical loading, the V diagram is always antisymmetrical and the M diagram is symmetrical.

Example 3.3

Draw the V and M diagrams for the beam shown in Figure 3.10a.

Solution:

1. The FBD, along with the proper reactions, is shown in Figure 3.10b. Beginning with the V diagram, imagine a cut a small distance to the right of the reaction at A. The 400-lb reaction would tend to move the member to the left of our cut upward, and hence a 400-lb shear force would be required at the cut to "hold" the cut portion down. This is positive shear having a magnitude of 400 lb. A +400-lb shear is, therefore, plotted at A as shown in Figure 3.10c. The only vertical force between A and B is the uniformly distributed load, and hence it is the only contributor to the change in shear between A and B. The change in shear between A and B is equal to the load between A and B, which in this case is the area under the loading curve between A and B. The change is $(-100 \text{ lb/ft}) \times (8 \text{ ft}) = -800$ lb. Recall that downward forces are negative by the adopted sign convention. Adding the −800-lb change in shear between A and B to the initial 400 lb at A gives a net shear of −400 lb at B.
2. We know that the shear ordinates at A and B are +400 lb and −400 lb respectively. It must be determined now how the shear varies between these two points. The slope of the V diagram at any point is equal to the intensity of the load at that point. The intensity of the load at any point between A and B is negative and constant, i.e., the load intensity is always −100 lb/ft. This indicates a constant negative slope at every point on the V diagram, or a straight line directed down and to the right from A to B. This is consistent with the previously determined magnitudes at A and B. The shear diagram is plotted as shown in Figure 3.10c. The upward 400-lb reaction at B closes the shear diagram to zero.

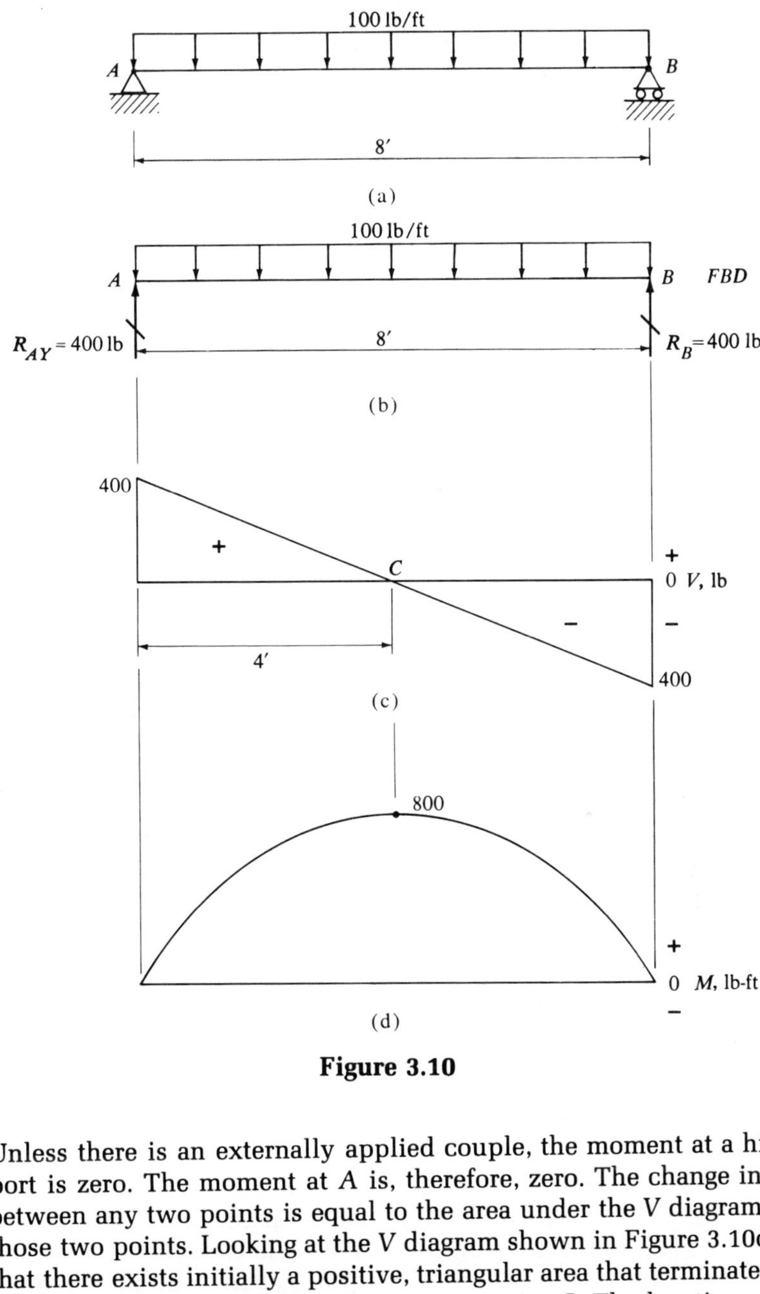

Figure 3.10

3. Unless there is an externally applied couple, the moment at a hinge support is zero. The moment at A is, therefore, zero. The change in moment between any two points is equal to the area under the V diagram between those two points. Looking at the V diagram shown in Figure 3.10c we note that there exists initially a positive, triangular area that terminates at some point of zero shear, which we denote as point C. The location of point C must be determined in order to calculate the area under the V diagram between A and C. To do this, we observe that the shear varies linearly between A and C, and, hence, we ask the question, **over how much of the span must we place a 100 lb-ft downward load to go from +400 lb at** A **to 0 lb at** C. The distance is (400 lb) ÷ (100 lb/ft) = 4 ft, or C is 4 ft from

point A. The location of point C may also be determined from similar triangles. For example, imagine a horizontal line drawn from the −400-lb ordinate (Figure 3.10c) at B to a vertical line drawn through the +400 lb ordinate at A. Using similar triangles the following relation is obtained:

$$\frac{400+400}{\overline{AB}} = \frac{400}{\overline{AC}}$$

$$\frac{800}{8} = \frac{400}{\overline{AC}}$$

$$\overline{AC} = 4 \text{ ft}$$

4. The moment at C is equal to the moment at A plus the change in moment from A to C, or $M_C = 0 + (400)(4)(\frac{1}{2}) = 800$ lb-ft. The shear ordinates between A and C are positive but decreasingly positive to zero at C. The slope of the moment starts off positively at A, but the slope gradually reduces to zero, i.e., a horizontal line, at C. This variance in slope gives rise to the curve shown in Figure 3.10d.
5. The change in moment from C to B is, again, equal to the area under the V diagram between C and B. This change is $(\frac{1}{2})(4)(-400) = -800$ lb-ft. Adding the change of −800 lb-ft from C to B to the +800 lb-ft at C gives a net of 0 lb-ft at B. This is consistent with the support at B since there is no moment introduced at a roller support.
6. The shear ordinates between C and B are negative and increasingly negative beginning with zero at C. The slope of the moment curve between C and B is, therefore, zero at C and increasingly negative up to point B. This is represented again by the curve shown in Figure 3.10d.
7. Characteristics of the shear and moment diagrams shown in Figure 3.10 are as follows:
   a. The following relations exist as evidenced by the plots:

   $w = C$, C is a constant
   $V = f(x)$* where x is distance along the beam
   $M = f(x^2)$

   b. For a uniformly distributed load on a simply supported beam the V diagram is antisymmetrical and the moment diagram is symmetrical.
   c. At points of zero shear, such as at point C in Figure 3.10c, the moment is a maximum or a minimum for that interval. Zero shear at a point indicates zero slope at that point on the M diagram and this gives rise to a horizontal tangent to the moment curve. In this case (Figure 3.10) the horizontal tangent, i.e., zero slope, at C indicates the peak of the concave downward moment curve.

---

* This is read, **Shear is a function of** x.

Example 3.4

Draw the V and M diagrams for the beam shown in Figure 3.11.

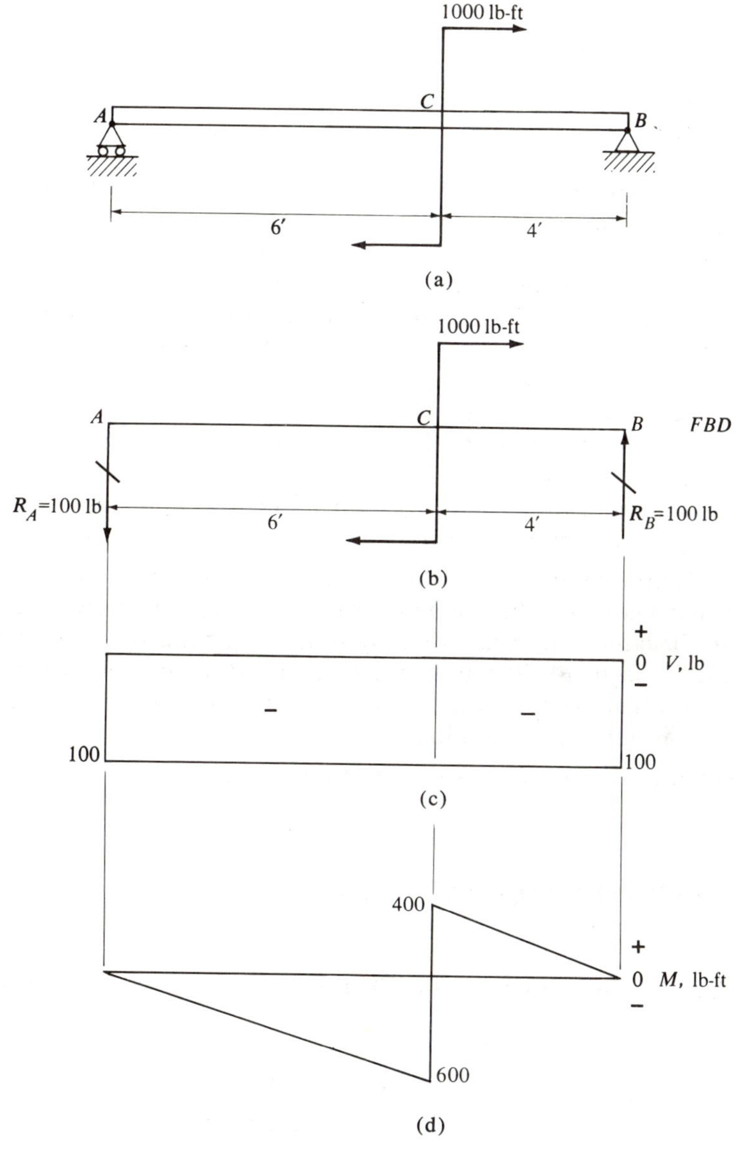

Figure 3.11

Solution:

1. The reactions shown on the FBD in Figure 3.11b were obtained as follows:

$$\Sigma M_A = 0 \quad (\curvearrowright +)$$

$$1000 - 10R_{BY} = 0$$
$$\underline{R_{BY} = 100 \text{ lb} \uparrow}$$

$$\Sigma F_Y = 0 \uparrow +$$
$$+100 - R_A = 0$$
$$\underline{R_A = 100 \text{ lb} \downarrow}$$

2. The construction of the V diagram shown in Figure 3.11c is straightforward. The moment diagram, however, needs some special explanation. Imagine a cut anywhere along the span between A and C (see Figure 3.11b). Considering that portion to the left of the cut, the moment anywhere along this segment is a function of the 100-lb reaction at A and the corresponding distance from A. This is shown by the M diagram in Figure 3.11d, which shows the moment between A and C as a linear variation of the reaction at A and the corresponding distance from A. If, however, an imaginary cut is taken just to the right of point C, again considering that portion to the left of the cut, we note that the moment is increased by the magnitude of the applied couple at C, i.e., an applied couple introduces an abrupt change in the moment at the point of application of the couple.
3. To determine the sign of the change in moment introduced by the couple, imagine grasping the beam just to the left of point C. Now let the couple act, i.e., imagine it to rotate thereby producing bending in segment CB. Note that the segment CB bends in the positive direction, or concave up. Hence, the applied couple introduces a positive change in the moment at point C of magnitude of 1000 lb-ft.
4. The moment diagram is then completed in the standard manner.

Example 3.5

Draw the V and M diagrams for the beam shown in Figure 3.12.

Solution:

1. The reactions shown on the FBD in Figure 3.12 were obtained as follows:

$$\Sigma F_Y = 0 \uparrow +$$
$$-(2)6 - 10 + R_A = 0$$
$$\underline{R_A = 22.0 \text{ kips} \uparrow}$$

$$\Sigma M_A = 0 \quad (\curvearrowright +)$$

$$+2(6)(3) + 10(10) - M_A = 0$$
$$\underline{M_A = 136 \text{ kip-ft} \circlearrowleft}$$

2. The construction of the V diagram shown in Figure 3.12c is straightforward. The moment diagram, on the other hand, deviates slightly from

those drawn previously. At the fixed support, there exists not only a vertical reaction but also a resisting moment of magnitude of 136 kip-ft. This moment introduced by the fixed support on the beam *AC* acts as, and is treated as, an applied couple. There is an abrupt change, therefore, in the moment at *A* caused by the 136 kip-ft moment. Letting the moment act on the beam would tend to introduce negative bending in the member. The change in moment at *A* caused by the resisting couple is negative and is plotted as shown in Figure 3.12d.

3. The completion of the moment diagram is straightforward (see Figure 3.12d).

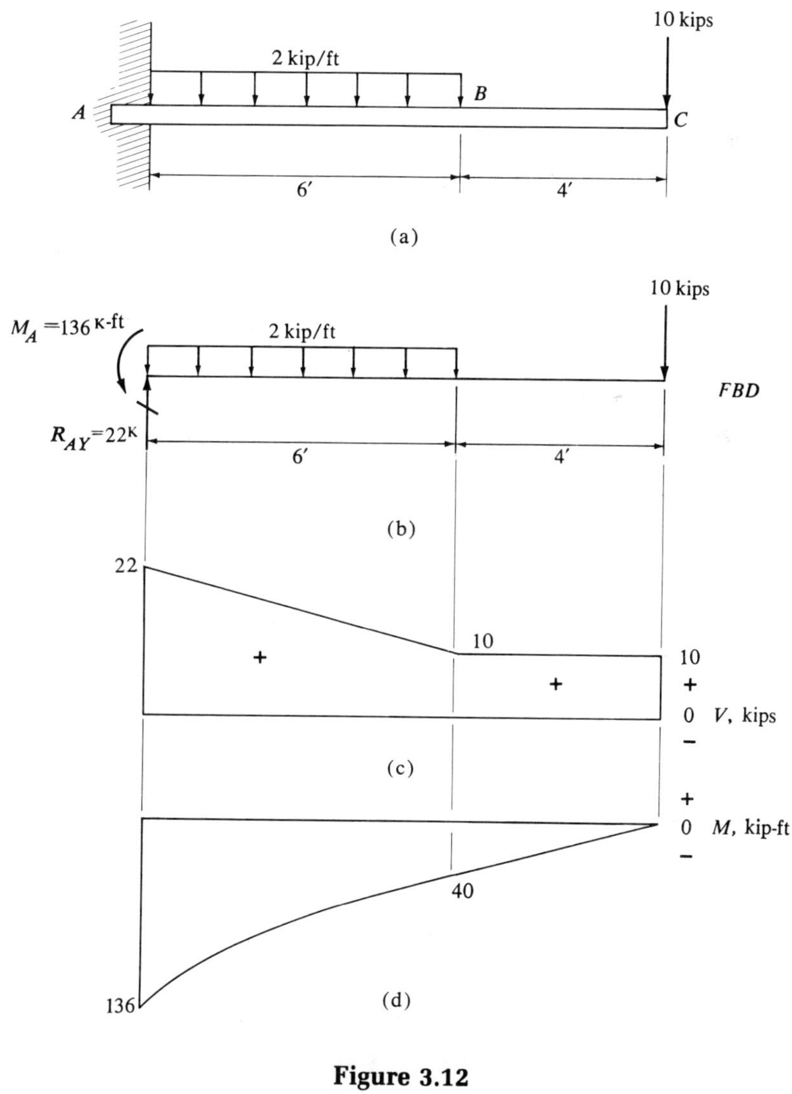

Figure 3.12

Example 3.6

Draw the V and M diagrams for the beam shown in Figure 3.13.

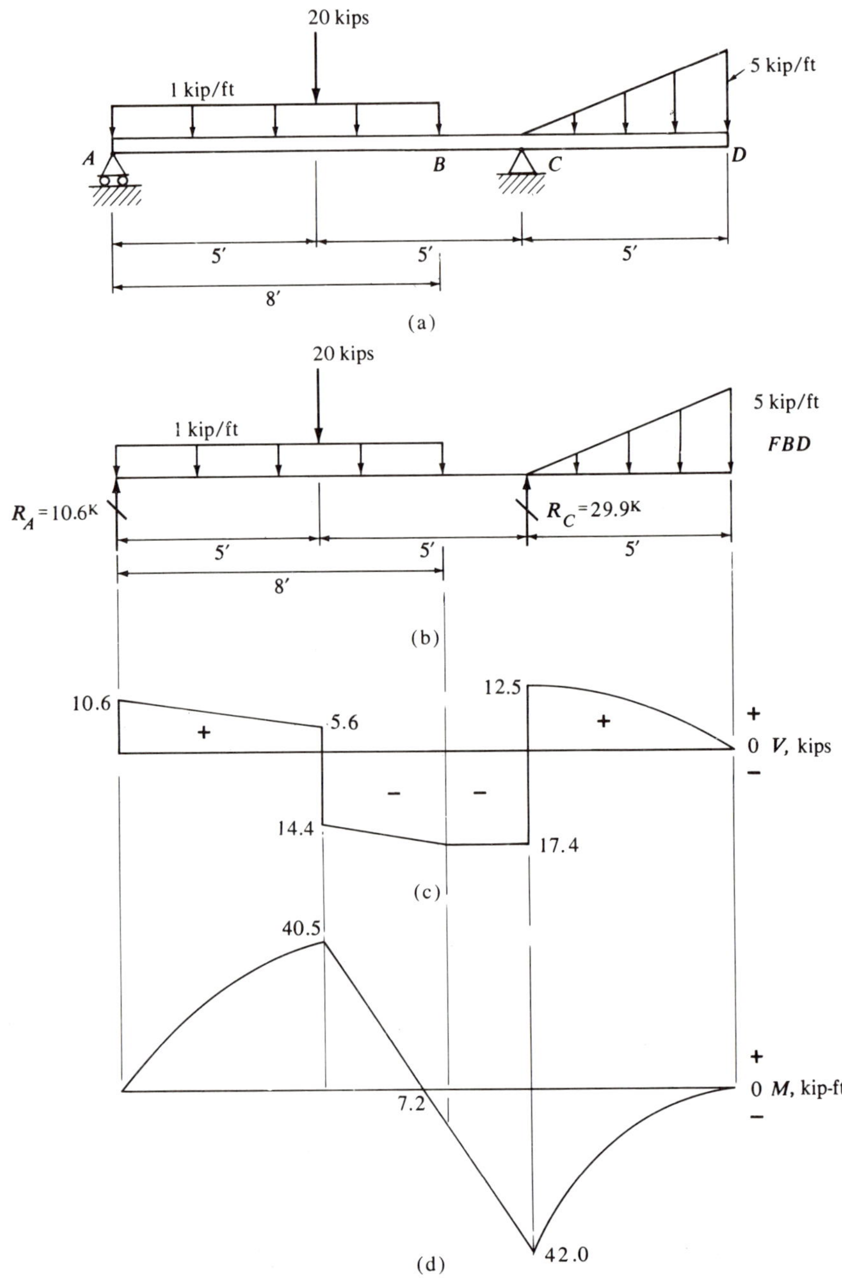

Figure 3.13

Solution:

1. The reactions shown on the FBD in Figure 3.13 were obtained as follows:

$$\Sigma M_A = 0 \; (+\curvearrowright)$$

$$(1)(8)(4) + 20(5) - R_C(10) + (\tfrac{1}{2})(5)(5)(15 - \tfrac{5}{3}) = 0$$

$$\underline{R_C = 29.9 \text{ kips} \uparrow}$$

$$\Sigma M_B = 0 \; (+\curvearrowright)$$

$$R_A(10) - (1)(8)(2+4) - (20)(5) + \tfrac{1}{2}(5)(5)(\tfrac{2}{3})5 = 0$$

$$\underline{R_A = 10.6 \text{ kips} \uparrow}$$

Check:

$$\Sigma F_Y = 0 \uparrow +$$

$$10.6 - (1)(8) - 20 + 29.9 - \tfrac{1}{2}(5)(5) = 0$$

$$10.6 - 28.0 + 29.9 - 12.5 = 0$$

$$40.5 - 40.5 = 0$$

$$0 = 0 \quad \text{Check}$$

2. The shear diagram, Figure 3.13c, needs no added explanation up to point *C*. The shear diagram for segment *CD* obeys all the rules stated previously for the construction of a shear diagram. The slope of the shear diagram at any point between *C* and *D* is given by the loading ordinates between *C* and *D*. Note that the loading ordinates acting on segment *CD* are negative (downward) with magnitudes initiating with zero at *C* and increasing linearly along the segment. The slope, therefore, of the shear diagram is zero at *C* and is negative and increasingly negative as we proceed from *C* to *D* as shown in Figure 3.13c. Since the loading ordinates vary linearly along the segment, i.e., $w = f(x)$ where x is distance along segment *CD*, the shear varies parabolically along the same segment, i.e., $V = f(x^2)$ where x is distance measured along segment *CD*.
3. The moment diagram up to point *C* is straightforward (see Figure 3.13d). The change in moment between *C* and *D* is equal to the area under the shear diagram between *C* and *D* which is, in this case, a semiparabolic area. Such an area is shown in Figure 3.14. The area is given by $A = (2/3)$

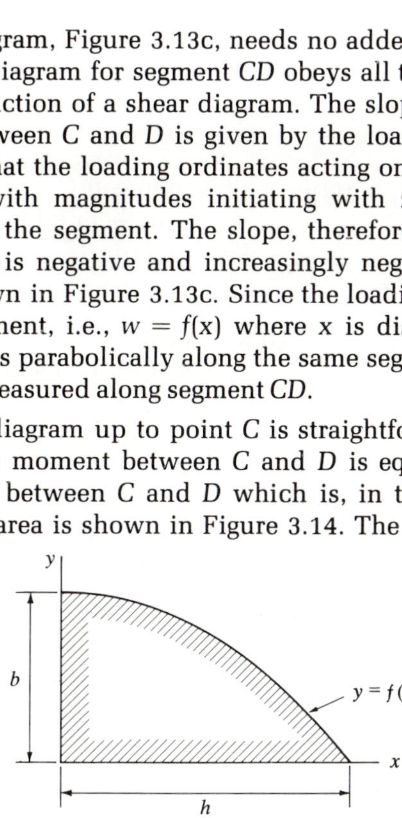

Figure 3.14

bh, or for the segment CD shown in Figure 3.13c, $A = (2/3)(12.5)(5) \simeq 42$ kip-ft. The shear ordinates between C and D are always positive with magnitudes decreasing along the segment. The slope of the moment diagram between C and D is positive at C but decreasingly positive as we proceed from C to D, eventually becoming zero at point D as indicated by a zero shear ordinate at point D. The complete M diagram is shown in Figure 3.13d. The moment diagram varies cubically along segment CD, i.e., $M = f(x^3)$.

4. In drawing V and M diagrams for beams subjected to loads which vary along the span, unconventional areas often appear under the V diagram. To calculate such areas, reference must be made to mathematical handbooks which tabulate properties of various areas. One such area is that shown in Figure 3.14. It appears so often that the student should memorize the equation for its area.

Example 3.7

Draw the V and M diagrams for the beam shown in Figure 3.15.

Solution:

1. The reactions shown on the FBD in Figure 3.15 were obtained as follows:

$$\Sigma M_E = 0 \; (\curvearrowright +)$$

$$R_A(10) - 4(4)(2+4) - 10(2) = 0$$

$$\underline{R_A = 11.6 \text{ kips} \uparrow}$$

$$\Sigma M_A = 0 \; (\curvearrowright +)$$

$$4(4)(2+2) + 10(8) - 10R_E = 0$$

$$\underline{R_E = 14.4 \text{ kips} \uparrow}$$

Check:

$$\Sigma F_Y = 0 \uparrow +$$

$$11.6 - 4(4) - 10 + 14.4 = 0$$

$$26 - 26 = 0 \quad \text{Check}$$

2. The shear diagram shown in Figure 3.15c was drawn in the conventional manner. The shear diagram crosses the 0-base line at an unknown location between points B and C. Calling the distance from point B to the point where the shear diagram crosses the 0-base line x, we may determine x using similar triangles as follows (see Figure 3.15):

$$\frac{(11.6 + 4.4)}{4} = \frac{11.6}{x}$$

$$x = 2.9 \text{ ft}$$

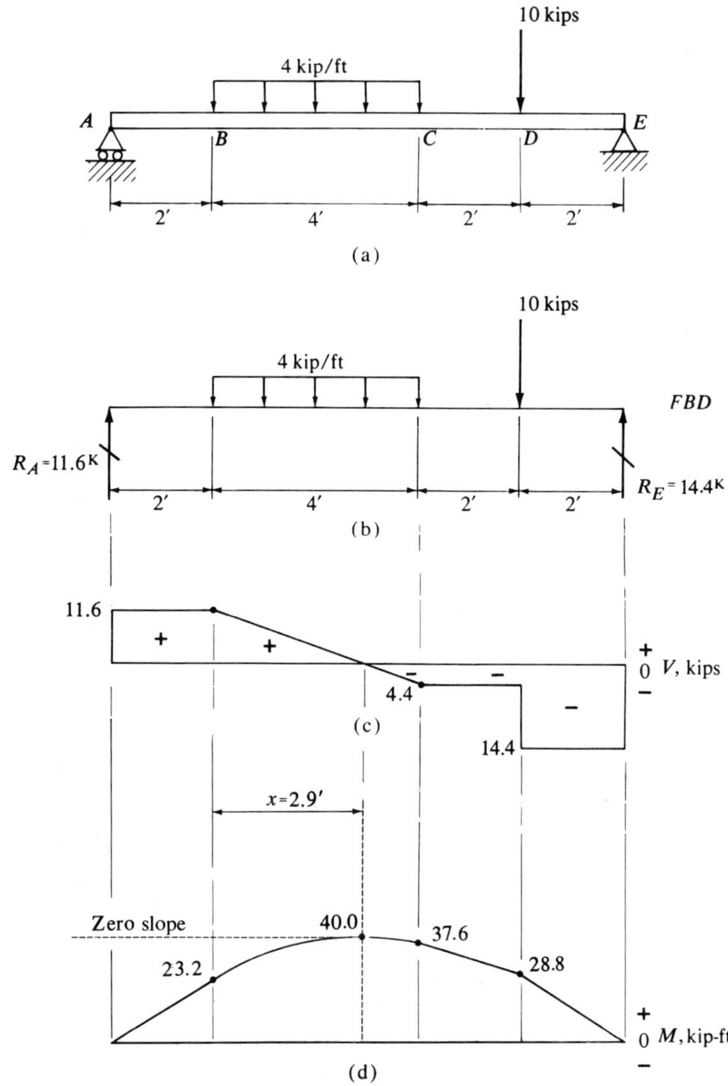

Figure 3.15

Knowing the distance x, we were able to calculate all the corresponding areas under the shear diagram.

3. Knowing where the shear diagram crosses the 0-base line is extremely important. At such a point, the shear is obviously zero. If the shear ordinate at that point is zero, then the slope of the moment diagram is also zero at that point. A zero slope at a point indicates a horizontal slope, or tangent, at that point. Horizontal tangents only appear at maximum or minimum points on a curve. (Maximum used here will indicate maximum positive moment and minimum will indicate maximum negative moment.) Referring to Figure 3.15d the maximum moment is 40.0 kip-ft and it occurs

2.9 ft from point B. Observe that the slope of the moment diagram is zero at the maximum point and that the shear ordinate is zero.

4. Points of maximum or minimum moment along a beam's span are, therefore, extremely important because these are the limits which eventually control the ultimate design.

Problems

3–1. Find the shear and the moment at sections just to the left and just to the right of the 1000-lb load.

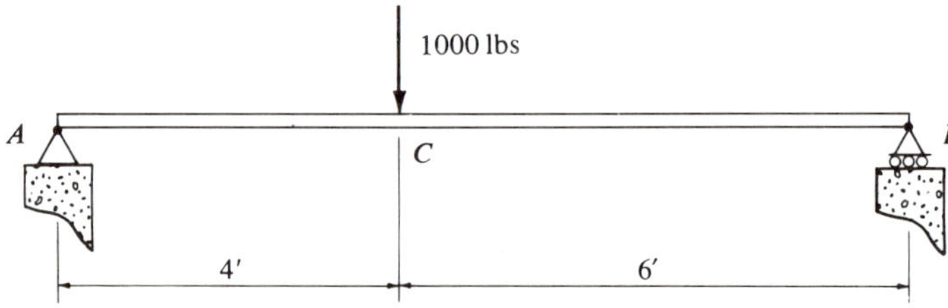

Figure P 3–1

3–2. Calculate the shear and the moment at point B for the cantilever beam shown.

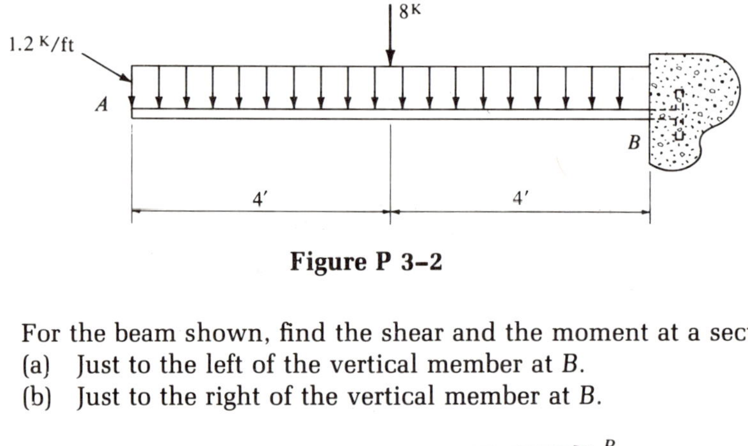

Figure P 3–2

3–3. For the beam shown, find the shear and the moment at a section:
(a) Just to the left of the vertical member at B.
(b) Just to the right of the vertical member at B.

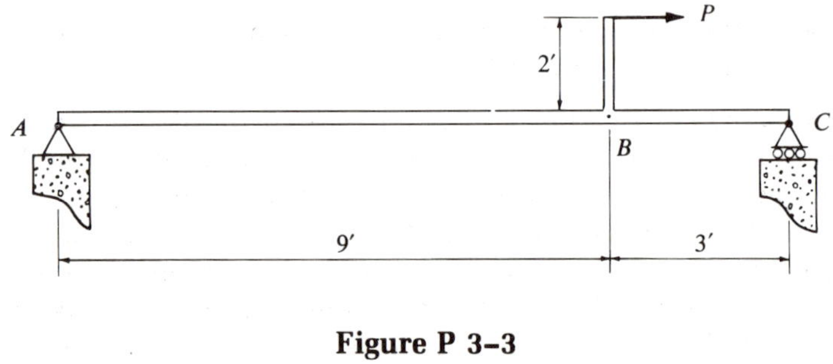

Figure P 3–3

3–4. Find the shear and the moment at the following sections for the beam shown:
(a) At cross section D.
(b) At a cross section just to the left of the support at B.
(c) At a cross section just to the right of the support at B.

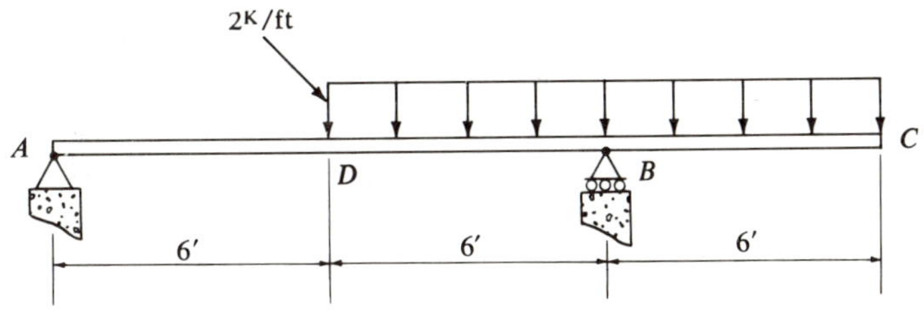

Figure P 3–4

3–5. For the beam shown, find the shear and the moment at the following cross sections:
(a) Just to the left of point B.
(b) At point C.

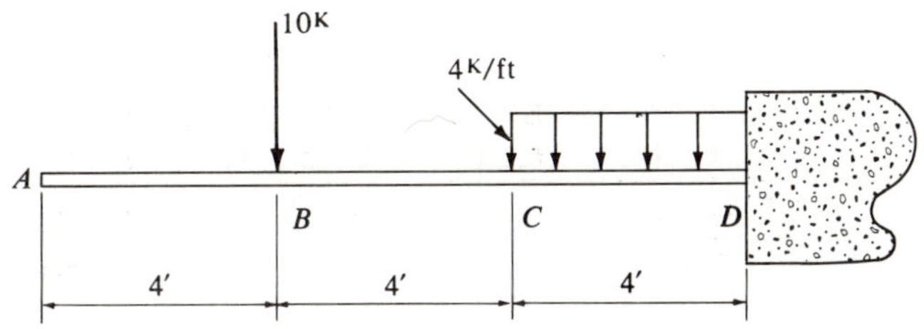

Figure P 3–5

3–6. For the beam shown, find:
(a) The shear and the moment at point C.
(b) The shear and the moment at point D.

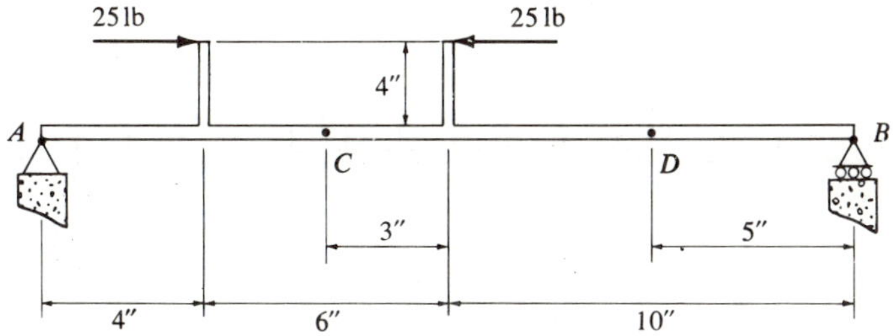

Figure P 3–6

3–7. For the beam shown in Figure P 3–7, and using the coordinate axis shown, write the equations for shear and moment at any section between B and C.

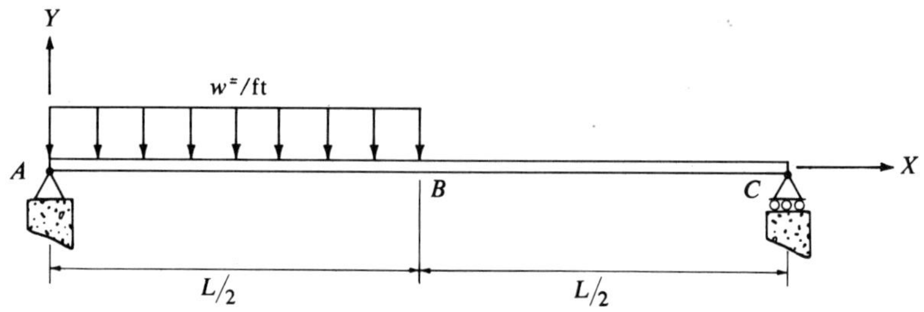

Figure P 3–7

3–8. For the beam shown in Figure P 3–8, and using the coordinate axis shown, write the equations for shear and moment at any section between the supports.

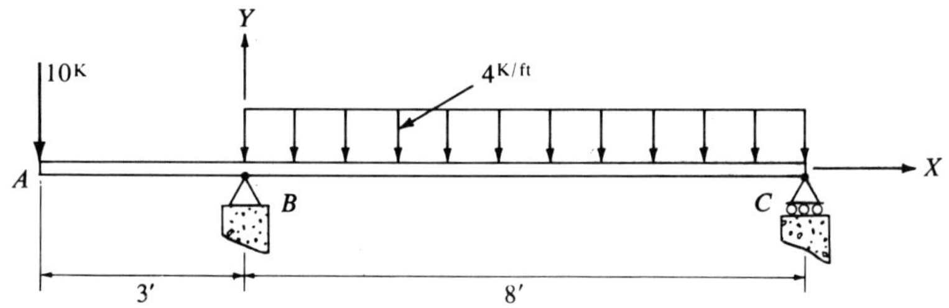

Figure P 3–8

3–9. For the beam shown in Figure P 3–9, and using the coordinate axis shown, write the equations for shear and moment at any section between the supports.

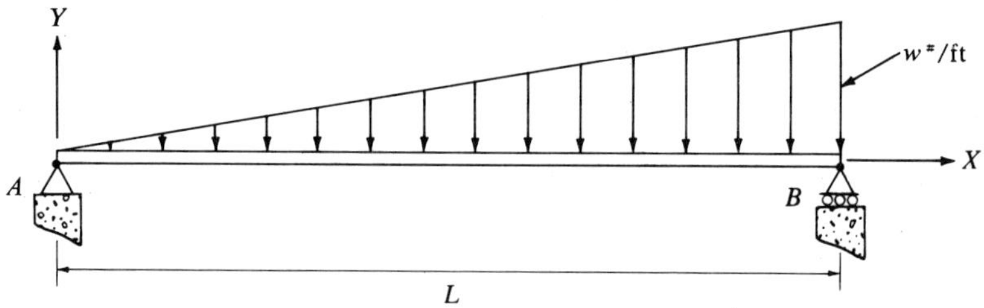

Figure P 3–9

3-10. For the beam shown in Figure P 3-10, and using the coordinate axis shown, write the equations for shear and moment at any section between the supports.

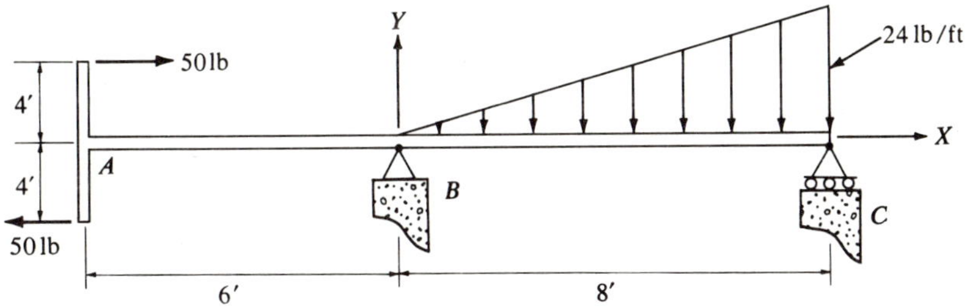

Figure P 3-10

3-11. For the cantilever type dam shown in Figure P 3-11, write the general equations for shear and moment using the coordinate axes shown.

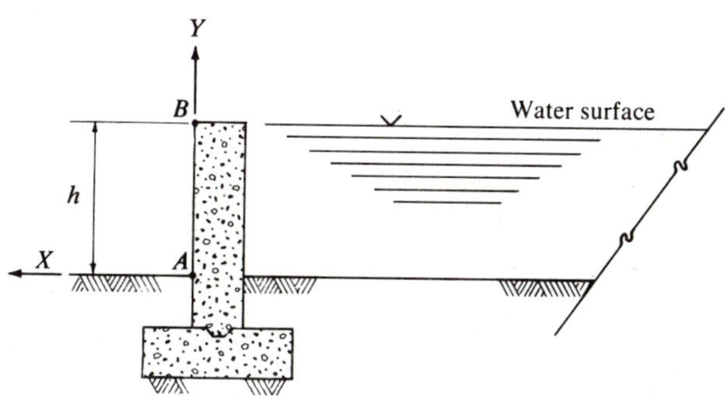

Figure P 3-11

3-12. For the beam shown in Figure P 3-12, write the general expressions for shear and moment for all segments of the span. Use the coordinate axes shown.

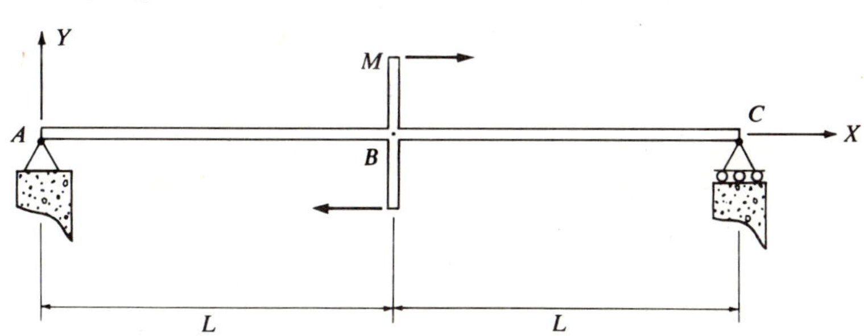

Figure P 3-12

3–13 to 3–21. Draw the complete shear and moment diagrams for the structures shown.

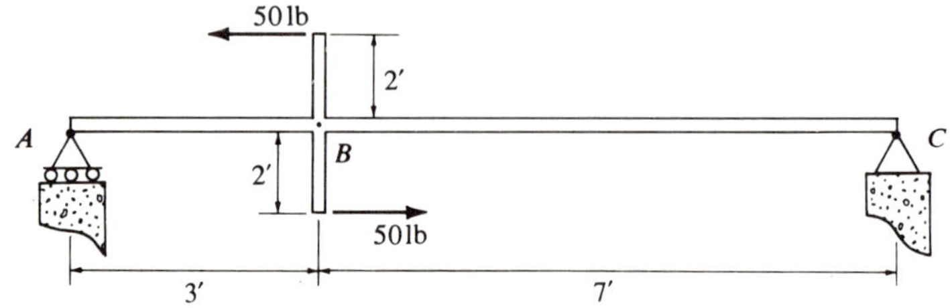

Figure P 3–13

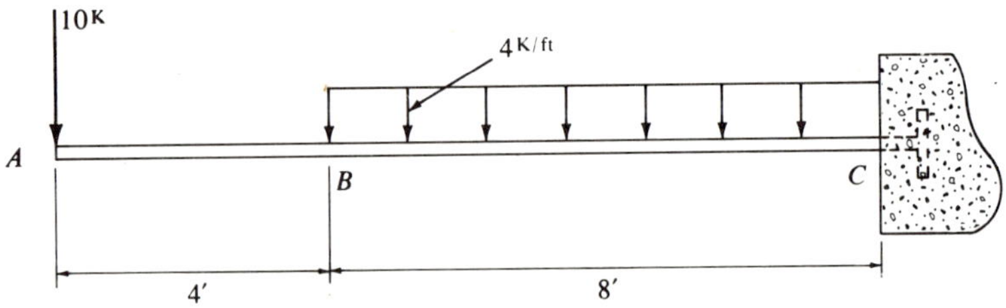

Figure P 3–14

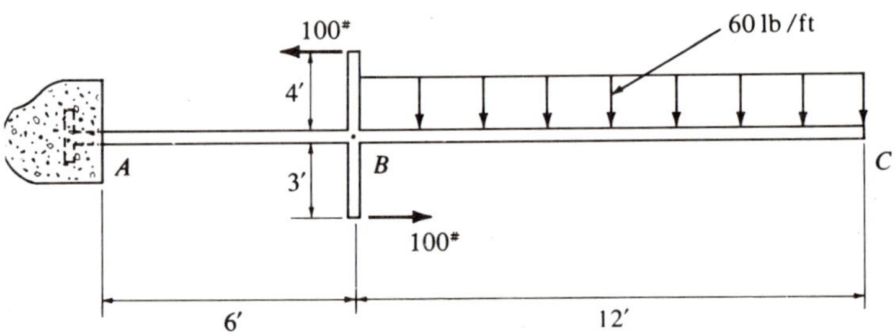

Figure P 3–15

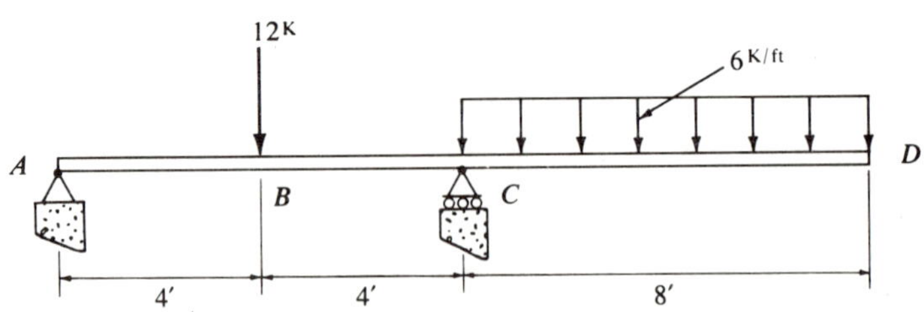

Figure P 3–16

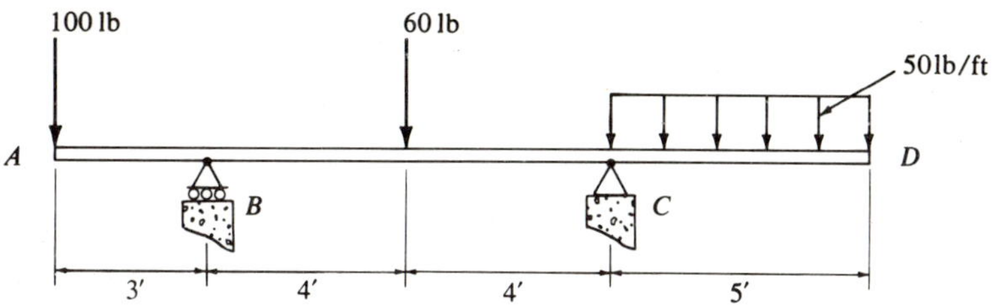

Figure P 3–17

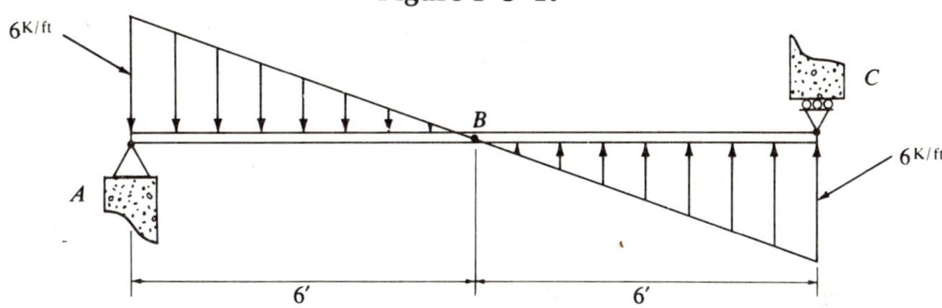

Figure P 3–18

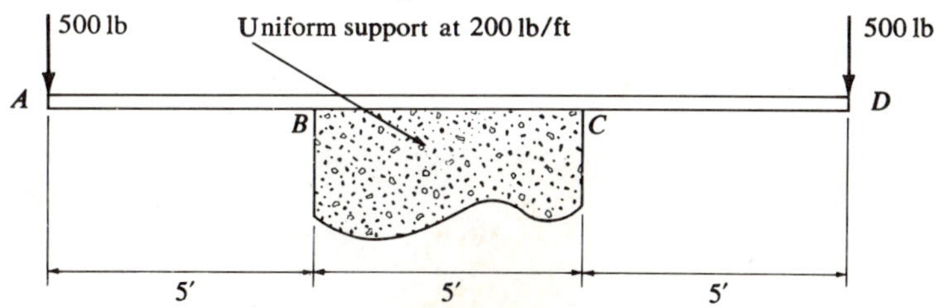

Figure P 3–19

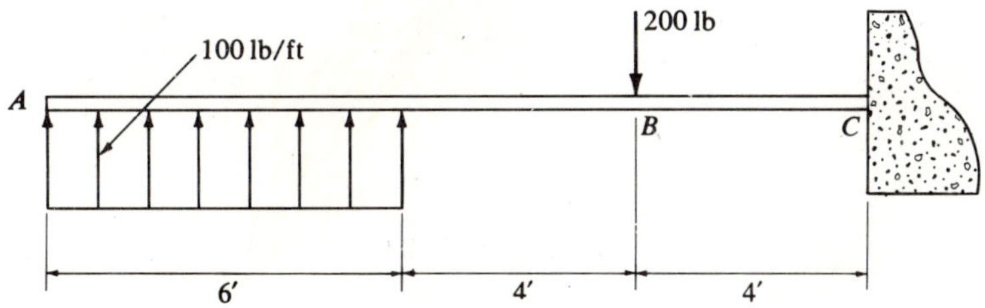

Figure P 3–20

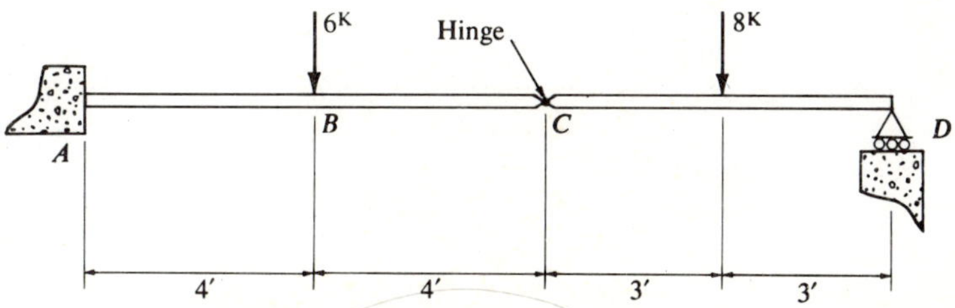

Figure P 3–21

3–22 to 3–30. Draw the complete shear and moment diagrams for the structures shown.

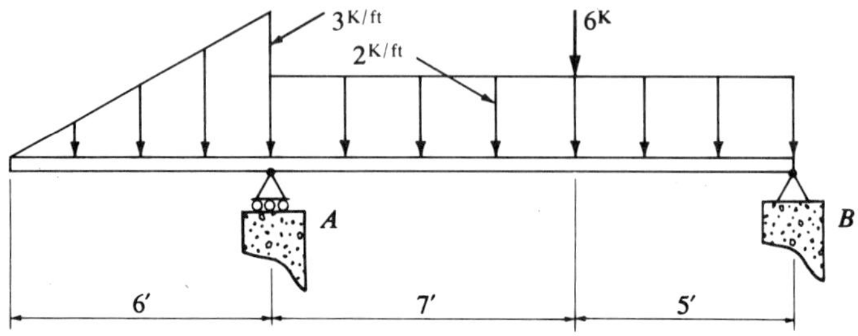

Figure P 3–22

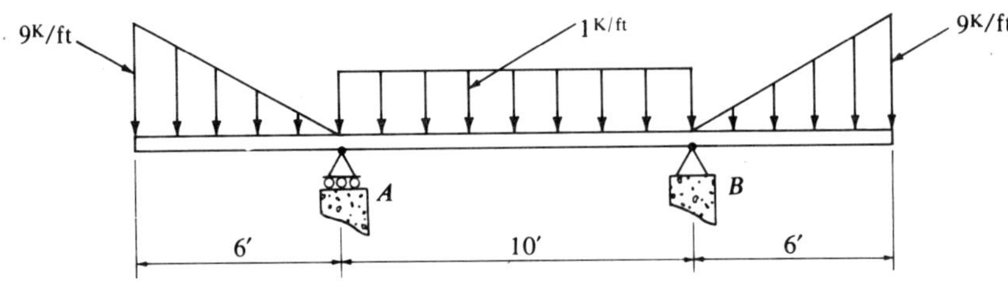

Figure P 3–23

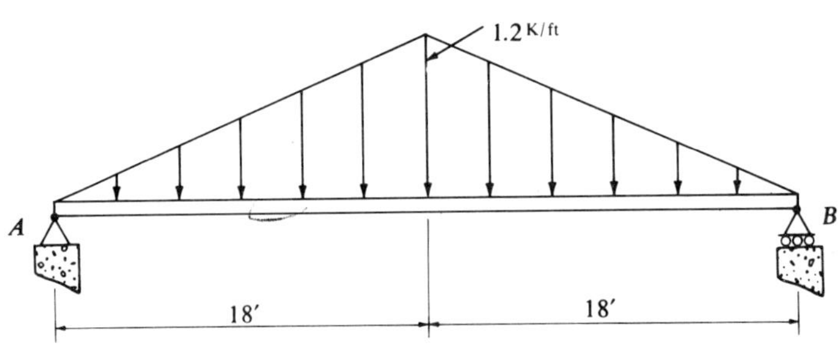

Figure P 3–24

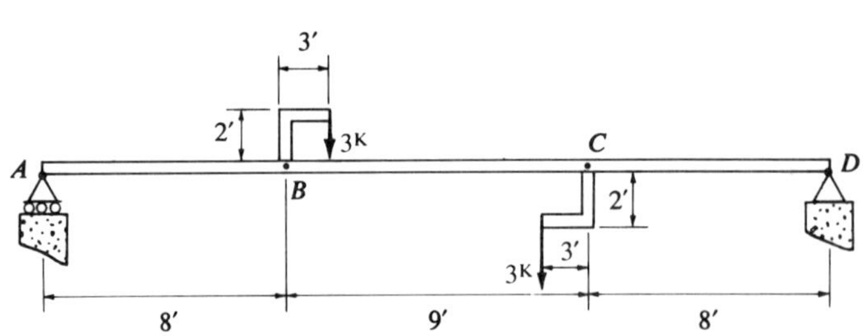

Figure P 3–25

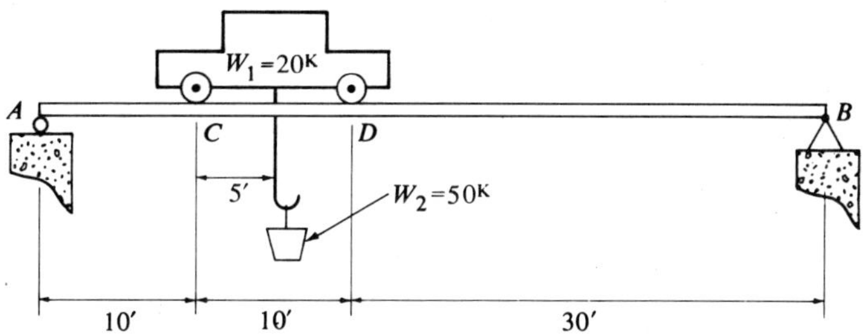

Figure P 3–26

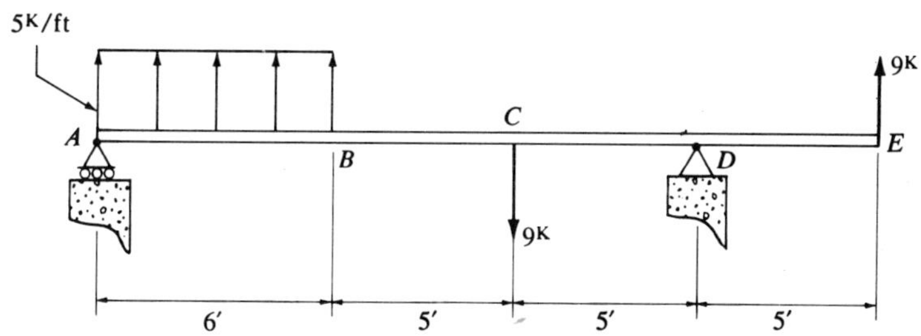

Figure P 3–27

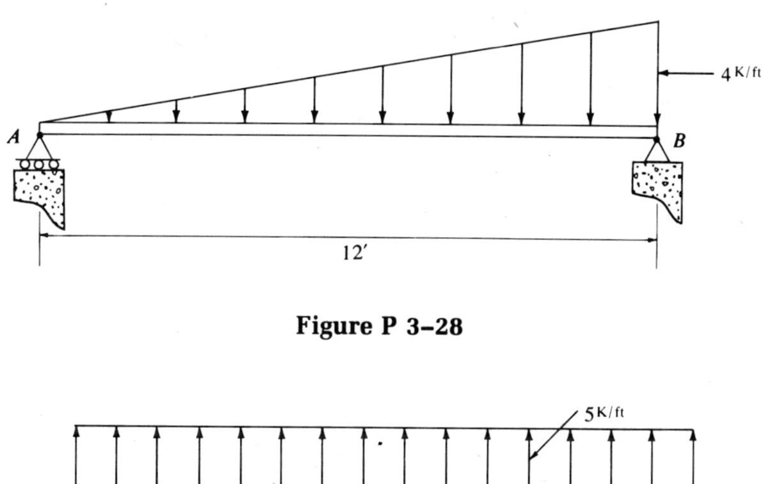

Figure P 3–28

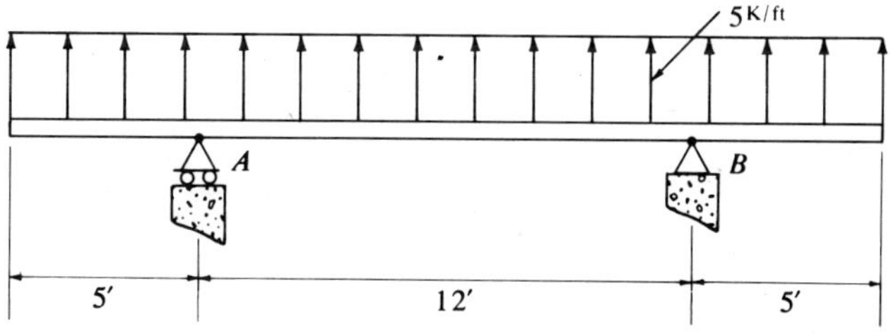

Figure P 3–29

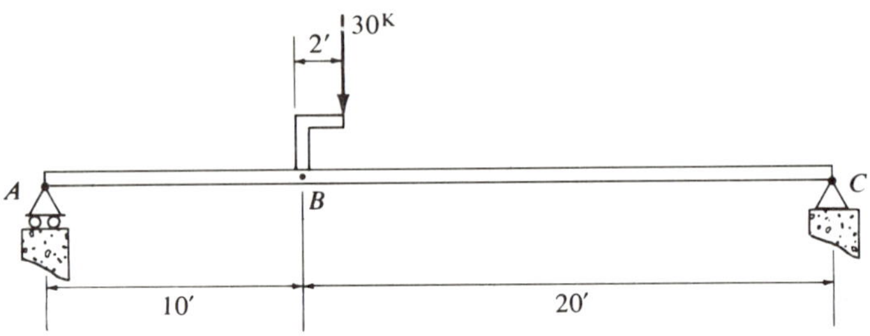

Figure P 3–30

3–31. A solid sign measuring 12 ft high by 30 ft long is supported by two steel poles as shown in Figure P 3–31. Draw the complete shear and moment diagrams for one steel pole due to a wind force of 25-lb per square foot. Neglect pole and sign weights.

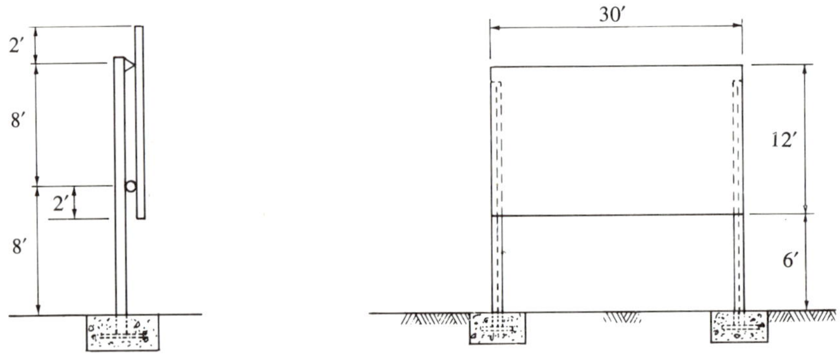

Figure P 3–31

3–32. A lifting rack for storage tanks is fabricated as shown in Figure P 3–32. Draw the complete shear and moment diagram for the member AB. The tank and its contents weigh 2500 lb.

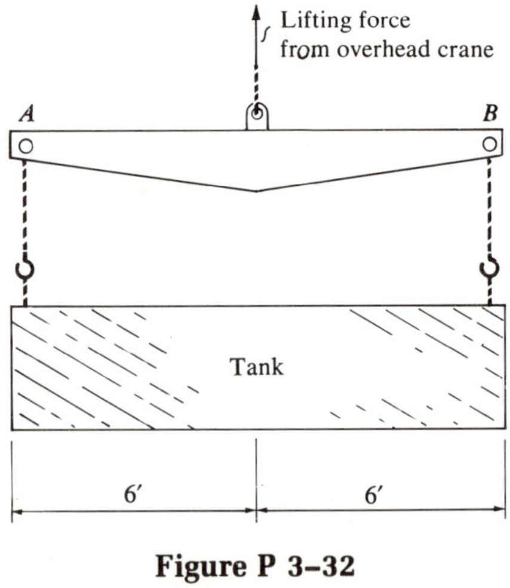

Figure P 3–32

3–33. A concrete footer measuring 14-ft long and 1-ft wide, as shown in Figure P 3–33, carries two 50-kip column loads. Assuming a uniform soil pressure, draw the complete shear and moment diagrams for the footer.

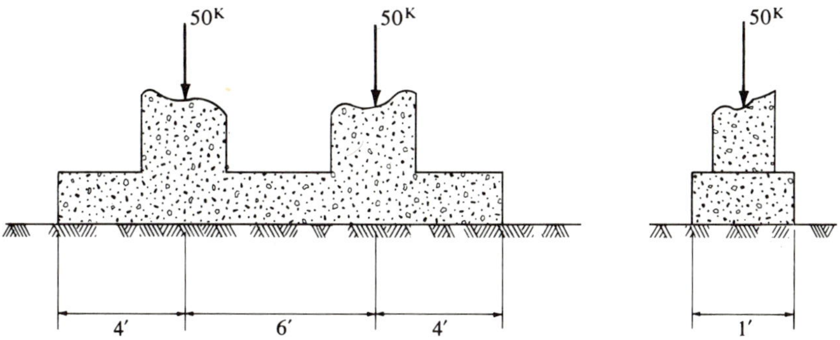

Figure P 3–33

3–34. For the transmission pole shown in Figure P 3–34, draw the complete shear and moment diagrams.

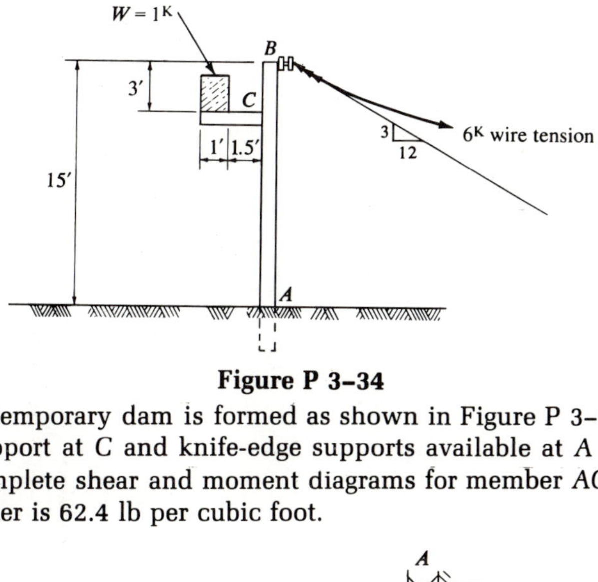

Figure P 3–34

3–35. A temporary dam is formed as shown in Figure P 3–35. Assuming no support at C and knife-edge supports available at A and B, draw the complete shear and moment diagrams for member AC. Unit weight of water is 62.4 lb per cubic foot.

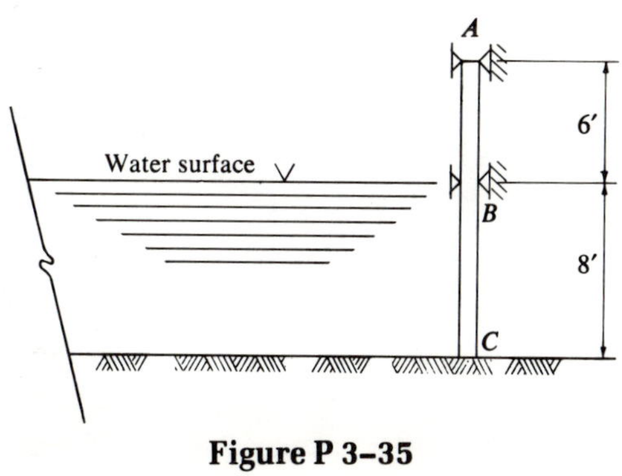

Figure P 3–35

CHAPTER FOUR

Trusses

4.1 Definition of a Truss

A truss is an engineering structure often utilized in the design of bridges, buildings, and towers. Trusses have consistently provided not only a practical, but also an economical, answer to many design problems. A truss may be simply defined as a rigid structure composed of a number of straight bars, all lying in the same plane and joined together at their ends by frictionless pins. Such a structure is shown in Figure 4.1. Since all the bars, or members, of the

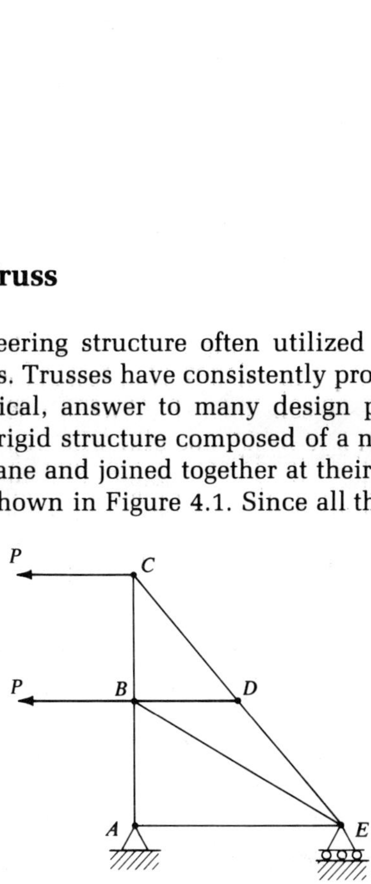

Figure 4.1

truss shown in Figure 4.1 are confined to the same plane, the truss is said to be a **plane truss.** Other pertinent characteristics of a plane truss are: (1) no member is continuous through a joint, e.g., there is no member *AC*, but there are instead two members *AB* and *BC*; (2) loads and reactions are applied to the

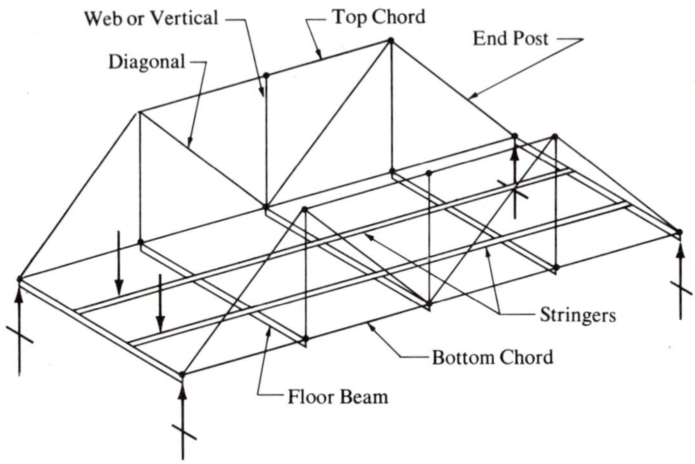

Figure 4.2

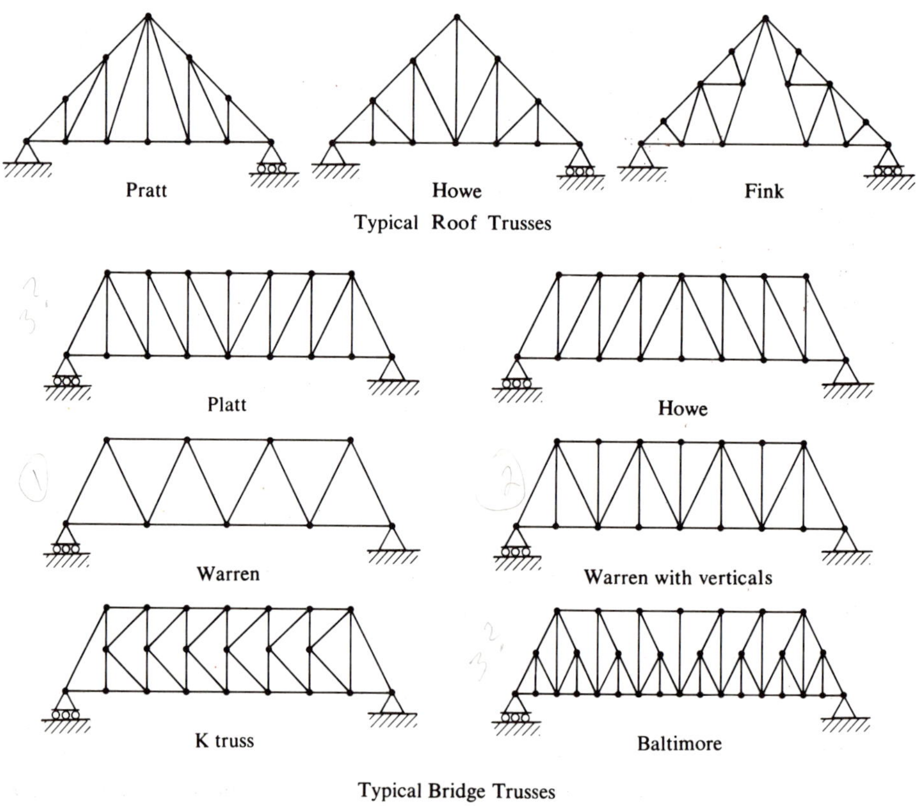

Figure 4.3

truss only at the joints. This last characteristic can be further explained by referring to Figure 4.2. Actual structures, such as the bridge truss shown in Figure 4.2, are actually comprised of several trusses to form a space framework. Notice, however, that as loads are applied to the bridge truss shown in Figure 4.2, the following load transfers are made: (1) the wheel loads are transmitted by the pavement (if the bridge is a highway type) to the stringers; (2) the stringers in turn transfer the loads to the floor beams through the reactions of the stringers with the floor beams; (3) the floor beams transfer these reaction forces to the trusses at the joints. Loads anywhere along the truss span, therefore, are transferred to the joints through the use of the floor system. Since the loads applied to the space framework are transmitted to the joints of an individual truss and act in its plane, the truss may be conveniently treated as a two-dimensional or plane structure. Figure 4.2 also points out members' names as they are commonly referred to in structural engineering. Figure 4.3 shows common types of roof and bridge trusses.

4.2 Simple Trusses

Suppose that a truss is formed with smooth (frictionless) pins at points *A*, *B*, *C*, and *D* as shown in Figure 4.4a. If a load, *P*, is applied at joint *B* as shown in Figure 4.4a, the truss will greatly deform and collapse as shown in Figure 4.4b. Imagine now removing members *BC* and *CD* of the structure shown in Figure 4.4a and adding a new member *BD* as shown in Figure 4.4c. This truss, comprised of three members joined together by smooth pins at *A*, *B*, and *D*, will not collapse under application of the load *P* and is said to be a **rigid truss,** i.e., the

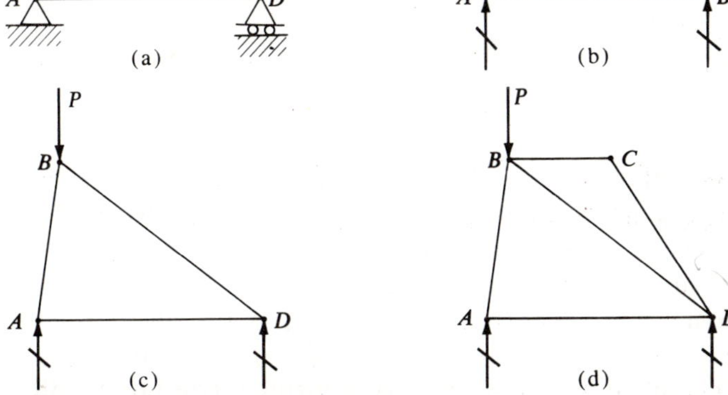

Figure 4.4

truss will not greatly deform and collapse, but will undergo only slight deformation caused by the small elastic deformations of the members comprising the truss. If we now add members *BC* and *CD* to the truss shown in Figure 4.4c, as shown in Figure 4.4d, we have constructed a rigid truss which is only a slight modification of the original unrigid truss shown in Figure 4.4a. A truss constructed in this manner is called a **simple truss.** A simple truss is defined as a system of bars arranged by joining together three bars at their extremities in the form of a triangle and then adding to this basic unit two more bars for each additional joint. To the rigid triangular truss shown in Figure 4.4c were added the two members *BC* and *CD* joined at joint *C* to obtain the truss shown in Figure 4.4d; hence, it meets all the criteria of a simple truss. Referring back to Figure 4.3, it can be seen that all of the typical trusses shown, with the exception of the Fink and Baltimore trusses, are simple trusses. The basic triangular unit is shaded for each case. The building onto the triangular unit in the prescribed fashion breaks down for the Fink and Baltimore trusses (see Figure 4.3) at the center line of their spans. These trusses are classified as **compound trusses.** A compound truss is defined as a rigid system formed by combining two or more simple trusses in such a manner so as to form one rigid framework. In Figure 4.3, the Fink roof truss is constructed by combining the two simple trusses *ABE* and *BCD* with member *ED* and hinge *B*. The Baltimore truss is constructed in a similar fashion. If a truss fails to meet the criteria necessary for either a simple or compound truss, it is classified as a **complex truss.** Such a truss is shown in Figure 4.5. Although a complex truss is statically

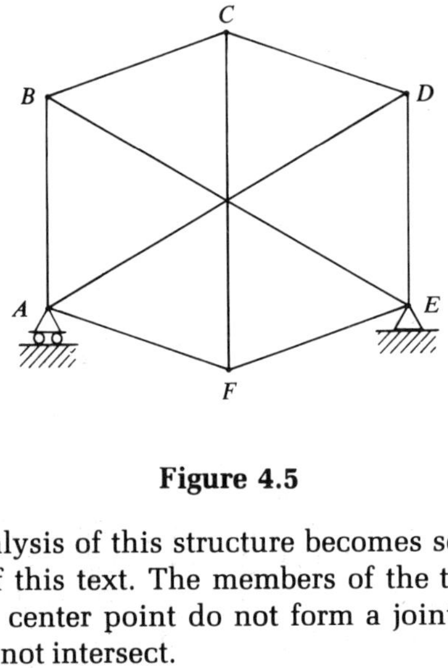

Figure 4.5

determinate, the analysis of this structure becomes somewhat involved and is beyond the scope of this text. The members of the truss shown in Figure 4.5 passing through the center point do not form a joint. These members simply cross and, hence, do not intersect.

Before moving on to the actual analysis of truss structures, the various classifications of bridge trusses shall be mentioned briefly. A bridge truss may be classified as a **deck truss** or a **through truss** depending upon how the floor system is connected to the truss itself. Both types are shown in Figure 4.6.

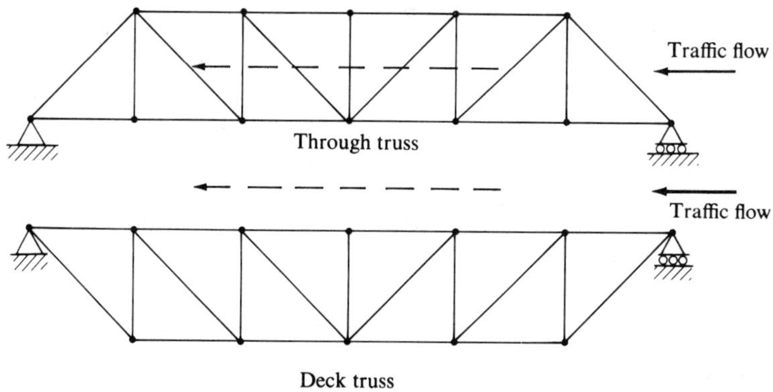

Figure 4.6

In some instances, the floor system is supported somewhere between the bottom and top chords. Such a truss is classified as a **pony truss.**

4.3 Bar Forces

The analysis of a truss simply involves finding the internal force acting in each member of the truss. Consider, for example, the simple truss shown in Figure 4.7a. Furthermore, let us remove member *AB* from the truss and draw a free-

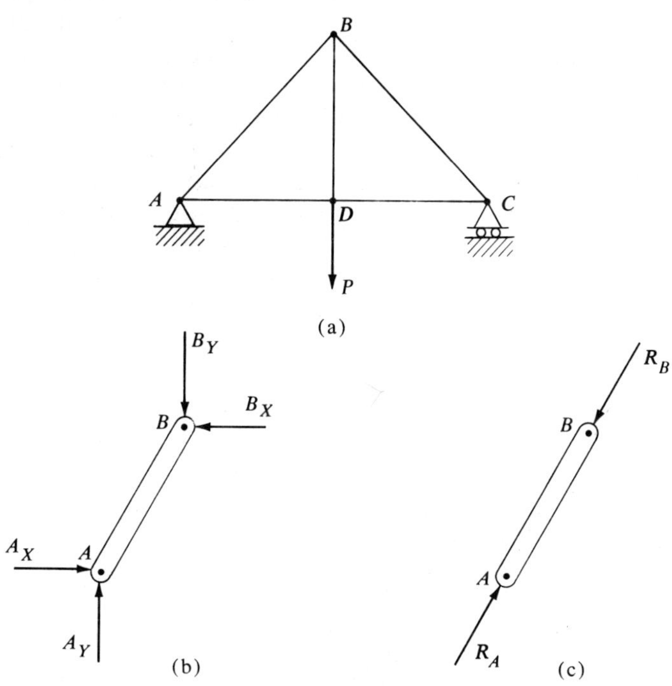

Figure 4.7

body diagram of the member as shown in Figure 4.7b. Recalling that all the members of a truss are joined by smooth pins, it is possible, therefore, to have a vertical and horizontal component acting at each end of the member AB as shown in Figure 4.7b. By applying the equations of static equilibrium to the FBD shown in Figure 4.7b, i.e., $\Sigma F_Y = 0$ and $\Sigma F_X = 0$, it is evident that $A_Y = B_Y$ and $A_X = B_X$. Since it is impossible for a smooth pin to transmit a moment, the sum of the moments about A in Figure 4.7b must be zero and likewise for point B. Hence, the moment produced by B_Y and B_X about point A and the moment produced by A_Y and A_X about point B both must be zero. Since any two component forces may be replaced by their resultant force without altering the conditions of the problem, the components A_Y and A_X and B_Y and B_X are shown replaced by R_A and R_B, their respective resultants, in Figure 4.7c. If the components B_Y and B_X do not produce moment about point A, then their resultant, R_B, must produce the same effect at point A. For a single force not to produce moment about a point, the line of action of the force must pass through the point in question. Therefore, if R_B does not produce moment about point A, its line of action must pass through point A. The line of action of R_A can be established using the same reasoning. The member AB is acted upon by two equal, but oppositely sensed, forces whose lines of action pass through the pins at the ends as shown in Figure 4.7c. For equilibrium, it is obvious that $R_A = R_B = R$. The truss member AB is subjected to an **internal axial force,** R, commonly referred to as a **bar force.** An internal axial force may be further defined as a force whose line of action coincides with the longitudinal axis of the member, i.e., since the members of the truss are straight, the longitudinal axis and a line joining the center of the pins at the ends are coincident. In a similar fashion, it could be shown that all the members of the truss shown in Figure 4.7a, or any plane truss for that matter, are subjected to internal axial forces. If we know that the bar force in a truss member acts along the longitudinal axis of the respective member, then the line of action of the bar force is known and we must simply determine its magnitude and sense. A truss member may be acted upon as shown in either of the two sketches of Figure 4.8. In the first case, Figure 4.8a, the forces tend to pull the member apart, or stretch it, and the member is said to be in **tension.** In Figure 4.8b, the forces tend to

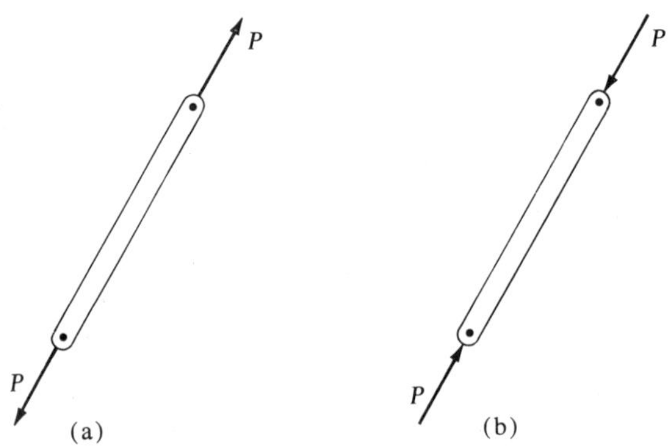

Figure 4.8

compress the member and the member is said to be in **compression.** The following sign convention will be adopted to distinguish between tension and compression forces: (1) bar forces producing tension in a member will be called positive; (2) bar forces producing compression in a member will be called negative.

It is obvious that in actual practice trusses are not fabricated with smooth pins at all the joints. In fact most trusses are made with riveted, bolted, or welded joints. In addition, the weight of the individual members themselves gives rise to a shear force and a bending moment in the truss members. The shear and bending introduced into the truss members by the joints which are somewhat restrained from rotation by nature of their fabrication and by the weight of the individual members themselves produce secondary effects which are small in comparison with the primary effects which are caused by considering only the bar forces. A truss analyzed by assuming that the members are subjected to axial forces only because there are smooth pins at the ends of each member is referred to as an **ideal truss.**

4.4 Method of Joints

As was stated previously, the analysis of truss structures involves finding the magnitude and sense of the bar forces. Since the bar forces are internal forces when considering the truss itself as a whole, we must somehow cut into the structure to expose these internal forces on a FBD for their eventual solution. Let us consider the truss shown in Figure 4.9a and its corresponding FBD in Figure 4.9b. As in the analysis of most structures, the FBD must be drawn first; and secondly, the reactions must be accurately determined. After the reactions have been determined, as shown in Figure 4.9b, the next step is to determine the bar forces in each of the truss members. If we remove an individual member and draw the corresponding FBD, all that is presented is a single member acted upon by an unknown force. Figure 4.10a shows such a FBD for member *AB* in the truss shown in Figure 4.9. From this FBD it is impossible to determine the force F_{AB}. Let us next draw a FBD for, say, the pin at *A* as shown in Figure 4.10b.

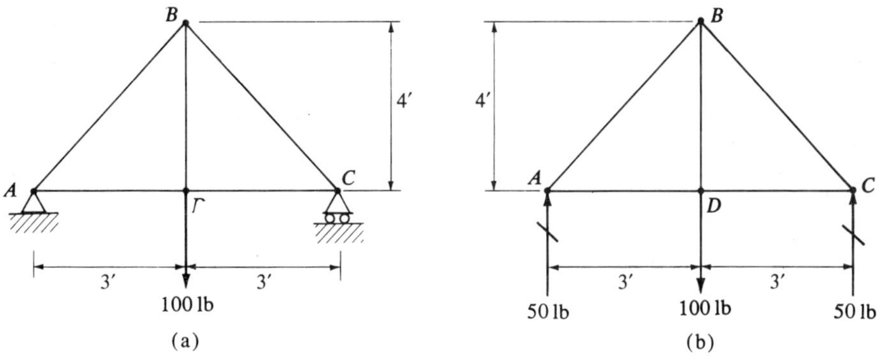

Figure 4.9

The FBD for the remaining truss is shown in Figure 4.10c. When we cut into a member we must show all the forces acting on the member at the cut section. For ideal truss members, the only forces present are axial forces. Referring again to Figure 4.10b, the force F_{AB} acting in member AB must also act on the pin at A. Recall that the pin and truss member are in contact; hence, forces of action and reactions between bodies in contact have the same magnitude, same line of action, and opposite sense. This law holds true as can be seen by observing the force F_{AB} in Figures 4.10a, 4.10b, and 4.10c. The line of action of F_{AB} and F_{BD} are known to be in the same direction as the longitudinal axis of the members in which they act. The senses, however, are not known and are assumed. As before, if a positive answer for the magnitude is obtained a correct sense was assumed, and if negative, the sense must be reversed. Notice that if the force tends to "push" down on the pin, as assumed for F_{AB} in Figure 4.10b, the corresponding effect on the member AB is a compression force as can be seen from Figures 4.10a and 4.10c. If the force tends to "pull" away from the pin, the corresponding effect is tension in the member. (See F_{AD} in Figure 4.10b and 4.10c.) The forces shown in Figure 4.10b form a concurrent force system, i.e., the lines of action of all the forces acting in the system pass through the same point. The resultant of a concurrent force system cannot be a

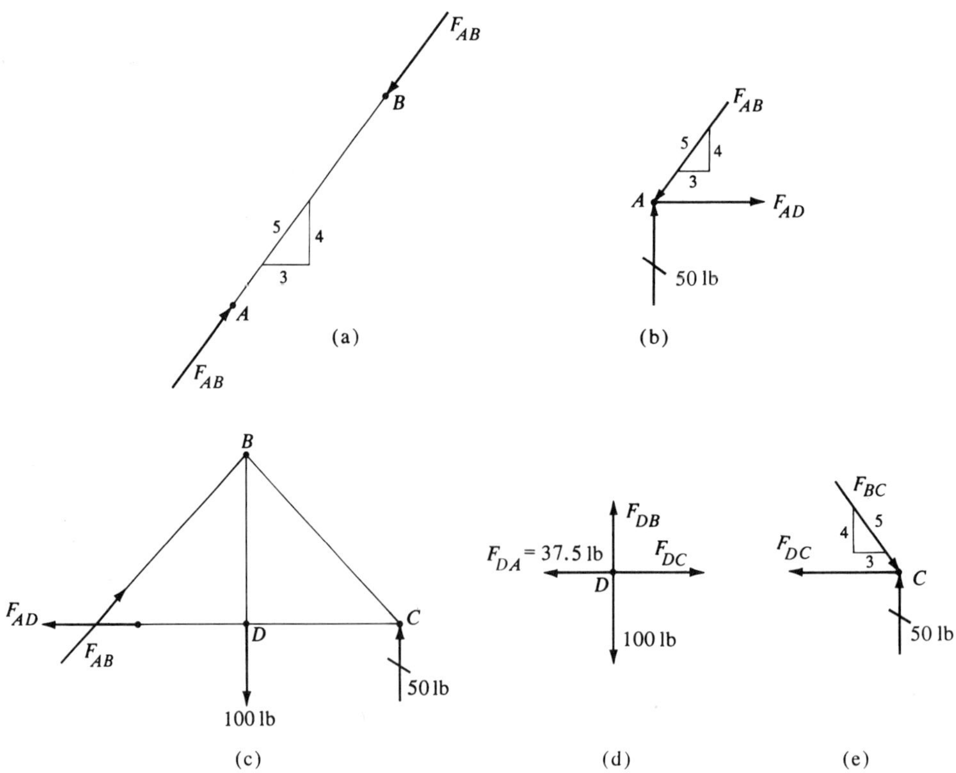

Figure 4.10

couple; therefore, the force system will be in equilibrium if $\Sigma F_X = 0$ and $\Sigma F_Y = 0$. With these two equations available at isolated joint A, Figure 4.10b, the two unknown forces F_{AB} and F_{AD} may be determined as follows:

$$\Sigma F_Y = 0 \uparrow +$$
$$50 - (\tfrac{4}{5})F_{AB} = 0$$
$$F_{AB} = +62.5 \text{ lb}$$
$$\therefore \underline{F_{AB} = 62.5 \text{ lb compression, } c}$$

$$\Sigma F_X = 0 \rightarrow +$$
$$-(\tfrac{3}{5})F_{AB} + F_{AD} = 0$$
$$F_{AD} = (\tfrac{3}{5})(62.5)$$
$$F_{AD} = +37.5 \text{ lb}$$
$$\therefore \underline{F_{AD} = 27.5 \text{ lb tension, } t}$$

Using the FBDs of joints D and C shown in Figures 4.10d and 4.10e, the remaining bar forces may be determined as follows:

$$@D\ \Sigma F_Y = 0 \uparrow + \qquad \Sigma F_X = 0 \rightarrow +$$
$$-100 + F_{DB} = 0 \qquad -F_{DA} + F_{DC} = 0$$
$$F_{DB} = +100 \text{ lb} \qquad F_{DC} = +37.5 \text{ lb}$$
$$\underline{F_{DB} = 100 \text{ lb } t} \qquad \underline{F_{DC} = 37.5 \text{ lb } t}$$

$$@C\ \Sigma F_Y = 0 \uparrow + \qquad \Sigma F_X = 0 \rightarrow +$$
$$-(\tfrac{4}{5})F_{BC} + 50 = 0 \qquad (\tfrac{3}{5})F_{BC} - F_{DC} = 0$$
$$F_{BC} = +62.5 \text{ lb} \qquad 37.5 - 37.5 = 0$$
$$\underline{F_{BC} = 62.5 \text{ lb } c} \qquad 0 = 0 \quad \text{Check}$$

A helpful hint in applying this method, which is commonly called the **method of joints**, is to isolate whenever possible only those joints acted upon by a maximum of two unknown bar forces. For example, if joint B of the previous example had been isolated initially rather than joint A, three unknown bar forces (F_{AB}, F_{BD}, and F_{BC}) would have appeared on the FBD of the isolated pin. As stated previously, only two equations of equilibrium are applicable; hence, the three unknown bar forces could not have been obtained from the single FBD at B. If joint A were isolated initially, F_{AB} could be obtained as was shown and then a FBD of pin B would have only F_{BD} and F_{BC} as unknowns. The two available equations of equilibrium could then be directly applied. It can be stated, therefore, that the order in which the joints are selected to be isolated can play an important role, lessening the amount of calculations.

If an entire truss is in equilibrium, then each pin must also be in equilibrium. FBDs of all the pins must include all the bar forces as well as all the reaction components. The number of unknowns encountered, therefore, must be equal to the total number of bar forces, which is equal to the number of truss members, plus the number of reaction components. There are two equations of static equilibrium available for each joint considered as a free body. If j denotes the number of joints including the points of reaction, then there is a total of $2j$ equations available. If m is the number of truss members and r is the number of reaction components, then the total number of unknowns is $m + r$. For a truss to be statically determinate, the number of unknowns must be equal to the number of available equations. In equation form, the criterion becomes:

$$m + r - 2j = 0$$

For the truss just solved (see Figure 4.9), $m = 5$, $r = 3$, and $j = 4$. Hence:

$$5 + 3 - 2(4) = 0$$
$$0 = 0$$

The truss was indeed statically determinate, which is now verified by the newly established criterion. If the sum $(m + r - 2j)$ is greater than zero, there are more unknowns than available equations and the structure is statically indeterminate. This could mean that there are too many unknown bar forces, too many unknown reaction components, or both. If the sum $(m + r - 2j)$ is less than zero, the truss lacks a sufficient number of members or a sufficient number of reactions. In such a case, the truss is said to be **statically unstable.** This criterion can be applied to the truss previously considered and shown in Figure 4.4a. For this truss, $m = 4$, $r = 3$, and $j = 4$ and the following is obtained:

$$m + r - 2j = 4 + 3 - 2(4) = -1 < 0$$

Since the criterion yields a result less than zero, the truss is classified as statically unstable. The collapse depicted by Figure 4.4b had earlier substantiated this fact. The numerical significance of the criterion is also important since it tells us that the truss lacks being statically stable by either one member or by one reaction component. Examination of the reactions shows that sufficient restraint has been provided and, hence, the truss lacks one member. The addition of member BD, as shown in Figure 4.4d, is sufficient to make the truss statically stable.

Example 4.1

Compute the bar forces in all the members of the simple truss in Figure 4.11.

Solution:

1. The reactions at A and G shown on the FBD in Figure 4.11a were obtained as follows:

$$\Sigma M_G = 0 \quad \curvearrowright +$$

$$120\,A_Y - 5(100 + 80 + 60 + 40 + 20) = 0$$

$$\underline{A_Y = 12.5^K \uparrow}$$

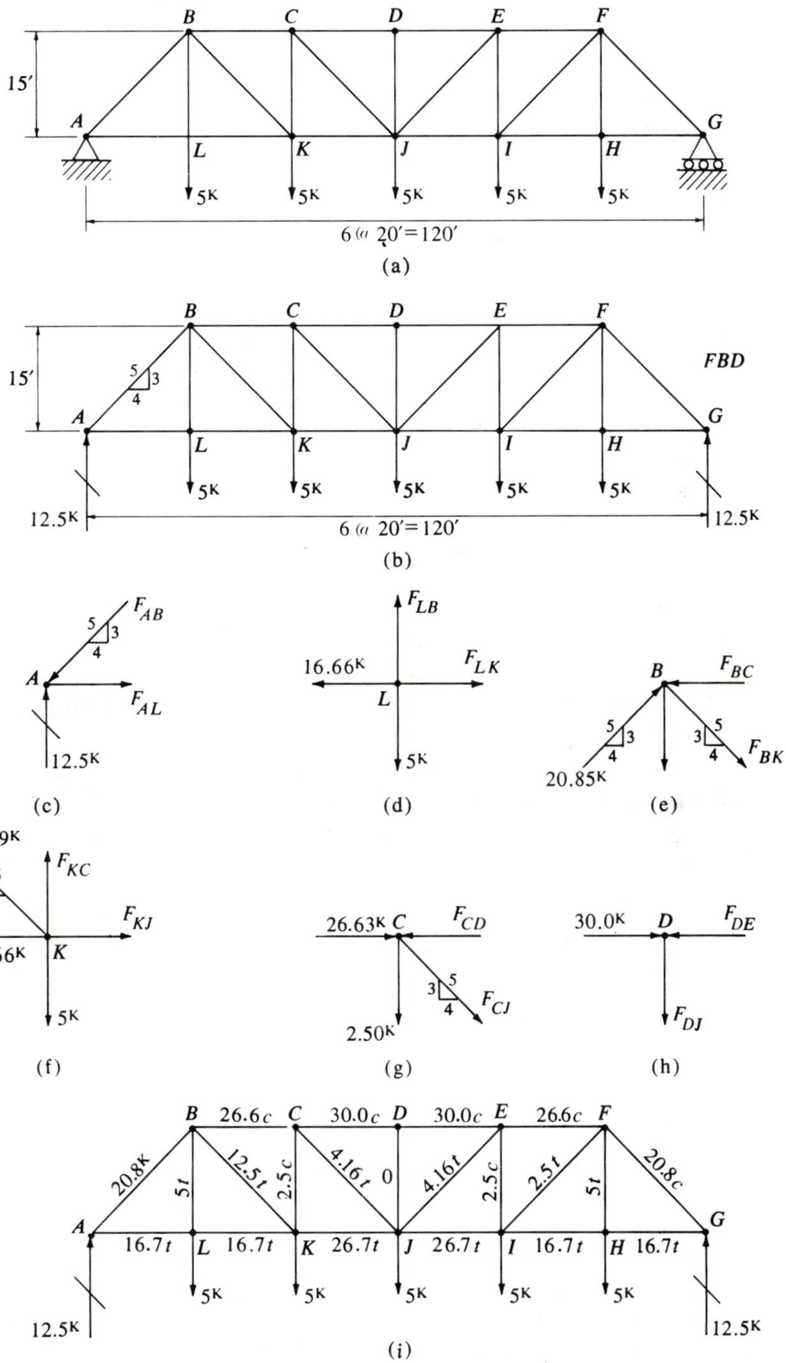
B C D E F
15′
A G
L K J I H
5K 5K 5K 5K 5K
6 @ 20′=120′
(a)
FBD
12.5K 12.5K
(b)
FAB
FAL
12.5K
(c)
FLB
16.66K
FLK
L
5K
(d)
FBC
FBK
20.85K
(e)
12.49K
FKC
FKJ
16.66K
K
5K
(f)
26.63K C
FCD
FCJ
2.50K
(g)
30.0K D
FDE
FDJ
(h)
26.6c 30.0c 30.0c 26.6c
20.8c 5t 12.5t 2.5c 4.16t 0 4.16t 2.5c 2.5t 5t 20.8c
16.7t 16.7t 26.7t 26.7t 16.7t 16.7t
(i)

Figure 4.11

$$\Sigma F_Y = 0 \uparrow +;\ 12.5 - 5(5) + G_Y = 0$$
$$\underline{G_Y = 12.5^K \uparrow}$$

$$\Sigma F_X = 0 \rightarrow +;\ \underline{A_X = 0}$$

2. The method of joints may be applied as follows:

Figure 4.11c:

$$\Sigma F_Y = 0 \uparrow +;\ 12.5 - \tfrac{3}{5}(F_{AB}) = 0$$
$$\underline{F_{AB} = 20.83^K\ c}$$

$$\Sigma F_X = 0 \rightarrow +;\ F_{AL} - (\tfrac{4}{5})\, 20.83 = 0$$
$$\underline{F_{AL} = 16.66^K\ t}$$

Figure 4.11d:

$$\Sigma F_Y = 0 \uparrow +;\ F_{LB} - 5 = 0$$
$$\underline{F_{LB} = 5^K\ t}$$

$$\Sigma F_X = 0 \rightarrow +;\ F_{LK} - 16.66 = 0$$
$$\underline{F_{LK} = 16.66^K\ t}$$

It should be noted that as a bar force is determined, its numerical magnitude and sense are always used thereafter. For example, F_{AL} was determined as 16.66^K t from the FBD shown in Figure 4.11c; therefore, F_{AL} was represented as a 16.66^K tension force in Figure 4.11d.

Figure 4.11e:

$$\Sigma F_Y = 0 \uparrow +;\ (\tfrac{3}{5})(20.83) - 5 - (\tfrac{3}{5})F_{BK} = 0$$
$$\underline{F_{BK} = 12.49^K\ t}$$

$$\Sigma F_X = 0 \rightarrow +;\ (\tfrac{4}{5})(20.83) + \tfrac{4}{5}(12.49) - F_{BC} = 0$$
$$\underline{F_{BC} = 26.63^K\ c}$$

Figure 4.11f:

$$\Sigma F_Y = 0 \uparrow +;\ \tfrac{3}{5}(12.49) + F_{KC} - 5 = 0$$
$$\underline{F_{KC} = -2.49^K\ t}$$
$$\therefore\ \underline{F_{KC} = 2.49^K\ c}$$

F_{KC} was assumed to pull away from joint K in the FBD shown in Figure 4.11f, which meant that F_{KC} was assumed to be a tension force. The negative result for its magnitude simply means that the sense of F_{KC} was incorrectly assumed and the force is actually a compressive force.

$$\Sigma F_X = 0 \rightarrow +; -16.66 - (\tfrac{4}{5})(12.49) + F_{KJ} = 0$$
$$\underline{F_{KJ} = 26.66^{K}\ t}$$

Figure 4.11g:

$$\Sigma F_Y = 0 \uparrow +; 2.49 - (\tfrac{3}{5})F_{CJ} = 0$$
$$\underline{F_{CJ} = 4.16^{K}\ t}$$

$$\Sigma F_X = 0 \rightarrow +; 26.63 + (\tfrac{4}{5})4.16 - F_{CD} = 0$$
$$\underline{F_{CD} = 30.0^{K}\ c}$$

Figure 4.11h:

$$\Sigma F_Y = 0 \uparrow +; \underline{F_{DJ} = 0}$$
$$\Sigma F_X = 0 \rightarrow +; 30.0 - F_{DE} = 0$$
$$\underline{F_{DE} = 30.0^{K}\ c}$$

3. Notice that the truss shown in Figure 4.11a is symmetrical about the center line with respect to both geometry and loading. When such symmetry exists, the remaining bar forces are identical to their counterpart on the opposite side of the center line. It must be emphasized that the condition of symmetry can only be applied, as in this case, when, and only when, symmetry exists with respect to both loading and geometry. Figure 4.11i shows the truss with all its corresponding bar forces.

Example 4.2

Compute the bar forces in all the members of the Howe roof truss shown in Figure 4.12.

Solution:

1. The reactions at A and E shown on the FBD in Figure 4.12b were obtained as follows:

$$\Sigma M_E = 0 \circlearrowright +$$

$$A_Y(40) - 30(\sin 60°)30 + 30(\cos 60°)(10)\tan 30° = 0$$
$$A_Y(40) - 779.4 + 86.5 = 0$$
$$\underline{A_Y = 17.32^{K} \uparrow}$$

$$\Sigma F_Y = 0 \uparrow +; 17.32 + E_Y - 30(\sin 60°) = 0$$
$$17.32 + E_Y - 25.98 = 0$$
$$\underline{E_Y = 8.66^{K} \uparrow}$$

$$\Sigma F_X = 0 \rightarrow +; 30(\cos 60°) - E_X = 0$$
$$\underline{E_X = 15.0^{K} \leftarrow}$$

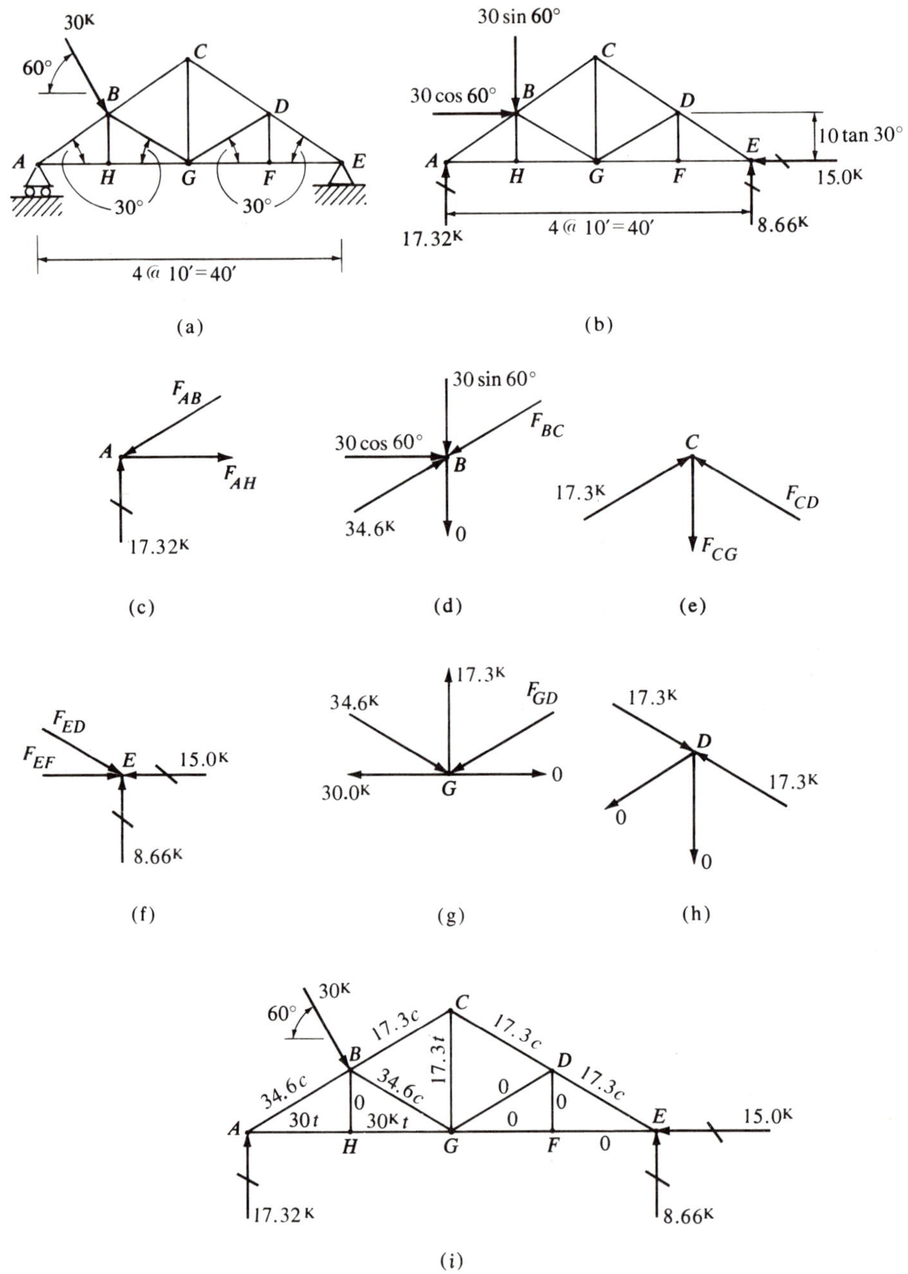

Figure 4.12

It should be noted that when dealing with **pitched roof trusses** (the ratio of rise, or height, to span length is called **pitch**), the geometry introduces additional complexity to the problem.

2. The method of joints may be applied as follows:

Figure 4.12c:

$$\Sigma F_Y = 0 \uparrow +;\ 17.32 - F_{AB}(\sin 30°) = 0$$
$$17.32 - F_{AB}(0.500) = 0$$
$$\underline{F_{AB} = 34.6^{K}\ c}$$

$$\Sigma F_X = 0 \rightarrow +;\ F_{AH} - F_{AB}(\cos 30°) = 0$$
$$F_{AH} - 34.6(0.866) = 0$$
$$\underline{F_{AH} = 30.0^{K}\ t}$$

After obtaining the bar forces F_{AB} and F_{AH}, the next logical step would be to isolate either joint B or joint H. At joint B, however, there are three unknown bar forces, F_{BC}, F_{BG}, and F_{BH}, and only two equations of equilibrium available. At joint H there are only two unknown bar forces, F_{BH} and F_{HG}. Impulse would dictate drawing a FBD of joint H, but a further examination of the joint reveals that this approach would be time-consuming. Imagining a FBD of joint H, it can be seen that the only possible vertical force acting is F_{BH} and it must be zero for equilibrium. Member BH is known as a **zero-force member.** Further examination of the joint at H indicates that the only possible horizontal forces are F_{AH} and F_{HG} and since they act alone and have the same line of action they must be equal to maintain equilibrium in the x direction at joint H. Therefore, the following may be concluded:

By observation at joint H,

$$F_{BH} = 0$$
$$\underline{F_{HG} = F_{AH} = 30.0^{K}\ t}$$

Examination of a truss prior to initiating the calculations should be done to seek out zero-force members as a time-saving measure. A closer look at the truss shown in Figure 4.12a reveals that member DF is also a zero-force member. Also, by mentally summing forces in the x-direction at joint F, $F_{EF} = F_{FG}$. We have, therefore, eliminated the need for FBDs of joints H and F and have obtained four bar forces by inspection. Continuing:

Figure 4.12d:

$$\Sigma F_Y = 0 \uparrow +$$
$$34.6(\sin 30°) + F_{BG}(\sin 30°) - F_{BC}(\sin 30°) - 30(\sin 60°) = 0$$
$$34.6 + F_{BG} - F_{BC} - 52.0 = 0$$
$$F_{BG} - F_{BC} = 17.4 \qquad \textbf{(4–1)}$$

$$\Sigma F_X = 0 \rightarrow +$$
$$30(\cos 60°) + 34.6(\cos 30°) - F_{BC}(\cos 30°) - F_{BG}(\cos 30°) = 0$$
$$17.3 + 34.6 - F_{BC} - F_{BG} = 0$$
$$F_{BG} + F_{BC} = 51.9 \qquad \textbf{(4–2)}$$

Solving Eqs. (4–1) and (4–2) simultaneously:

$$F_{BG} - F_{BC} = 17.4 \qquad (4\text{–}1)$$
$$F_{BG} + F_{BC} = 51.9 \qquad (4\text{–}2)$$
$$2F_{BG} = 69.3$$
$$\underline{F_{BG} = 34.6^{K}\ c}$$
$$34.6 + F_{BC} = 51.9$$
$$\underline{F_{BC} = 17.3^{K}\ c}$$

Figure 4.12e:

$$\Sigma F_X = 0 \rightarrow +;\ 17.3(\cos 30°) - F_{CD}(\cos 30°) = 0$$
$$\underline{F_{CD} = 17.3^{K}\ c}$$

$$\Sigma F_Y = 0 \uparrow +;\ 17.3(\sin 30°) + 17.3(\sin 30°)\ - F_{CG} = 0$$
$$\underline{F_{CG} = 17.3^{K}\ t}$$

Figure 4.12f:

$$\Sigma F_Y = 0 \uparrow +;\ 8.66 - F_{ED}(\sin 30°) = 0$$
$$8.66 - F_{ED}(0.500) = 0$$
$$\underline{F_{ED} = 17.3^{K}\ c}$$

$$\Sigma F_X = 0 \rightarrow +;\ 17.3(\cos 30°) + F_{EF} - 15 = 0$$
$$15 + F_{EF} - 15 = 0$$
$$\underline{F_{EF} = 0}$$

As previously observed (by inspection at F),

$$\underline{F_{DF} = 0}$$
$$\underline{F_{EF} = F_{FG} = 0}$$

Figure 4.12g:

$$\Sigma F_Y = 0 \uparrow +;\ 17.3 - 34.6(\sin 30°) - F_{GD}(\sin 30°) = 0$$
$$17.3 - 17.3 - F_{GD}(0.500) = 0$$
$$\underline{F_{GD} = 0}$$

All the bar forces are obtained, but, as a check, let us isolate the last remaining joint, joint D, and see if equilibrium is maintained:

Figure 4.12h:

Inspection of the FBD shows that joint D is in equilibrium as the only forces acting are two equal, but oppositely sensed forces F_{DE} and F_{CD}. This serves as a check on the computations.

3. Figure 4.12i shows the truss along with all the corresponding bar forces.

4.5 Method of Sections

The method of joints was utilized in the previous section to determine all the forces in the members of a truss. It may be convenient sometimes to determine the bar forces in only a few certain members as a check on the results obtained by the method of joints. If the truss is, for example, unsymmetrically loaded and has many joints and members, the method of joints may become a laborious task. A second method, **the method of sections,** has proven to be a useful supplement, or alternative in some cases, to the method of joints. The method of sections consists basically of isolating a portion of the truss by an imaginary cut passing through the members for which the bar forces are desired. The unknown bar forces at a cut section can be determined from the equations of equilibrium applied to the FBD isolated by the cut section. Consider the truss shown in Figure 4.13. All the bar forces in this truss were determined in Example 4.1. Suppose, as a check on the previous solution, the forces in members *CD, CJ,* and *KJ* are to be determined using the method of sections. An imaginary cutting plane is passed through the members *CD, CJ,* and *KJ* and a FBD is drawn for that portion of the truss to the left of the cut as shown in Figure 4.13b. (A FBD for that portion of the truss to the right of the cut could be drawn and similar results would be obtained.) The bar forces in the cut members were internal forces in the truss shown in Figure 4.13a. The cut, however, has ex-

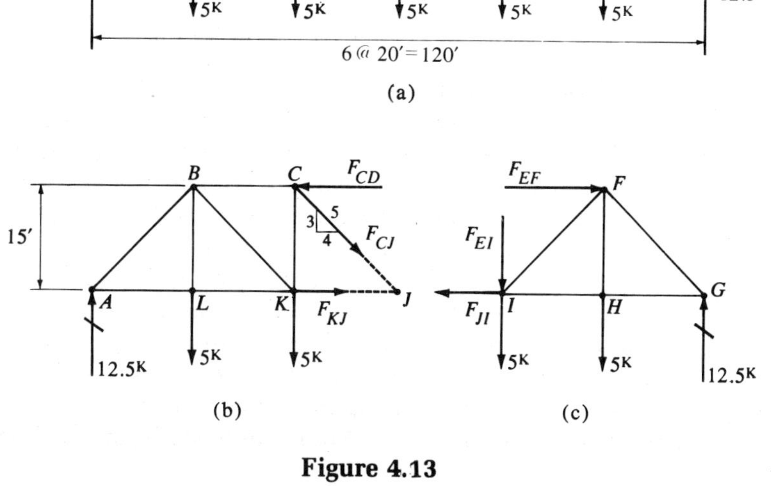

Figure 4.13

posed these internal forces and they are called upon to keep the free body shown in Figure 4.13b in equilibrium. The senses of the bar forces are assumed as indicated in the figure. Three unknown bar forces have been released by the cut and, hence, three equations must be available. The force system acting on the FBD shown in Figure 4.13b is coplanar but not concurrent, i.e., all the forces lie in the same plane but their lines of action do not pass through the same point. All three equations of equilibrium are applicable to this force system. The bar forces may be obtained as follows:

Figure 4.13b:

$$\Sigma F_Y = 0 \uparrow +;\ 12.5 - 5 - 5 - (\tfrac{3}{5})F_{CJ} = 0$$

$$\underline{F_{CJ} = 4.16^K\ t}$$

$$\Sigma M_C = 0 \quad (+\curvearrowright)$$

$$12.5(40) - 5(20) - F_{KJ}(15) = 0$$

$$\underline{F_{KJ} = 26.66^K\ t}$$

It should be noted that point C was selected as a moment center since the lines of action of both F_{CD} and F_{CJ} pass through point C, thus eliminating both from the resulting moment equation. Continuing with the solution,

$$\Sigma M_J = 0 \quad (+\curvearrowright)$$

$$12.5(60) - 5(40) - 5(20) - F_{CD}(15) = 0$$

$$\underline{F_{CD} = 30.0^K\ c}$$

Check:

$$\Sigma F_X = 0 \rightarrow +$$

$$-30.0 + \tfrac{4}{5}(4.16) + 26.66 = 0$$

$$-30.0 + 30.0 = 0$$

$$0 = 0$$

Point J was selected as a moment center again to eliminate F_{CJ} and F_{KC} from the resulting moment equation. After F_{CJ} and F_{KJ} were determined, the force in member CD could have been determined by using $\Sigma F_X = 0$. Since a moment equation was used instead, $\Sigma F_X = 0$ was used as a check. The results obtained using the method of sections check with those obtained in Example 4.1.

When using the method of sections, a decision must be made as to where the cutting section must be passed. Obviously it must cut through the member, or members, for which the bar forces are desired. At the same time, however, the number of additional or undesired members cut should be kept at a minimum in order that the application of one equilibrium equation be sufficient to yield the desired bar force. For example, suppose that the force in member EI of the truss shown in Figure 4.13a is desired. After careful study, an imaginary cutting plane is passed through members EF, EI, and JI and a FBD of that portion of the truss to the right of the cut was drawn as shown in Figure 4.13c. It should be noted that the selected cutting plane released only one unknown bar force with a vertical component and that bar force (F_{EI}) is the only bar force that is to be determined. By summing the forces in the vertical direction, the desired unknown bar force may be determined as follows:

Figure 4.13c:

$$\Sigma F_Y = 0 \uparrow +; -F_{EI} - 5 - 5 + 12.5 = 0$$

$$\underline{F_{EI} = 2.5^K\ C}$$

Careful selection of the cutting plane, therefore, provided an easy and direct solution for the desired bar force. Although a FBD for that portion of the truss to the right of the cut was drawn, as shown in Figure 4.13c, and used in the solution, a FBD of that portion of the truss to the left of the cut could have been utilized to yield a similar result. The reason for isolating the right portion rather than the left portion is convenience. The left portion involves more truss to draw and more forces to consider and, hence, is more time-consuming.

There can be no set rules to follow when using the method of sections. The keynote to applying the method effectively is to carefully study the truss before passing the cutting plane. Select a cutting plane that will yield a desired bar force with each application of an equation of equilibrium. This was accomplished for the bar forces obtained earlier in this section using the method of sections.

Example 4.3

Determine the force in members CD, KJ, ND, and NJ of the K truss shown in Figure 4.14.

Solution:

1. The reactions at A and G shown on the FBD in Figure 4.14b were obtained as follows:

$$\Sigma M_A = 0 \ \circlearrowright +$$

$$10(20) + 10(40) - 120G_Y = 0$$

$$\underline{G_Y = 5^K \uparrow}$$

$$\Sigma F_Y = 0 \uparrow +$$

$$A_Y - 10 - 10 + 5 = 0$$

$$\underline{A_Y = 15^K \uparrow}$$

2. In order to determine the bar forces in CD and KJ, cutting section 1–1 (Figure 4.14b) was passed through the truss. A FBD of that portion of the truss to the left of the cutting section is shown in Figure 4.14c along with

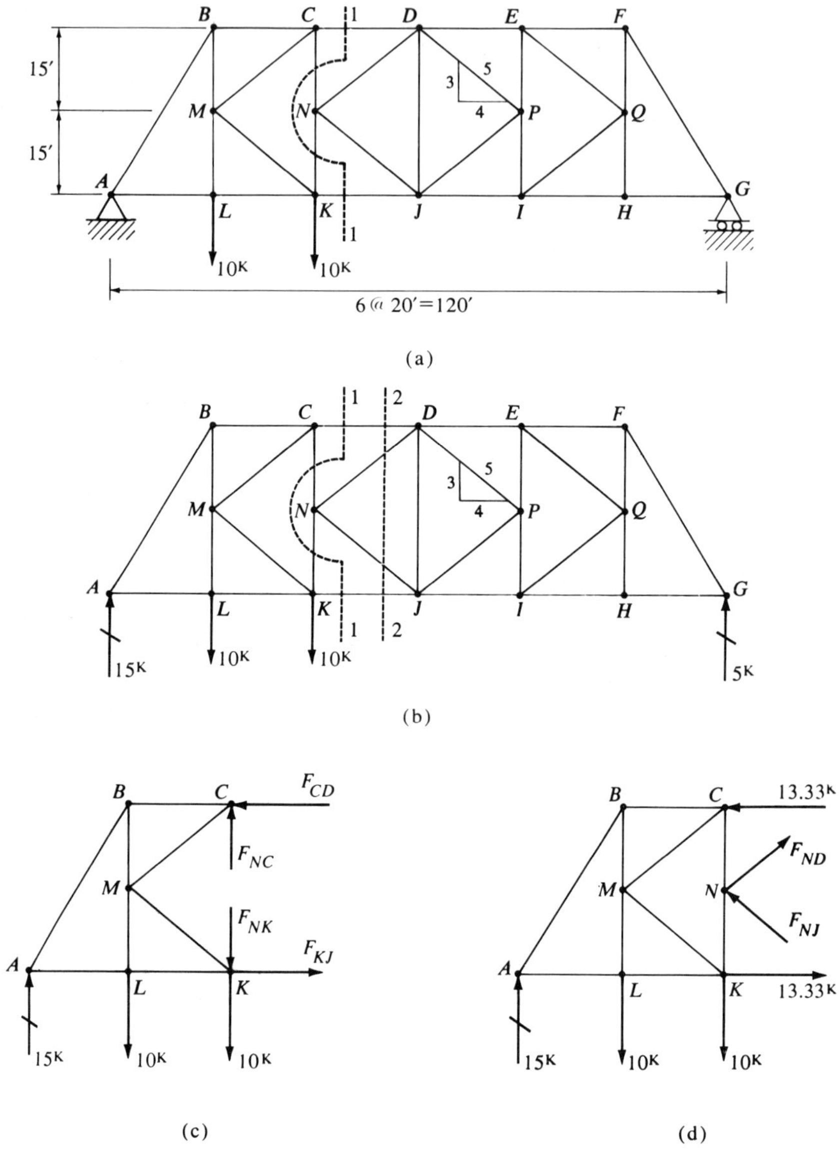

Figure 4.14

the assumed senses for the released bar forces. This particular cutting section was selected because summing moments about point C in Figure 4.14c will eliminate all the unknown bar forces with the exception of F_{KJ} from the resulting moment equation. The desired force F_{KJ} can be obtained with the application of a single equilibrium equation. The same could be said about summing moments about point K for F_{CD}. In this case either point could be selected initially and then the remaining chord force can be determined by summing forces in the x-direction. Or:

Figure 4.14c:

$$\Sigma M_C = 0 \quad (+\curvearrowright)$$

$$15(40) - 10(20) - F_{KJ}(30) = 0$$

$$\underline{F_{KJ} = 13.33^{K}\ t}$$

$$\Sigma F_X = 0 \rightarrow +;\ 13.33 - F_{CD} = 0$$

$$\underline{F_{CD} = 13.33^{K}\ c}$$

A similar cut could be used to determine all the forces in the chords for the truss shown in Figure 4.14a and, hence, could provide a very efficient supplement to the method of joints if all the forces in the members of the truss were to be determined. This cut is typical for the analysis of *K* trusses.

3. A careful study of the truss will reveal that there is no cut possible that will yield either F_{ND} or F_{NJ} with an application of a single equation of equilibrium. After determining the forces in members *CD* and *KJ*, therefore, cutting section 2–2 (Figure 4.14b) is passed through the truss and the corresponding FBD is shown in Figure 4.14d. The desired bar forces may be obtained as follows:

Figure 4.14d:

$$\Sigma F_Y = 0 \uparrow +$$

$$15 - 10 - 10 + (\tfrac{3}{5})F_{ND} + (\tfrac{3}{5})F_{NJ} = 0$$

$$F_{ND} + F_{NJ} = 8.33 \qquad \textbf{(4–3)}$$

$$\Sigma F_X = 0 \rightarrow +$$

$$13.33 - 13.33 + (\tfrac{4}{5})F_{ND} - (\tfrac{4}{5})F_{NJ} = 0$$

$$F_{ND} = F_{NJ} \qquad \textbf{(4–4)}$$

or substituting into Eq. (4–3),

$$2F_{ND} = 8.33$$

$$\underline{F_{ND} = 4.16^{K}\ t}$$

and, $\underline{F_{NJ} = 4.16^{K}\ c}$

Example 4.4

Determine the force in members *CE* and *BF* of the truss shown in Figure 4.15.

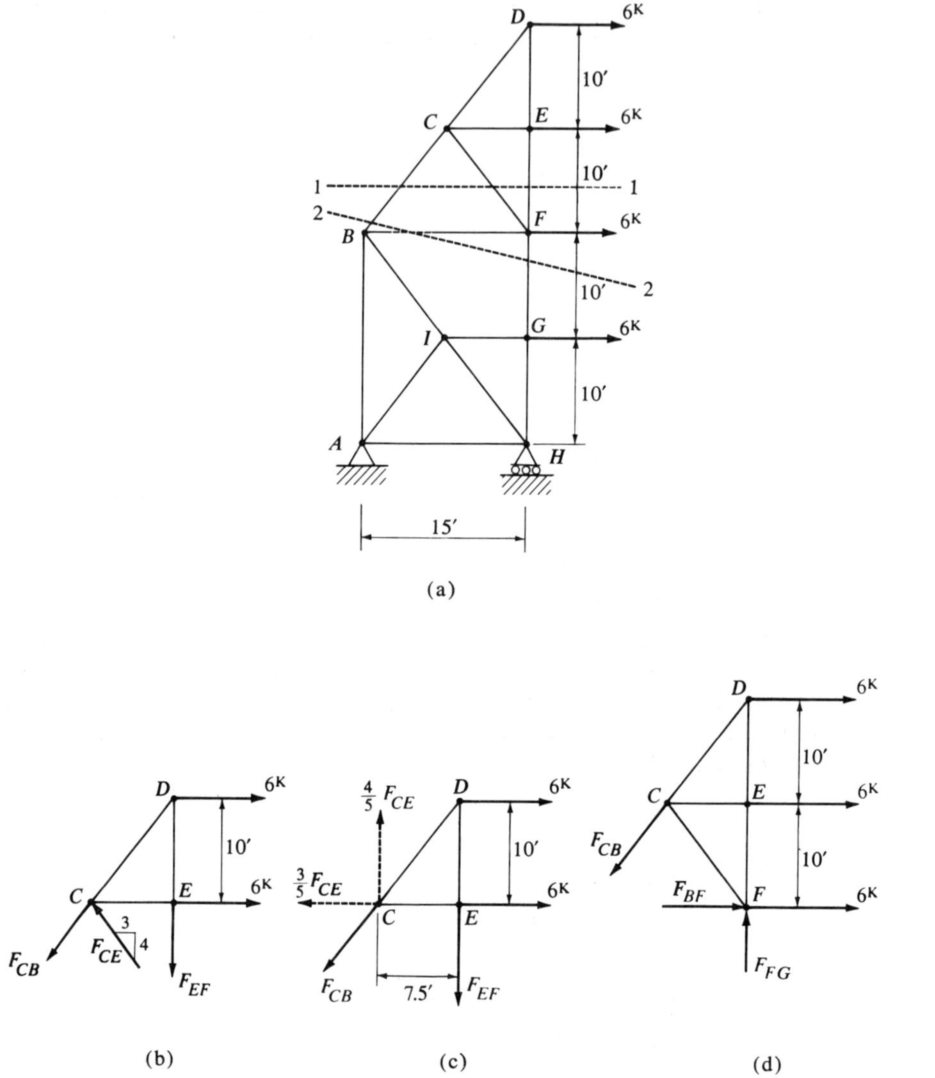

Figure 4.15

Solution:

1. The truss shown in Figure 4.15a is not a typical or commonly used truss. The criterion established in Section 4.4 to determine statical determinancy and statical stability will, therefore, be applied here as follows:

$$m = \text{number of truss members} = 15$$
$$r = \text{number of reaction components} = 3$$
$$j = \text{number of joints (including reaction points)} = 9$$
$$m + r - 2j = 15 + 3 - 2(9) = 18 - 18 = 0$$

The truss is both statically determinate and stable.

2. It would seem that the next step in the solution would be to solve for the reactions at A and H. If, however, a horizontal cutting plane is passed through the truss and that portion of the truss above the cut is isolated as a FBD, there would be no need for the reactions to fulfill the requirements of this particular example. Considering member CE, cutting section 1–1 (Figure 4.15a) is passed through the truss and the top portion of the truss is chosen as a free body as shown in Figure 4.15b. Three unknown forces are involved: F_{CE} (the desired bar force) and two additional forces. If point D is selected as a moment center, the lines of action of the forces F_{CB} and F_{EF} pass through point D and, hence, both will be eliminated from the moment equation. F_{CE} is resolved into x and y components at point C as shown in Figure 4.15c. The direction of the line of action of F_{CE} was determined from triangle CEF in Figure 4.15a, which is a 3:4:5 triangle. Summing moments, the solution proceeds as follows:

Figure 4.15c:

$$\Sigma M_D = 0 \quad (\curvearrowright +)$$

$$\left(\tfrac{4}{5}\right)F_{CE}(7.5) + \left(\tfrac{3}{5}\right)F_{CE}(10) - 6(10) = 0$$

$$\underline{F_{CE} = 5.0^{K}\ c}$$

It should be noted that F_{CE} was determined from the application of a single moment equation because of the proper cutting section and corresponding moment center used.

3. Considering member BF, cutting section 2–2 (Figure 4.15a) is passed through the truss and the top portion of the truss is chosen as a free body as shown in Figure 4.15d. Again, three unknown forces are involved. Two of these may be eliminated by summing moments about point D, which yields a moment equation containing only one unknown force (F_{BF}), the desired force. The solution is:

$$\Sigma M_D = 0 \quad (\curvearrowright +)$$

$$-6(10) - F_{BF}(20) - 6(20) = 0$$

$$F_{BF} = -9.0^{K}\ c$$

$$\therefore \underline{F_{BF} = 9.0^{K}\ t}$$

The sense of F_{BF} was chosen assuming member BF to be in compression; the negative sign obtained indicated that the member is in tension.

4. This example was chosen to demonstrate that the proper choice of a cutting section can eliminate time-consuming calculations. In this case, even the reactions were not necessary and both required bar forces were obtained with the application of a single moment equation. The selection of the proper cutting section is not obvious and, hence, careful study should precede its selection.

Example 4.5

Compute the bar forces in members UR, US, RQ, CD, UQ, DQ, and DV of the subdivided truss shown in Figure 4.16a.

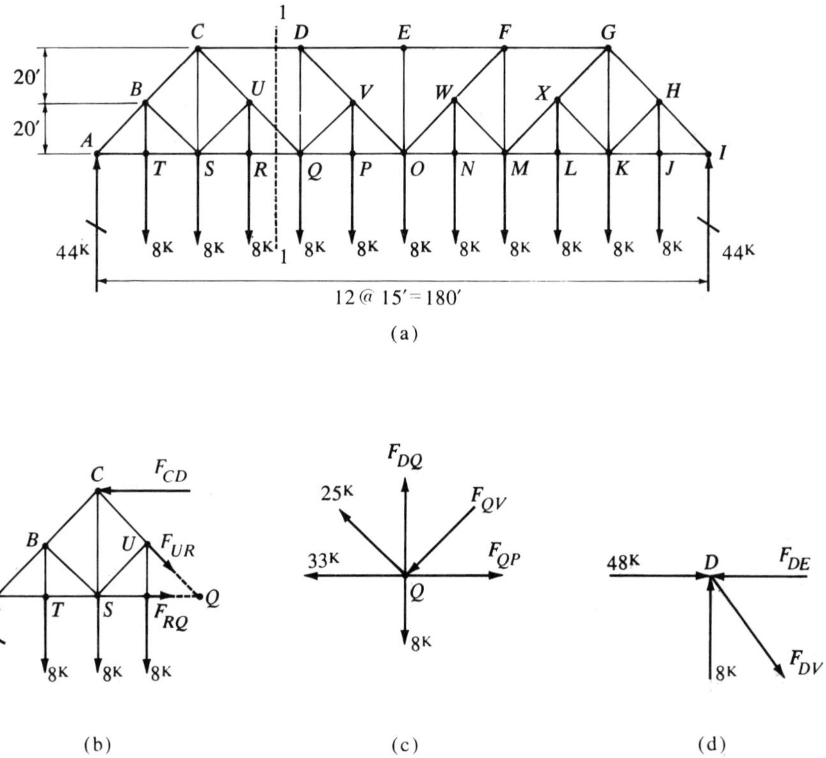

Figure 4.16

Solution:

1. The lengths AS, SQ, QO, OM, MK, and KI in Figure 4.16a are generally referred to as **panel lengths.** A panel length is also the distance between floor beams (see Figure 4.2). The area, say $CDQS$ in Figure 4.16a, is called a **panel.** Since the panel lengths of the truss shown in Figure 4.16a are further divided into two lengths (panel length AS, for example, is divided into lengths AT and TS), the truss is said to be **subdivided,** or has **subdivided panels.** Trusses with subdivided panels are used to counteract heavy floor systems, i.e., the loads acting at the panel points are reduced with the additional panel subdivision. The short diagonals, such as US in Figure 4.16a, are usually referred to as **substruts** and the small verticals, such as UR in Figure 4.16a, are known as **hangers.**
2. To solve for the force in hanger UR, imagine a FBD of the joint at R of the truss shown in Figure 4.16a. The only vertical forces acting at the joint are the 8-kip applied load and the force in UR. By inspection, therefore, F_{UR}

must be 8-kips tension. The following rule can be used next: **the substrut has a vertical component equal to one-half of the hanger force.** Considering member *US*, therefore,

$$\text{Vertical component of } US = (\tfrac{1}{2})F_{UR} = 4^{\mathrm{K}}$$
$$(\tfrac{4}{5})F_{US} = 4^{\mathrm{K}}$$
$$F_{US} = 5^{\mathrm{K}}$$

A substrut, by definition, is a compression-carrying member. Hence,

$$\underline{F_{US} = 5^{\mathrm{K}}\ c}$$

If the truss in Figure 4.16a were to be used as a deck truss (see Figure 4.6) rather than a through truss as shown, the force in the hanger *UR* would be 8-kips compression and member *US* would carry a 5-kip tension force. Member *US* would then be called a **subtie,** which is by definition a tension member. The classification of substrut or subtie is, therefore, a function of whether the hanger force is compression or tension.

3. To obtain the forces in members *RQ*, *CD*, and *UR*, cutting section 1–1 is passed through the truss and the left portion of the truss is shown as a free body in Figure 4.16b. The solution proceeds as follows:

Figure 4.16b:

$$\Sigma M_Q = 0 \quad (+\curvearrowright)$$

$$44(60) - 8(45 + 30 + 15) - (40)F_{CD} = 0$$
$$\underline{F_{CD} = 48.0^{\mathrm{K}}\ c}$$

$$\Sigma M_C = 0 \quad (+\curvearrowright)$$

$$44(30) - 8(15) + 8(15) - (40)\,F_{RQ} = 0$$
$$\underline{F_{RQ} = 33.0^{\mathrm{K}}\ t}$$

$$\Sigma F_Y = 0 \uparrow +$$
$$44 - 8 - 8 - 8 - (\tfrac{4}{5})F_{UR} = 0$$
$$\underline{F_{UR} = 25.0^{\mathrm{K}}\ t}$$

It should be noted again that the proper selection of the cutting section and moment center leads to a simple and direct solution for the required bar forces.

4. To find the force in vertical *DQ*, joint *Q* will be isolated as shown in Figure 4.16c. The method of joints is applied as follows:

Figure 4.16c:

$$\underline{F_{QV} = 5^{\mathrm{K}}\ c} \quad \text{(substrut rule)}$$

$$\Sigma F_Y = 0 \uparrow +$$
$$\tfrac{4}{5}(25) + F_{DQ} - \tfrac{4}{5}(5) - 8 = 0$$
$$F_{DQ} = -8.0^K\ c$$

The sense of F_{DQ} was incorrectly assumed as tension (see Figure 4.16b) since a negative answer was obtained. Hence:

$$\underline{F_{DQ} = 8.0^K\ t}$$

5. The force in DV, which is the upper portion of diagonal DO, is determined by applying the method joints to joint D as follows:

   Figure 4.16d:

$$\Sigma F_Y = 0 \uparrow +$$
$$8 - (\tfrac{4}{5})F_{DV} = 0$$
$$\underline{F_{DV} = 10.0^K\ t}$$

6. The members selected for solution in the example consisted of one of each type of member contained in the truss or, more specifically:
   1) UR—hanger
   2) US—substrut
   3) RQ—lower chord
   4) CD—upper chord
   5) UQ—lower portion of a diagonal
   6) DQ—vertical
   7) DV—upper portion of a diagonal

   Both the method of joints and the method of sections were utilized effectively for the determination of the forces in the members listed. If this order of solution is repeated again and again, eventually the forces in all the members will be determined. For a subdivided truss, however, an order of solution can be followed for an effective determination of all the bar forces. The order of solution is basically the same as followed in this example. The order of solution and method to be used (as in this example) are as follows:

   1) All hangers; by inspection
   2) All substruts; substrut rule
   3) All chords; method of sections, sum moments
   4) Lower portion of diagonal; sum vertical forces at cuts used in step 3
   5) All verticals; method of joints, ΣF_Y at joint
   6) Upper portion of diagonal; method of joints, ΣF_Y at joint.

4.6 Method of Shears and Moments

The **method of shears and moments** is a special method of solution used primarily for all types of parallel chord trusses, i.e., the top and bottom chords are parallel. The method, as presented here, will be further limited to parallel chord trusses with equal panel lengths. Consider the parallel chord truss shown in Figure 4.17. Furthermore, let us cut the truss at sections 1–1 and 2–2. The corresponding FBDs are shown in Figures 4.17b and 4.17c. Considering Figure

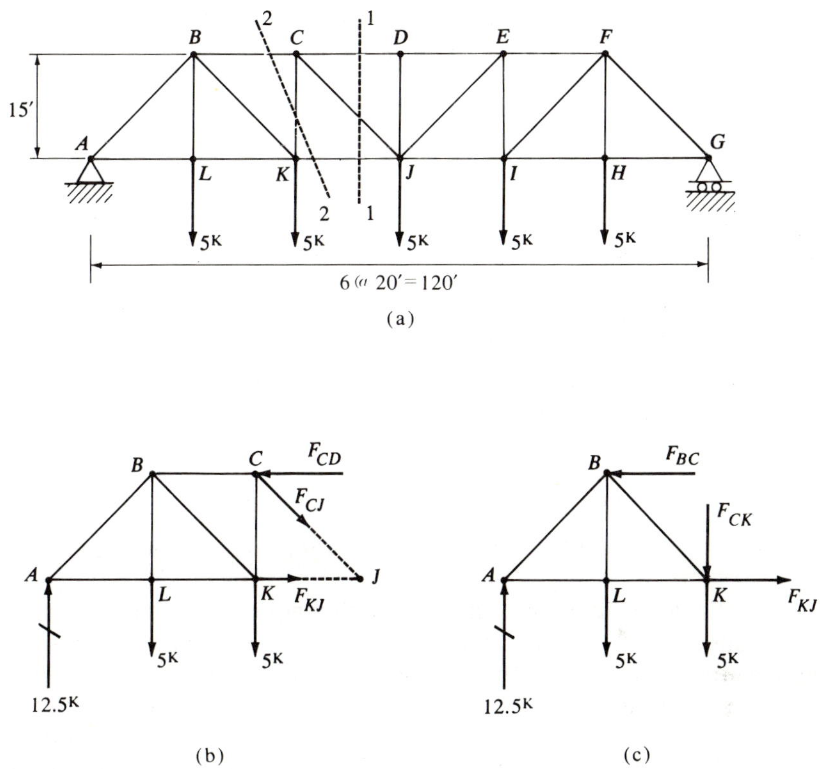

Figure 4.17

4.17b, there is an unbalanced vertical force acting at the cut section, or in the panel *KJ*. This unbalanced force is equal to the sum of the vertical forces to the left of the cut, or to the left of the panel *KJ*. This force has been previously defined as shear. Thus, the shear *V* in the panel *KJ* can be obtained as follows:

Figure 4.17b:

$$V = 12.5 - 5 - 5 = 2.5^{K} \uparrow$$

For equilibrium, this shear force must be resisted in the panel. The only resistance available in the vertical direction in the panel *KJ* is provided by the vertical component of the diagonal *CJ*. Thus:

$$(\tfrac{3}{5})F_{CJ} = V = 2.5^K$$
$$\underline{F_{CJ} = 4.16^K\ t}$$

The diagonals, therefore, carry the shear in the panel. The shear force acting in the panel is upward and, hence, the vertical component of F_{CJ} must be directed downward giving rise to a tension force in *CJ*. Similarly, the force in vertical *CK* can be obtained from Figure 4.17c. Thus:

$$F_{CK} = V = 2.5^K$$
$$\underline{F_{CK} = 2.5^K\ c}$$

Again, the force in *CK* must be directed downward to balance the upward shear force; hence, the bar force is compression. The shear in any panel may be determined very conveniently by drawing the shear diagram and the bar forces in the diagonals and verticals may be obtained without the aid of FBDs.

Referring again to the FBD shown in Figure 4.17b, there is a net moment produced by the applied loads and reaction at the section that must be resisted, or balanced, by the couple produced by the bar forces in members *CD* and *KJ*. In equation form this becomes:

$$12.5(40) - 5(20) = (15)F_{KJ};\ \underline{F_{KJ} = 26.6^K\ t}$$
$$12.5(60) - 5(40) - 5(20) = 15F_{CD};\ \underline{F_{CD} = 30.0^K\ c}$$

Two equations are involved since in the first equation moments were summed about panel point *K*, and, in the second equation, moments were summed about panel point *C*. The two moment equations are required because there are two unknowns to be determined: F_{KJ} and F_{CD}. Since the net moment produced by the applied loads and the reaction is clockwise at both points *K* and *C*, a tension force in *KJ* and a compression force in *CD* are required to produce a balancing counterclockwise moment. Notice that the bending moment produced by the external loads acting on the truss in Figure 4.17a about any bottom panel point is the same as that about any top panel point. The bending moments at the bottom panel points may be easily determined by drawing the moment diagram. The shear and moment diagrams for the truss being considered are shown in Figure 4.18. Suppose the bar forces in members *BL*, *BC*, and *LK* are to be determined using the methods of shears and moments. From the shear diagram in Figure 4.18b, the shear in panel *LK* is 7.5 kips. This must be carried by the vertical component of diagonal *BK*. Thus:

$$7.5 = \tfrac{3}{5}F_{BK}$$
$$\underline{F_{BK} = 12.5^K\ t}$$

The shear in the panel is positive and, therefore, would tend to push the left portion of the truss upward. The vertical component of F_{BK} must be directed downward to counteract this upward movement; hence, *BK* is in tension. To obtain F_{LK} we would cut through panel *LK* and sum moments about point *B*.

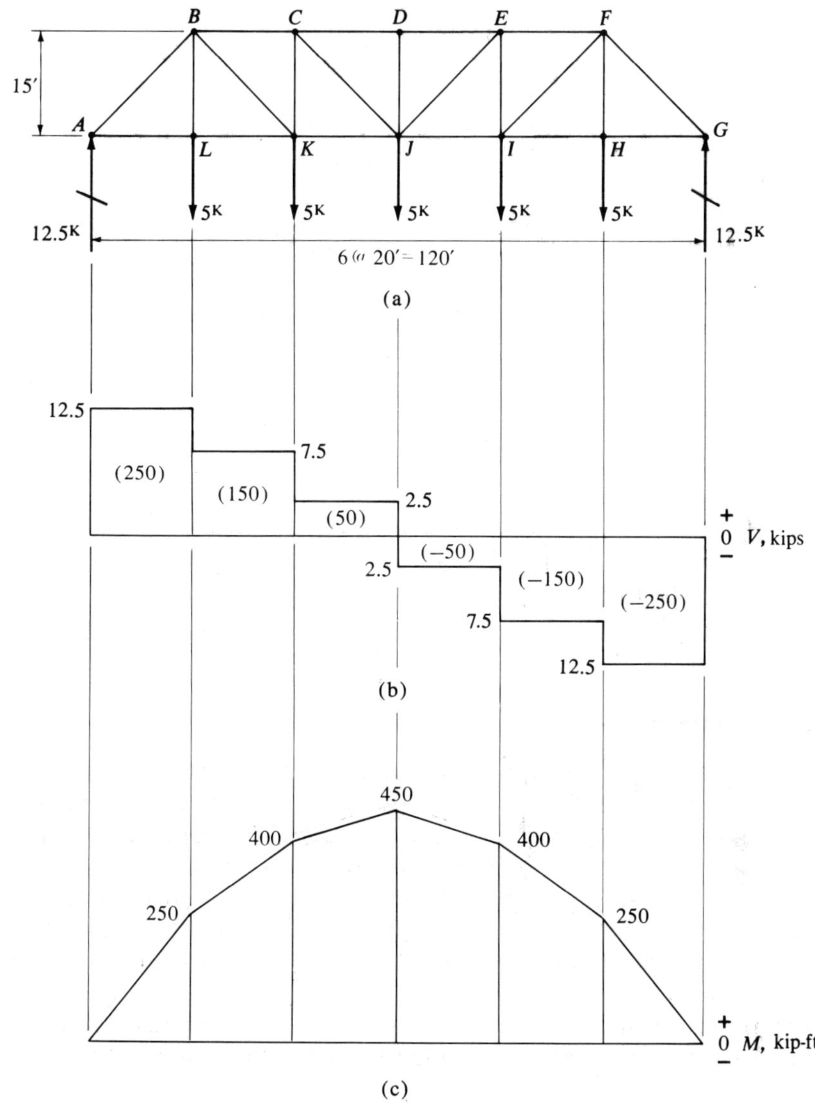

Figure 4.18

From the moment diagram shown in Figure 4.18c, the moment about B is 250 kip-ft. The bar force in LK is obtained as follows:

$$250 = 15\,F_{LK}$$
$$F_{LK} = 16.66^{K}\ t$$

To obtain F_{BC}, we would again cut through panel LK and sum moments about point K. From the moment diagram in Figure 4.18c, the moment about K is 400 kip-ft. Thus:

$$400 = 15\,F_{BC}$$
$$F_{BC} = 26.6^{K}\ C$$

The sense of the chord bar forces, i.e., whether tension or compression, is determined from the fact that the chords must produce a counterclockwise couple to resist the bending moment about each panel point which is clockwise as seen from the moment diagram.

The method of shears and moments is limited primarily to parallel chord trusses with equal panel lengths. It is presented here not so much as an alternative method but as a visual explanation of the behavior of trusses under loads and how those loads are carried by this important structure. Bending moment along the truss length is resisted by couples produced by the chord bar forces. Hence, this practical and economical observation can be made: moment-carrying capacity can be increased in a truss by increasing the distance between the chords up to a certain limit. Small forces in the chords with a large distance between the chords still produce a large couple available to resist bending moments along the span. Small chord forces mean an economical design since less chord material will be required to carry the small loads. Increased truss heights eventually, however, increase fabrication and construction costs. Shear along the span length is carried by the diagonals and verticals. Although analyzed in entirely different methods, both beams and trusses must be designed to resist shear and moment.

4.7 Space Trusses

When several straight members are joined together at their ends to form a three-dimensional configuration, the structure so obtained is called a **space truss.** Such a structure is shown in Figure 4.19. The ideal assumptions made for the plane trusses, such as no member being continuous through a joint and all loads and reactions being applied at the joints, are also applicable to the space trusses. Rather than smooth (frictionless) pins joining the members, members of a space truss are joined together by means of ball-and-socket connections.

Although the principles for the analysis of space trusses are basically similar to those used in the analysis of plane trusses, the solution of a space truss

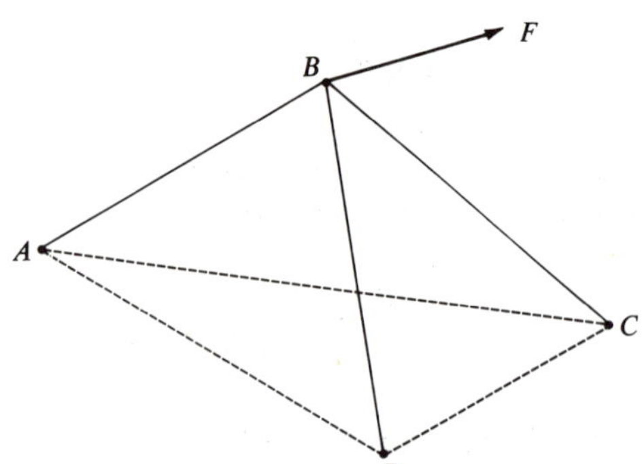

Figure 4.19

becomes somewhat more complex and laborious due to the three-dimensional geometry. Since a ball-and-socket joint provides three components of constraint, i.e., components in the x, y, and z direction, the conditions of equilibrium for each joint must be satisfied by the three equations, $\Sigma F_X = 0$, $\Sigma F_Y = 0$, $\Sigma F_Z = 0$. Hence, when applying the method of joints to such a truss, care should be exercised to select joints in such an order that no selected joint will contain more than three unknown bar forces.

Problems

4–1. Classify the following trusses as to statical determinancy and stability.

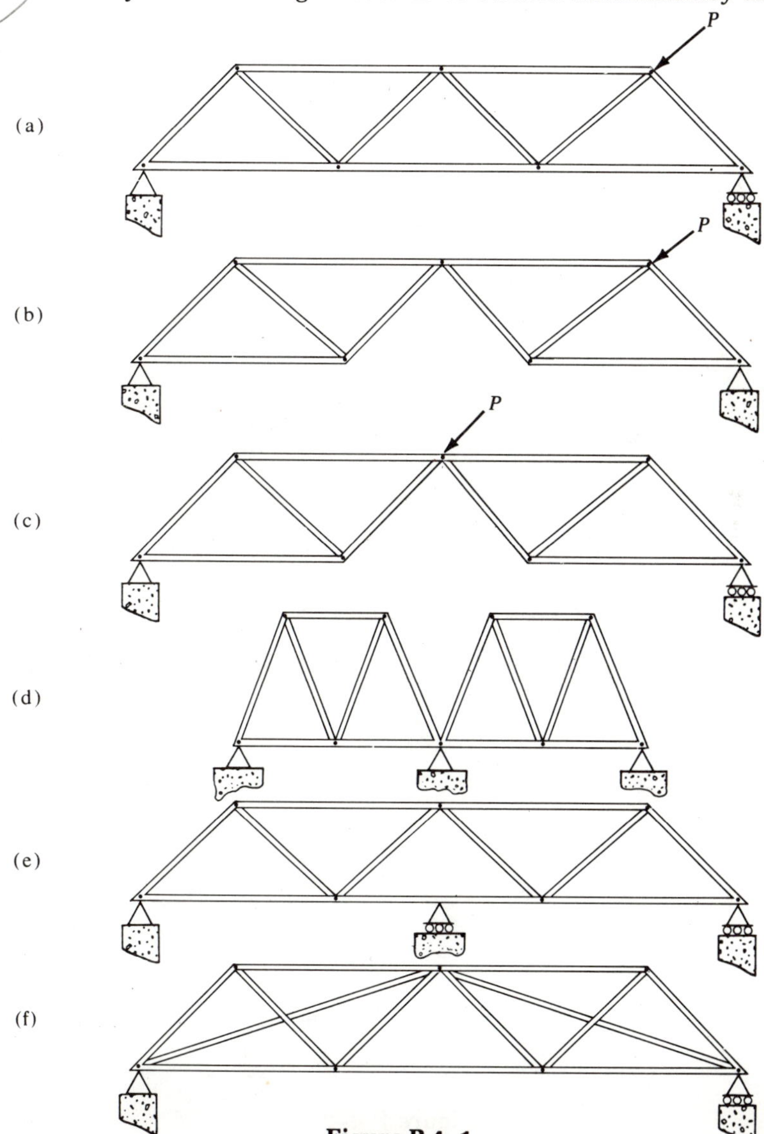

Figure P 4–1

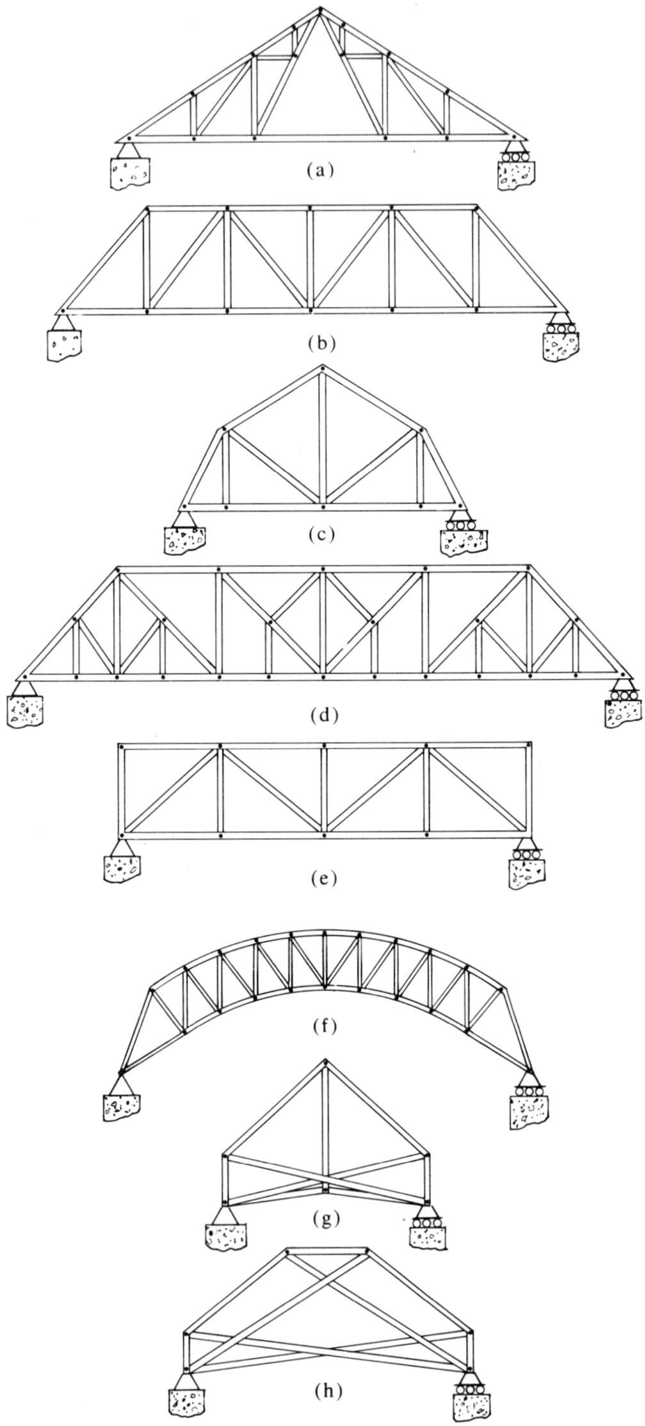

Figure P 4–2

4–2. Classify the trusses shown as being simple, compound, or complex.

4–3. A solid sign measuring 10-ft high by 30-ft long is supported by two steel trusses as shown. One truss is located at each end of the sign. Determine the bar forces due to a wind pressure of 30 lb per square foot.

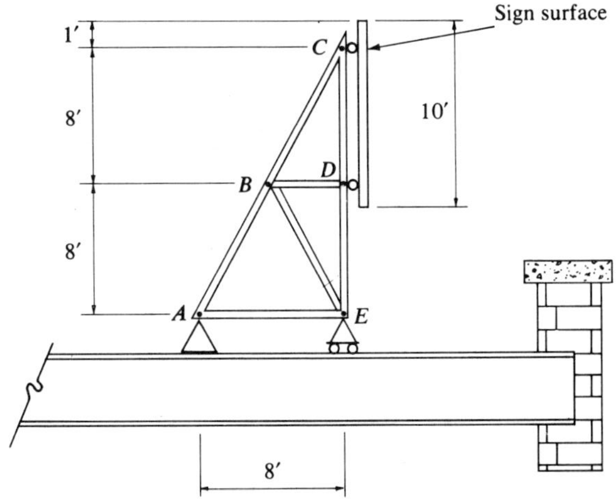

Figure P 4–3

4–4. Compute the bar forces in all members of one truss of the transmission tower shown.

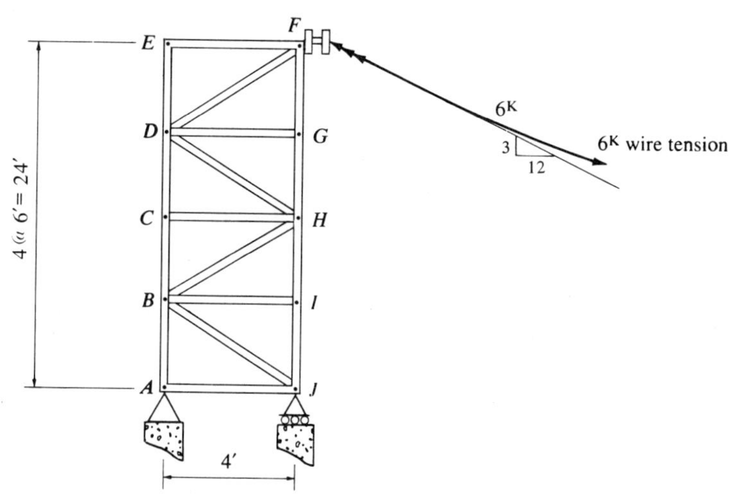

Figure P 4–4

4–5. A lifting rack for storage is fabricated as shown. Assume the truss weight to be negligible, and determine all the bar forces. The storage bin and its contents weigh 2500 lb.

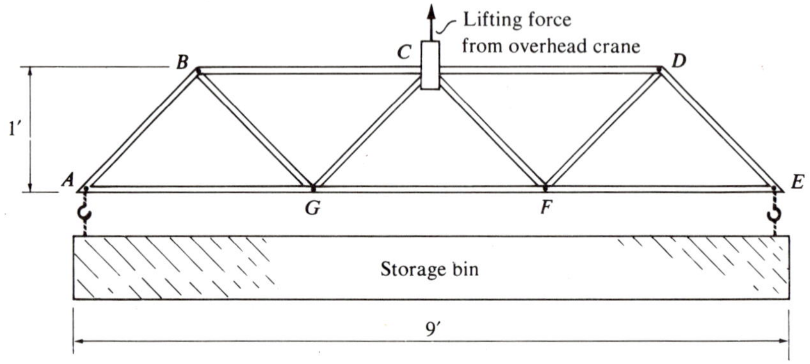

Figure P 4–5

4–6. Determine the force in all the members of the canopy truss shown.

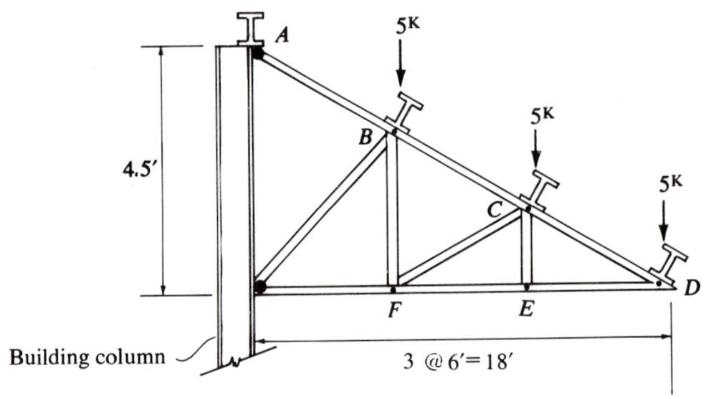

Figure P 4–6

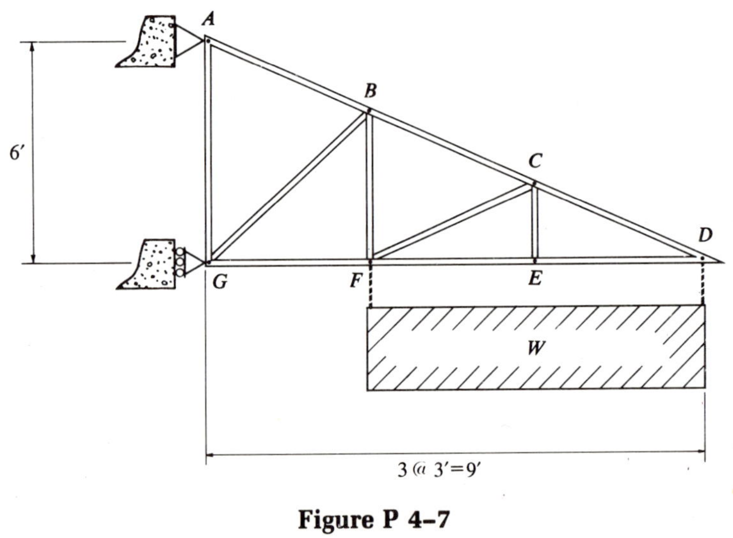

Figure P 4–7

4–7. Determine the maximum allowable magnitude of the weight W for the sign supported by the truss shown in Figure P–47 if the maximum allowable bar force that can be tolerated is 6 kips.

4–8 to 4–10. For the trusses shown, determine all the bar forces.

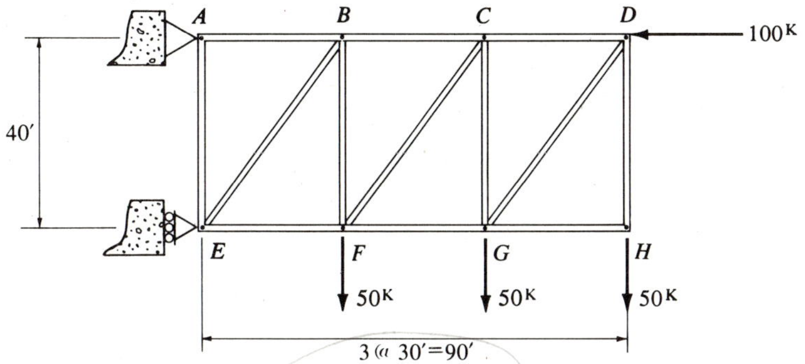

Figure P 4–8

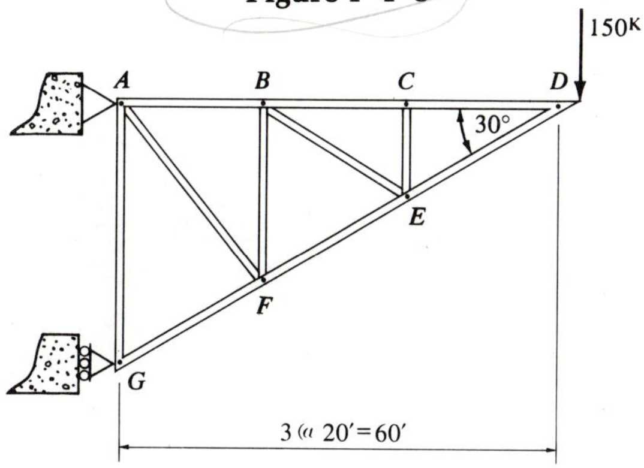

Figure P 4–9

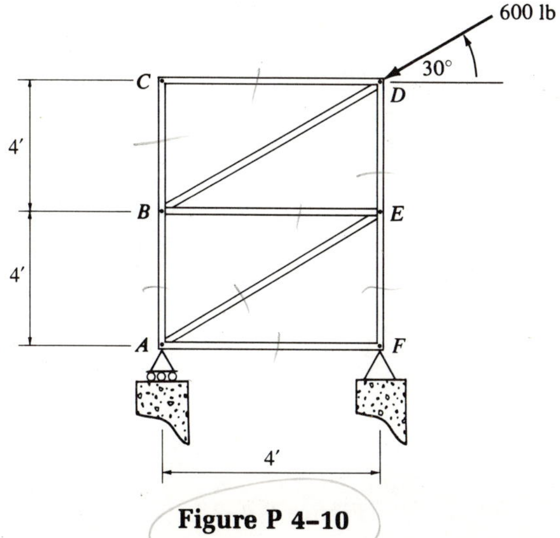

Figure P 4–10

4–11. Determine the bar forces in members *BC*, *CL*, and *HL*.

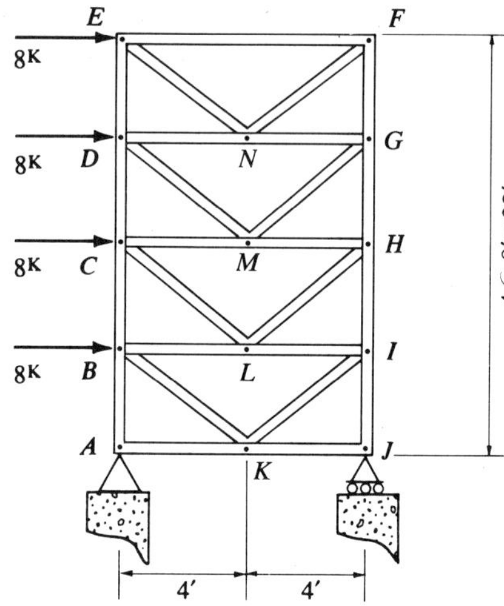

Figure P 4–11

4–12. Determine the forces in members *AB*, *BH*, and *DF* of the truss shown.

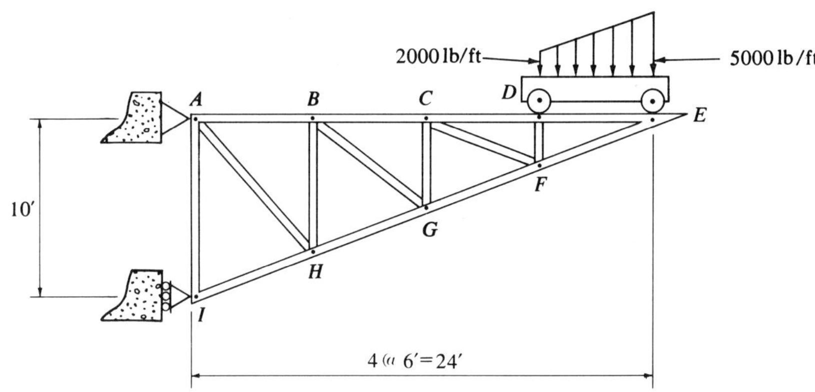

Figure P 4–12

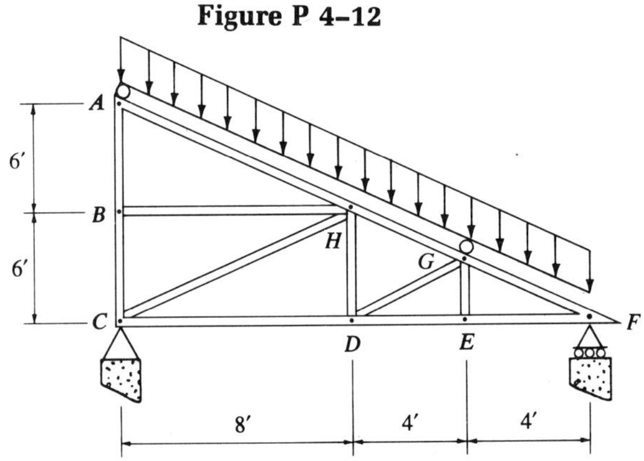

Figure P 4–13

4–13. The aggregate chute shown is supported by a truss at points A and G. Determine the force in member HD if the chute and its contents when full weigh 150-lb per running foot.

4–14. Determine the forces in members DE, EJ, KP, and OP of the truss shown.

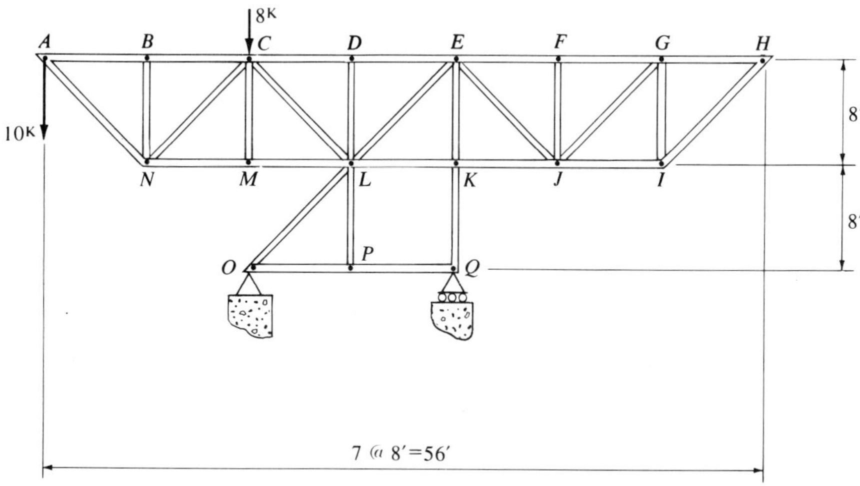

Figure P 4–14

4–15. Determine the force in member MI of the truss shown.

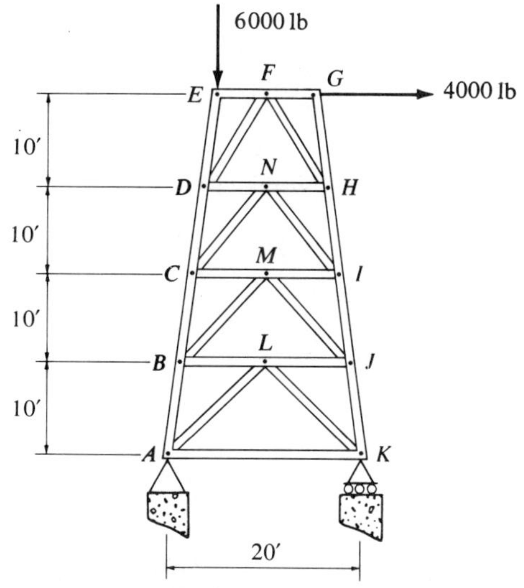

Figure P 4–15

4–16. Find the forces in members CD, JI, LD, and LI of the truss shown.

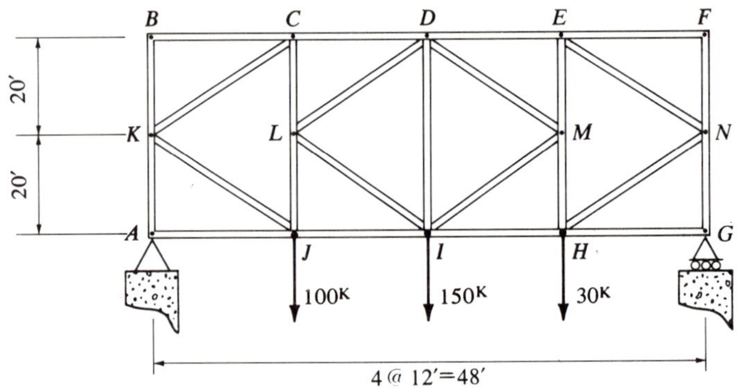

Figure P 4–16

4–17. Find the forces in members CD, ND, NJ, and KJ of the K truss shown.

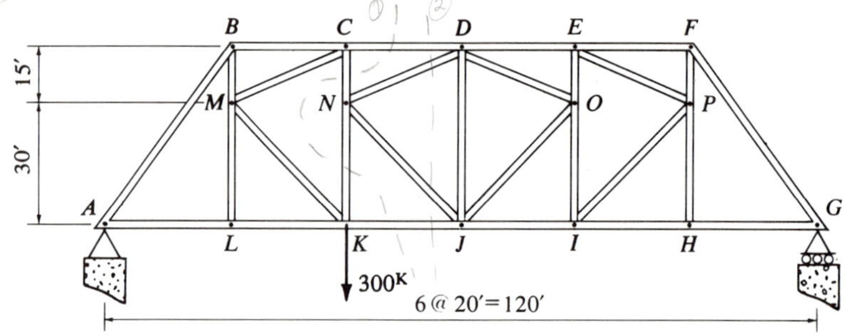

Figure P 4–17

4–18. Find the forces in members BN, CN, ML, OC, and CM of the truss shown.

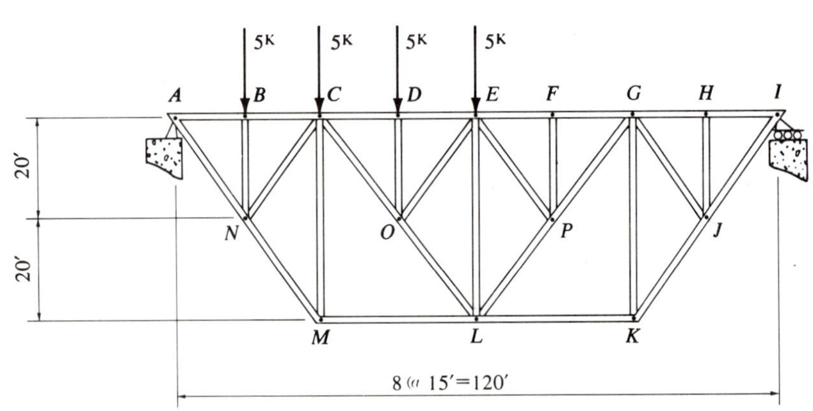

Figure P 4–18

4–19. Solve Problem 4–16 by the method of shears and moments.

4–20. Solve for *BC*, *HG*, *BG*, and *CG* by the method of shears and moments.

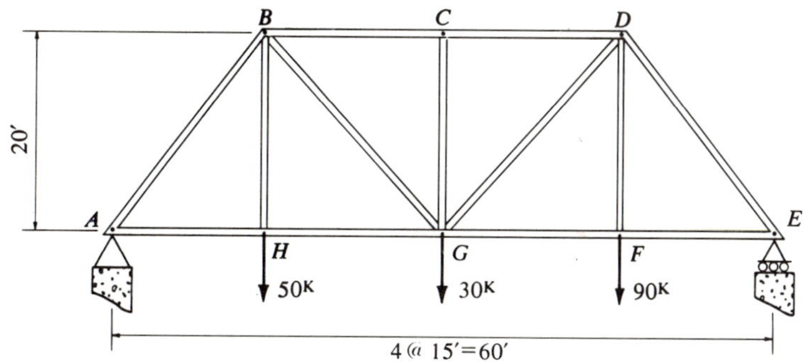

Figure P 4–20

4–21. Find the forces in all members of the roof truss shown.

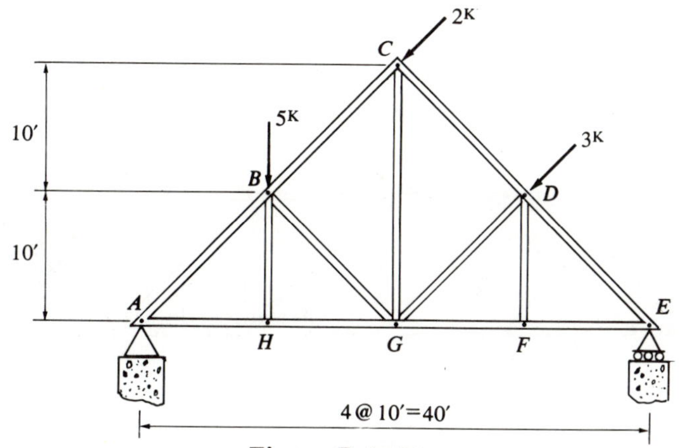

Figure P 4–21

4–22. Find the forces in members *DE*, *AE*, and *AL* of the parallel chord truss shown.

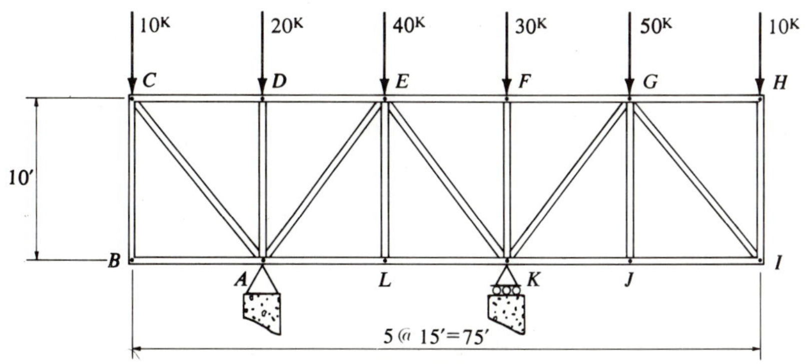

Figure P 4–22

4–23. Determine the forces in members *DI*, *CD*, and *CI* of the truss shown.

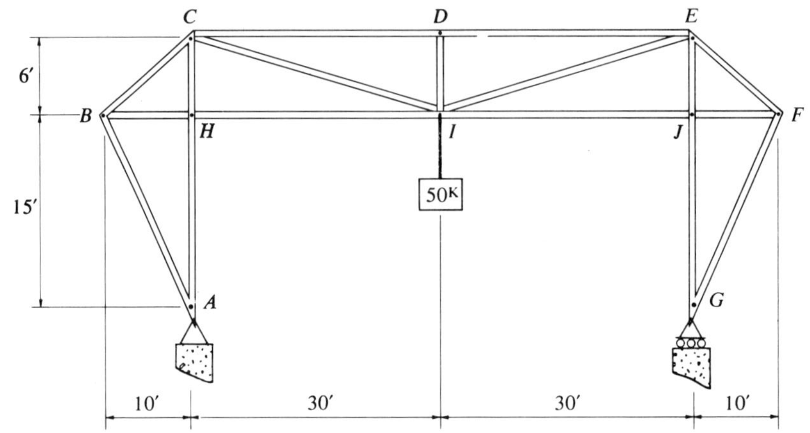

Figure P 4–23

4–24. Solve for the bar forces in members *CD*, *DV*, *UQ*, *SU*, and *UR* of the subdivided truss shown.

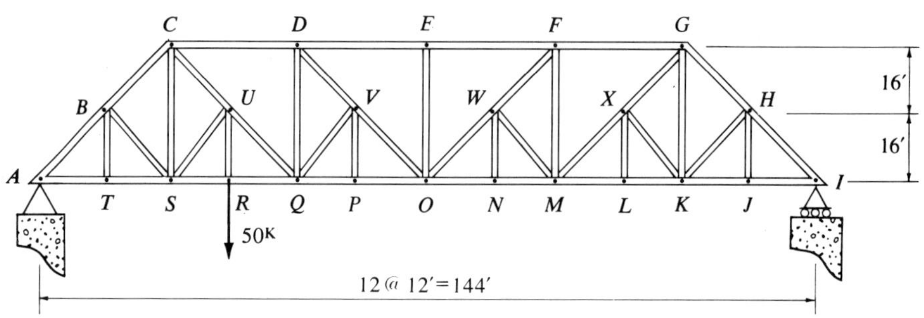

Figure P 4–24

CHAPTER FIVE

Rigid Frames, Arches, and Composite Structures

5.1 Rigid Frames

A **rigid frame** is a basic structural unit, just as are the beam and truss. Recall that the members of a truss are joined together by smooth pins incapable of transmitting moment. The members of a rigid frame are rigidly joined together by joints that are capable of resisting both force and moment. Thus, a moment-resistant joint can transmit both a force and couple from one member to the other members forming the joint. The difference between a pin-connected joint and a rigid joint is illustrated by the A-frame and corresponding FBDs shown in Figures 5.1a and b.

The rigid frame can take on a great variety of forms and, hence, has become one of the commonest and most important structures used in engineering practice. This is true particularly in building construction where the rigid frame forms the skeleton of the building. These rigid frameworks are three-dimensional assemblages of members, but for purposes of analysis and design, planar portions are isolated, thereby simplifying the problem. Some typical isolated rigid frames are shown in Figure 5.2. Such frames may be of steel or reinforced concrete construction. For steel rigid frames, the rigid joints may be formed by riveted, bolted, or welded connections. In reinforced concrete structures, rigid joints are obtained by **monolithic construction,** i.e., the joining members are formed and poured simultaneously to form one rigid unit.

Rigid frames are generally highly statically indeterminate. The scope of this chapter will be limited to statically determinate rigid frames. A frame, statically determinate or indeterminate, is completely analyzed when the variation in axial force, shear, and moment is known along the length of each member. These variations will be treated in this chapter and will provide a

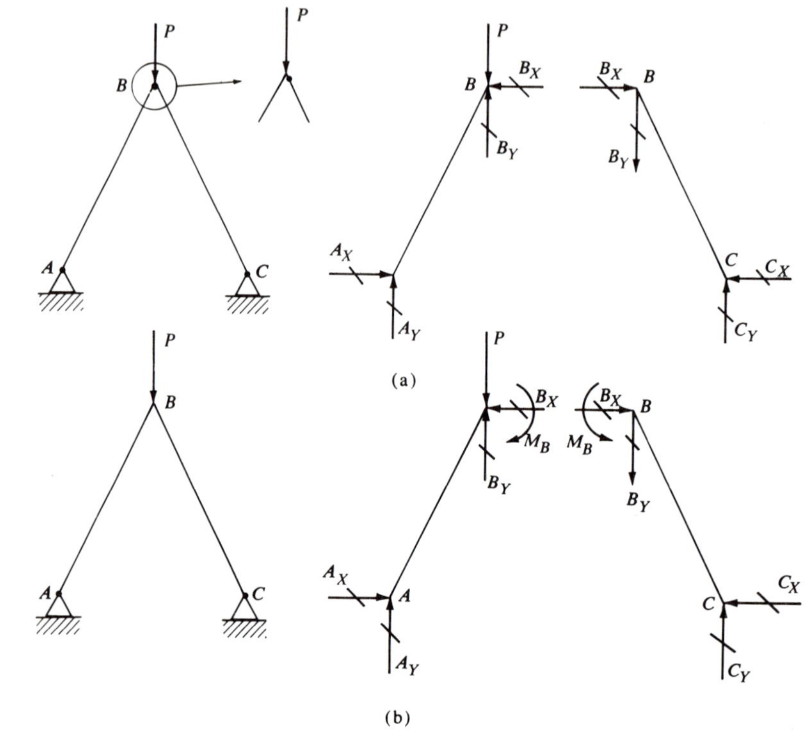

Figure 5.1

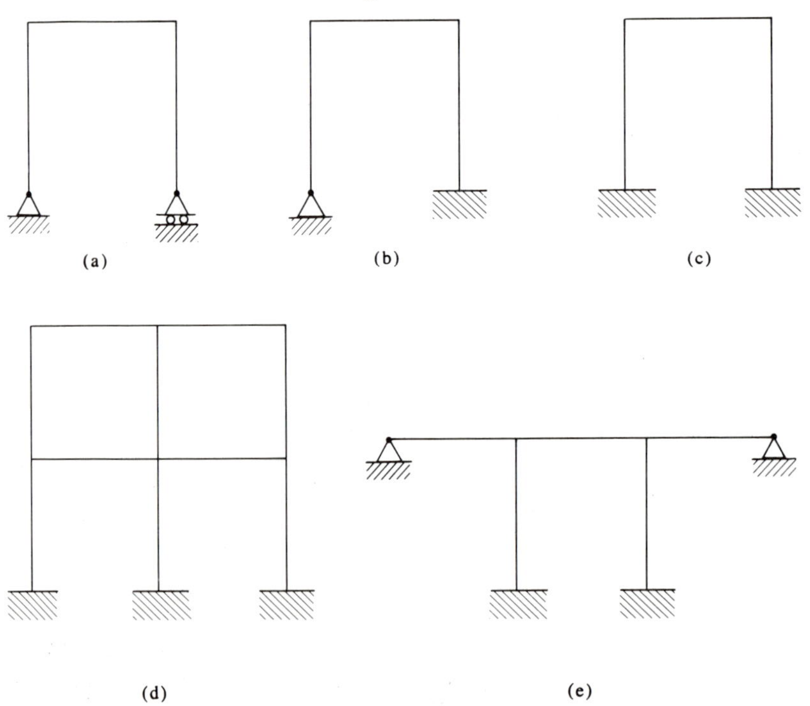

Figure 5.2

basis for the treatment of statically indeterminate frames in a later chapter. The analysis of statically determinate rigid frames will be illustrated by the following examples.

Example 5.1

Completely analyze the simply supported rigid frame shown in Figure 5.3a.

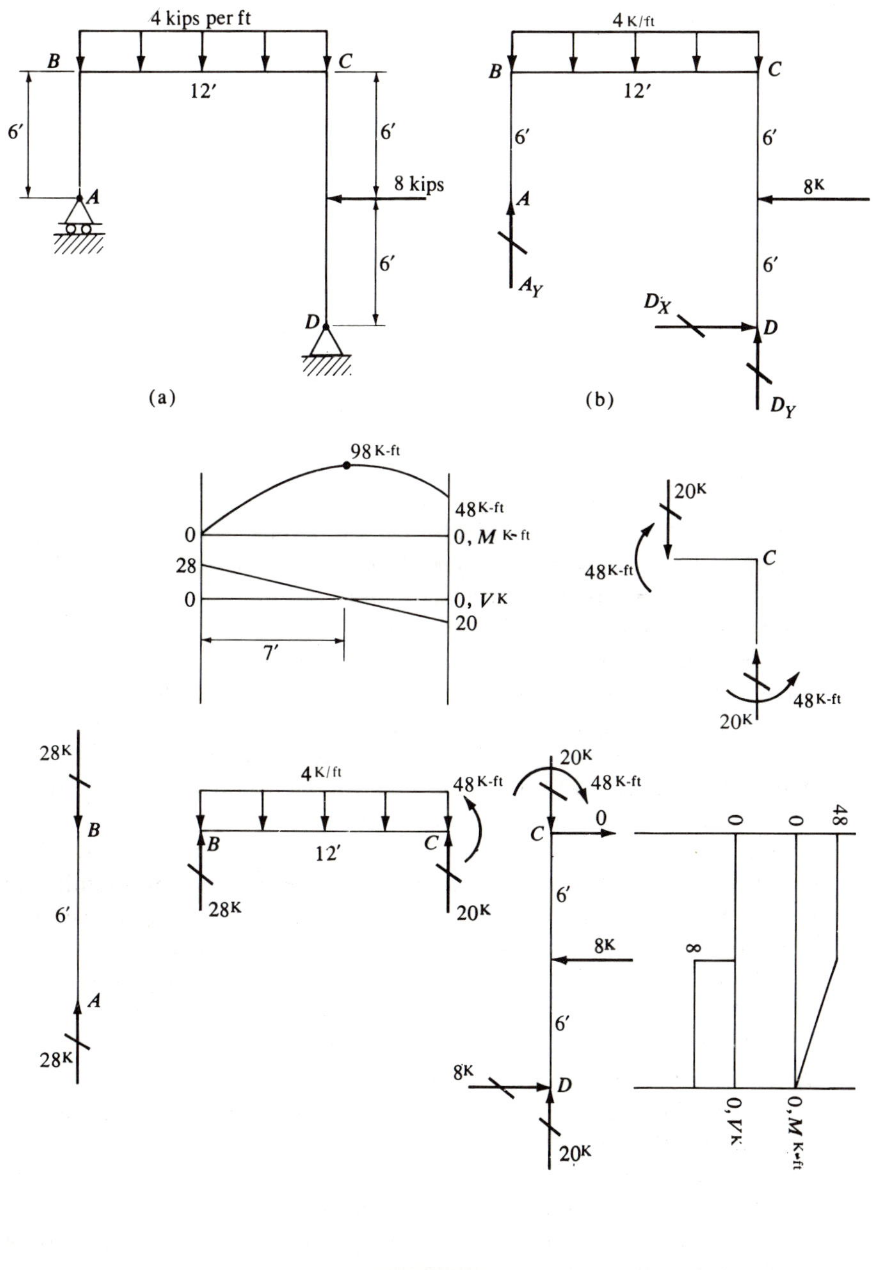

Figure 5.3

Solution:

1. The analysis of this frame can be accomplished by drawing the V and M diagrams for each member of the frame. As the initial step, the reactions must be determined. A FBD of the frame is shown in Figure 5.3b. There are three unknown reactions and, hence, the three equations of equilibrium are sufficient to provide a solution for the reactions. Referring to Figure 5.3b:

$$\Sigma F_X = 0 \rightarrow +$$
$$D_X - 8 = 0$$
$$\underline{D_X = 8.0 \text{ kips} \rightarrow}$$

$$\Sigma M_D = 0 \quad \circlearrowright +$$
$$A_Y(12) - (4)(12)\left(\tfrac{12}{2}\right) - 8(6) = 0$$
$$\underline{A_Y = 28 \text{ kips} \uparrow}$$

$$\Sigma F_Y = 0 \uparrow +$$
$$28 - 4(12) + D_Y = 0$$
$$\underline{D_Y = 20 \text{ kips} \uparrow}$$

2. With all the external forces acting on the frame known, each member can now be isolated and the forces acting on it determined. With these forces known, the V and M diagrams for each member may be determined. As an example, let us look at member CD. If we isolate CD by passing a cutting plane just below the joint at C, the internal forces acting at the cut section are released as shown in Figure 5.3c. These forces acting at the C end of member CD are required to keep the member in equilibrium. The shear force at the C end is zero by applying $\Sigma F_X = 0$ for the FBD of member CD. The axial force is 20 kips (downward) by $\Sigma F_Y = 0$ and the bending moment is 48 kip-ft (clockwise) by $\Sigma M_C = 0$. End reactions for the remaining members can be obtained in a similar fashion. Note, however, the FBD of joint C shown in Figure 5.3c. The joint C transmits both the 20-kip force and the 48 kip-ft moment to member BC as can be seen from the FBD of member BC. The internal end reactions once determined for one member may be transferred to the joining member as equal but opposite reactions. For member BA, the shear force and moment at the B end are zero from $\Sigma F_X = 0$ and $\Sigma M_B = 0$ respectively. The axial force at B is 28 kips (downward) by $\Sigma F_Y = 0$. Since for every action there is an equal but opposite reaction, the 28-kip force acts upward at the B end of member BC (see Figure 5.3c).
3. With all the forces acting on each isolated member known, the V and M diagrams can then be drawn. In order to be consistent with the methods and principles outlined for drawing V and M diagrams in an earlier chapter, a set procedure must be followed when applying these methods and principles to framed structures. The procedure is to simply view each member from the "inside" of the frame and then proceed to draw the V and M diagrams as for any beam. Take member CD for instance. Viewing member CD

from the inside would put end C on our left; hence, the V and M diagrams could be sketched considering CD a beam with C as the left end. Observing member AB from the inside would put end A on our left and we could proceed from left to right when drawing the V and M diagrams for AB. Since member BC is already a horizontal member, we may treat it as any conventional beam. The corresponding V and M diagrams for the given frame are shown in Figure 5.3c.

4. For each member to be completely analyzed, the variation in the axial force must also be known. Some analysts prefer to plot an axial force diagram to depict this variation. In a vast majority of cases, the axial force in a member will be constant along its length thereby making such a plot unnecessary.

Example 5.2

Analyze the simply supported frame shown in Figure 5.4a.

Solution:

1. The frame shown in Figure 5.4a is commonly referred to as a **gable frame.** The analysis of a gable frame differs somewhat from the procedure followed when analyzing the conventional rectangular rigid frames. A FBD of the entire frame is shown in Figure 5.4b. Applying the equations of equilibrium to the FBD:

$$\Sigma M_A = 0 \quad (+\curvearrowright)$$

$$(0.5)(18)(\tfrac{18}{2}) + (1)(24)(12) - 24E_Y = 0$$

$$\underline{E_Y = +15.375 = 15.37 \text{ kips} \uparrow}$$

$$\Sigma F_Y = 0 \uparrow +$$

$$A_Y + 15.37 - (1)(24) = 0$$

$$\underline{A_Y = +8.63 = 8.63 \text{ kips} \uparrow}$$

$$\Sigma F_X = 0 \rightarrow +$$

$$(0.5)(18) - A_X = 0$$

$$\underline{A_X = +9.00 = 9 \text{ kips} \leftarrow}$$

2. If each member of the frame is isolated and all the forces acting on it are determined, the V and M diagrams can be drawn for each member thereby providing the required analysis for the entire frame. The vertical members AB and ED are isolated as shown by the FBDs in Figure 5.4c. The forces or moments acting at the B and D ends are obtained by applying the equations of equilibrium to the FBDs. Thus, at end B

$$V = 0 \qquad M = 81 \text{ kip-ft} \qquad N = 8.63 \text{ kips}$$

acting as shown in Figure 5.4c. Similarly, at end D we have

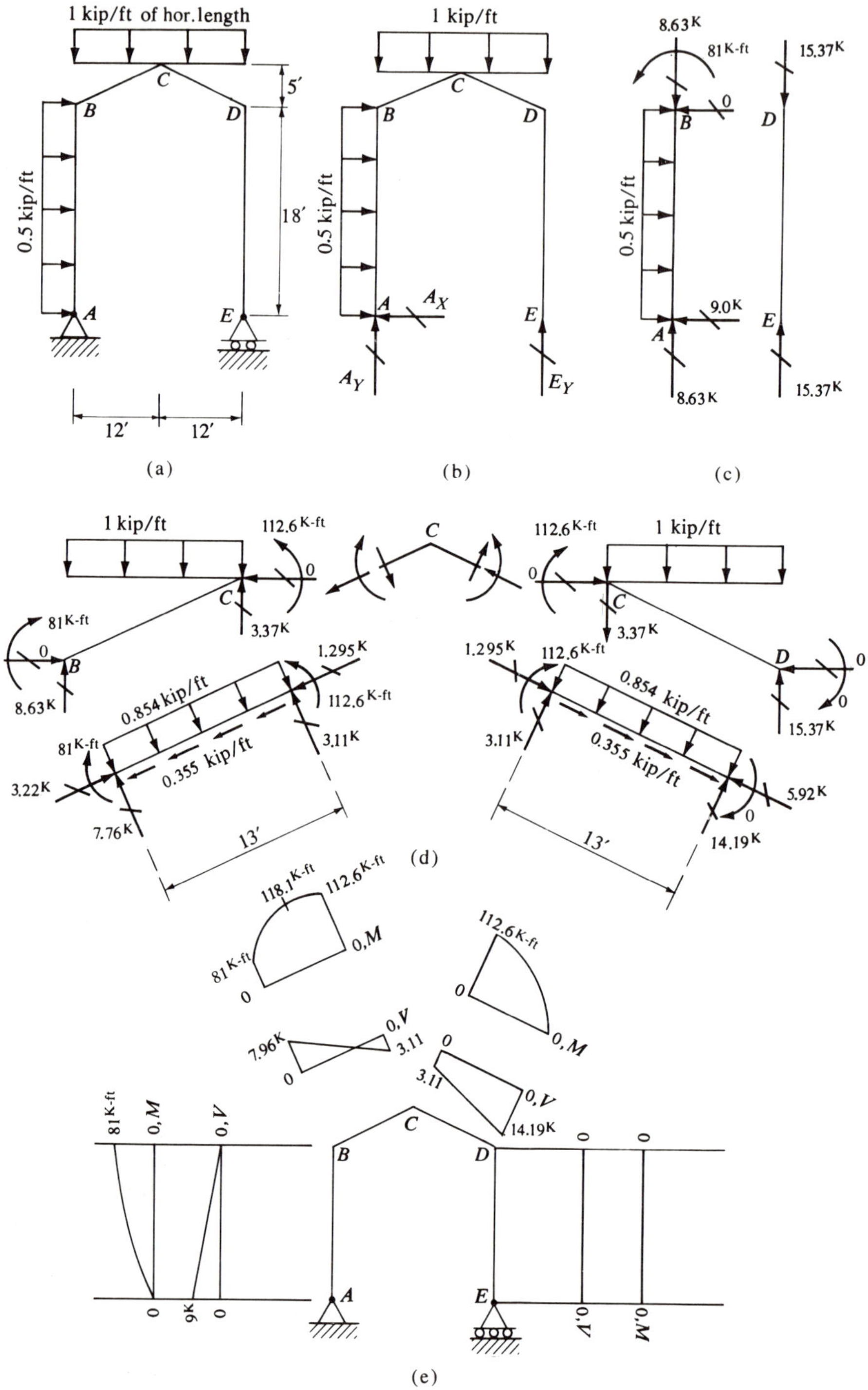
1 kip/ft of hor. length
5′
18′
0.5 kip/ft
12′
12′
(a)
1 kip/ft
A_X
A_Y
E_Y
(b)
8.63K
81K-ft
15.37K
9.0K
8.63K
15.37K
(c)
1 kip/ft
112.6K-ft
81K-ft
8.63K
3.37K
1.295K
0.854 kip/ft
0.355 kip/ft
3.11K
3.22K
7.76K
13′
(d)
5.92K
14.19K
118.1K-ft
112.6K-ft
0,M
7.96K
0,V
3.11
14.19K
(e)

Figure 5.4

$$V = 0 \qquad M = 0 \qquad N = 15.37 \text{ kips}$$

acting as shown in Figure 5.4c. Next let us isolate member BC as a free body as shown in Figure 5.4d. The forces acting at the B end are equal but opposite to those acting at the B end of member AB. With the forces known at the B end, we can apply $\Sigma F_X = 0$, $\Sigma F_Y = 0$, and $\Sigma M_C = 0$ to obtain the forces at the C end. Thus:

$$F_{CX} = 0 \qquad F_{CY} = 3.37 \text{ kips} \qquad M_C = 112.6 \text{ kip-ft}$$

acting as shown in Figure 5.4d. In order to determine the shear and axial force acting at any section of the member, we must resolve all the acting forces into components parallel and normal to the longitudinal axis of the member. For example, at end B we have:

$$N \text{ (axial force)} = (8.63)(\tfrac{5}{13}) = 3.22 \text{ kips}$$
$$V \text{ (shear force)} = (8.63)(\tfrac{12}{13}) = 7.96 \text{ kips}$$

In a similar fashion at end C we have:

$$N = (3.37)(\tfrac{5}{13}) = 1.295 \text{ kips}$$
$$V = (3.37)(\tfrac{12}{13}) = 3.11 \text{ kips}$$

These act as shown in the FBD of Figure 5.4d. No modification is required for the end moments, or couples, since a couple is independent of the orientation of the member. Next, we must consider the uniformly distributed load of 1 kip per foot of horizontal length. We may not treat this distributed load as a force, i.e., we may not resolve the loading intensity of 1 kip per foot into intensities normal and parallel to the longitudinal axis of the member BC. Recall that a force is completely defined by a magnitude, a direction, and a sense. The distributed load, however, does not meet this criterion since a length is also associated with the magnitude, direction, and sense. We may replace the distributed load, however, with an equivalent load equal in magnitude to the area under the loading curve and acting through the centroid of the area. This equivalent force may then be resolved into components parallel and normal to the longitudinal axis of the member and then these components can be converted back into uniformly distributed loads acting normal and parallel to the member. For member BC in Figure 5.4c we have:

$$\text{Equivalent Force} = (1)(12) = 12 \text{ kips} \downarrow$$
$$\text{Force normal to } BC = (12)(\tfrac{12}{13}) = 11.09 \text{ kips}$$
$$\text{Force parallel to } BC = (12)(\tfrac{5}{13}) = 4.61 \text{ kips}$$

Converting these components to uniformly distributed loads:

$$\frac{11.09 \text{ kips}}{13 \text{ ft}} = 0.854 \text{ kips/ft acting normal to } BC$$

$$\frac{4.61 \text{ kips}}{13 \text{ ft}} = 0.355 \text{ kips/ft acting parallel to } BC$$

acting as indicated in Figure 5.4c. As a check, the forces acting normal and parallel to BC can be summed as follows:

$$\begin{aligned} \Sigma F_{\text{Normal}} &= 0 \\ 7.96 + 3.11 - (0.854)(13) &= 0 \\ 0 &= 0 \end{aligned}$$

$$\begin{aligned} \Sigma F_{\text{Parallel}} &= 0 \\ 3.22 + 1.295 - (0.355)(13) &= 0 \\ 0 &= 0 \end{aligned}$$

In a similar manner, the FBD for member CD is obtained as shown in Figure 5.4d. The forces and moments acting at the C and D ends of member CD were conveniently obtained from the C and D ends of members CB and DE respectively.

3. With each frame member isolated as a free body showing all end moments and all applied forces resolved into components normal and parallel to the member, the V and M diagrams can readily be plotted using conventional procedures. It is customary to plot the V and M diagrams for the whole frame as shown in Figure 5.4e utilizing the FBDs of Figures 5.4c and 5.4d.

5.2 Three-Hinged Arches

An **arch** is a fundamental structural unit and may be defined as a structure which increases its ability to carry applied loads by the development of horizontal reaction components at the two end supports. Arches are classified as hingeless (or fixed), two-hinged, or three-hinged as shown in Figures 5.5a, b, and c respectively. Arches may also be classified on the basis of the building materials, the most common being reinforced concrete, steel, and glue-lami-

(a) (b) (c)

Figure 5.5

nated timber. Reinforced concrete arches are almost always of the hingeless type since these fixed supports are easily obtainable by the monolithic construction employed for reinforced concrete structures. Steel and glue-laminated timber arches are generally of the two- and three-hinged type. The hingeless and two-hinged arches are statically indeterminate; we will limit our discussion to the statically determinate three-hinged arch.

The arch is structurally advantageous because the horizontal forces developed at the supports create internal moments opposite to those resulting from the applied loads, thereby decreasing the maximum net moment at any section. Consider the simply supported beam and three-hinged arch shown in Figure 5.6a. Both structures have identical span lengths and are subjected to

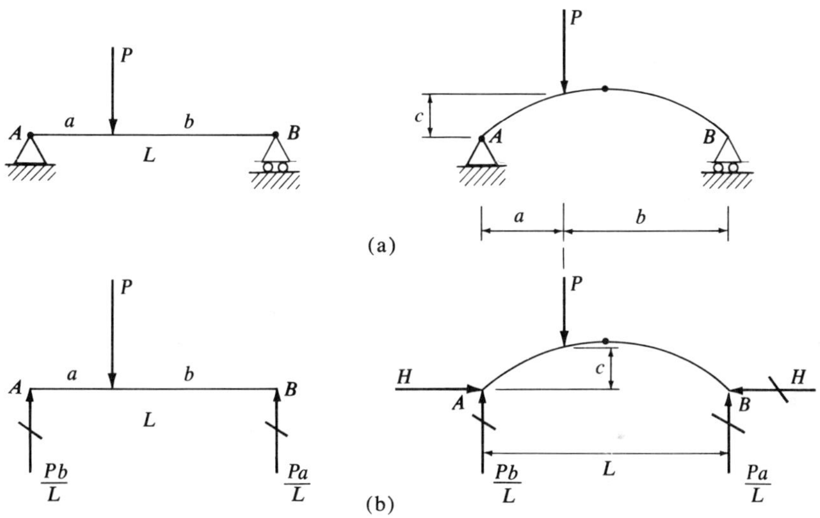

Figure 5.6

identical loads. Let us compare the positive moments at a section beneath the applied load for both structures. Referring to the FBDs shown in Figure 5.6b we have:

$$M_C = \frac{Pba}{L} \text{ for the beam}$$

$$M_C = \frac{Pba}{L} - Hc \text{ for the arch}$$

The maximum positive moment is substantially reduced by the arch action. At any given section of the beam, there exist only shear and bending moment. A corresponding section of the arch, however, is also subjected to an axial force. An arch is completely analyzed, therefore, when the axial force, shear, and moment at any section perpendicular to the arch axis are known.

The curve of an arch axis can take on a wide variety of forms, the most common being parabolic or circular as shown in Figures 5.7a and 5.7b respectively.

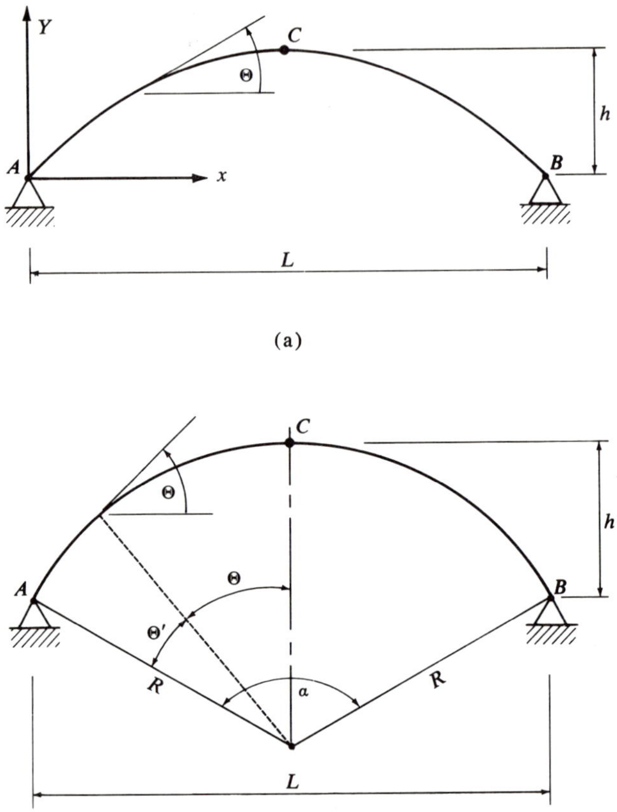

(a)

(b)

Figure 5.7

The curves may be symmetrical as shown or unsymmetrical. The internal hinge may or may not be located at the span centerline. The height of an arch, designated as h in Figure 5.7, is called the **rise,** while the ratio h/L is referred to as the **rise-to-span ratio.** This ratio is generally determined by the function of the arch, i.e., is the structure to be a bridge, an aircraft hangar, etc. The equation of the parabolic arch axis, with point A as the origin, is:

$$y = \frac{4h}{L^2}(Lx - x^2)$$

and the slope of the axis at any point is given by:

$$\tan \Theta = \frac{4h}{L^2}(L - 2x)$$

The circular arch is usually distinguished by the subtended angle α shown in Figure 5.7b. The slope at any point on the circular axis may be determined utilizing the angular relationship shown in Figure 5.7b.

Examination of the three-hinged arches shown in Figure 5.7 reveals that there are two available reaction components at each support for a total of four. In addition to the three equations of equilibrium, we also have the condition that the moment at C is zero, thereby rendering the structures statically determinate. A problem arises, unfortunately, when we try to draw the V and M diagrams for the arch member. The shear and axial force at any section of the arch depends upon the slope of the tangent to the curve at that section. Thus, the shear and axial force are not only functions of the applied loads but also of the direction of the arch axis. To compound the problem, the base line for the V and M diagrams is a curve rather than a straight line. **The conventional techniques followed for sketching V and M diagrams are, therefore, not applicable for curved members.**

Example 5.3

For the three-hinged parabolic arch shown in Figure 5.8a, determine the axial force, shear, and moment at points D and E of the arch.

Solution:

1. The reactions can be determined by referring to the FBDs shown in Figures 5.8b and c. Thus:

Figure 5.8b:

$$\Sigma M_B = 0 \quad \circlearrowright +$$

$$R_A(40) - 100(30) = 0$$

$$\underline{R_A = 75 \text{ kips} \uparrow}$$

$$\Sigma F_Y = 0 \uparrow +$$

$$75 - 100 + R_B = 0$$

$$\underline{R_B = 25 \text{ kips} \uparrow}$$

Figure 5.8c:

$$\Sigma M_C = 0 \quad \circlearrowright +$$

$$-25(20) + H_B(10) = 0$$

$$\underline{H_B = 50 \text{ kips} \leftarrow}$$

Figure 5.8b:

$$\Sigma F_X = 0 \rightarrow +$$

$$H_A - 50 = 0$$

$$\underline{H_A = 50 \text{ kips} \rightarrow}$$

2. With the reactions known, the axial force, shear, and moment at any section can now be determined. If a cutting plane is passed through the arch

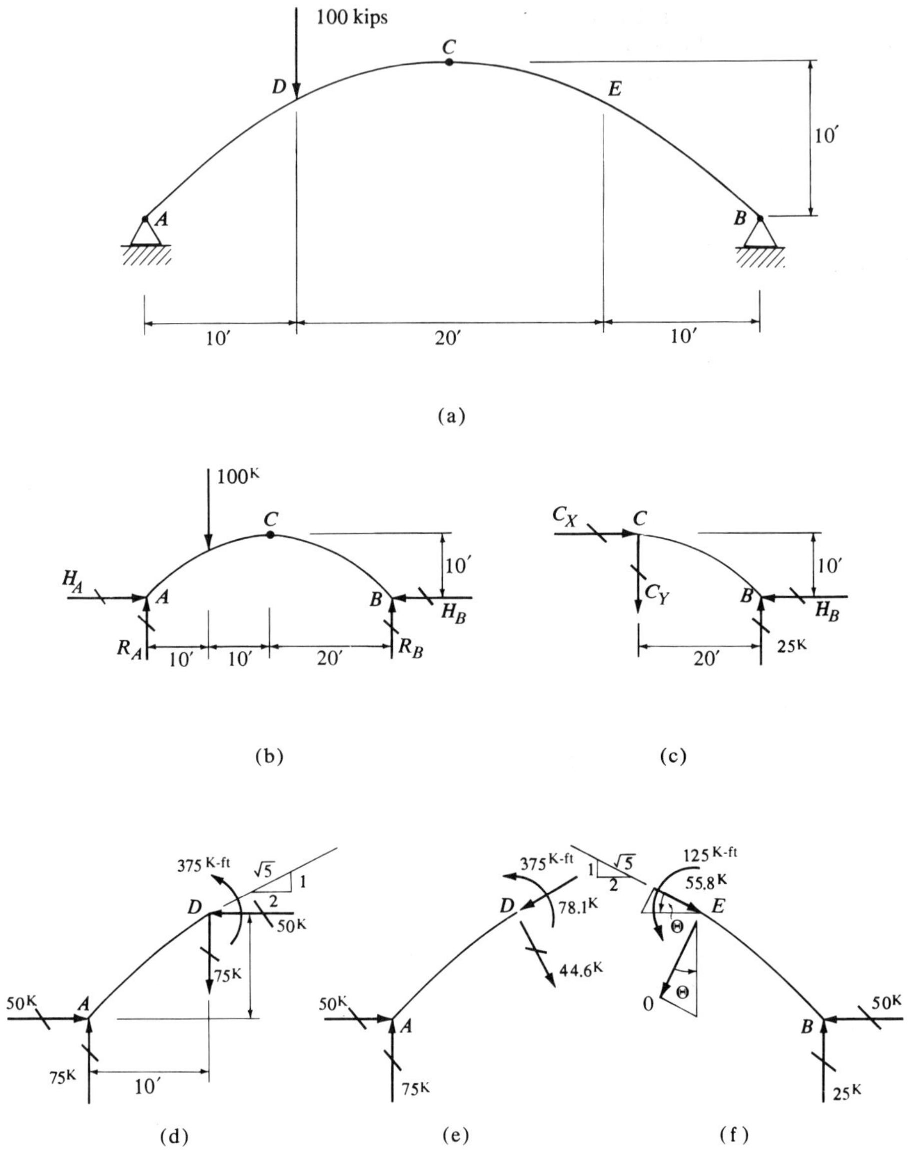

Figure 5.8

just to the left of point D, that portion of the arch to the left of the cut may be isolated as a free body as shown in Figure 5.8d. By applying the equations of equilibrium to the free body we have at end D:

$$D_X = 50 \text{ kips} \qquad D_Y = 75 \text{ kips} \qquad M_D = 375 \text{ kips-ft}$$

acting as indicated in Figure 5.8d. In order to determine the axial force and shear acting at D, we must resolve D_Y and D_X into components normal and parallel to the section at D. The slope of the tangent to the parabola at point D is given by:

$$\tan \Theta = \frac{4h}{L^2}(L - 2x) \quad (\text{origin @ } A)$$

$$@\ D,\ x = 10 \text{ ft}$$

$$\tan \Theta_D = \frac{(4)(10)}{(40)^2}(40 - 20) = \frac{1}{2};\ (\Theta_D = 26° - 34')$$

Resolving D_Y and D_X into normal and tangential components we have at D

$$\begin{aligned}
V_D &= 75 \cos \Theta - 50 \sin \Theta \\
&= 75\left(\frac{2}{\sqrt{5}}\right) - 50\left(\frac{1}{\sqrt{5}}\right) \\
V_D &= 44.6 \text{ kips} \\
N_D &= 75 \sin \Theta + 50 \cos \Theta \\
&= 75\left(\frac{1}{\sqrt{5}}\right) + 50\left(\frac{2}{\sqrt{5}}\right) \\
N_D &= 78.1 \text{ kips} \\
M_D &= 375 \text{ kip-ft}
\end{aligned}$$

acting as shown in Figure 5.8e. Similarly at E we have:

$$\tan \Theta_E = \frac{4h}{L^2}(L - 2x) = \frac{(4)(10)}{(40)^2}(40 - 60) = -\frac{1}{2}$$

Figure 5.8f:

$$\begin{aligned}
\Sigma F_X &= 0; & E_X &= 50 \text{ kips} \rightarrow \\
\Sigma F_Y &= 0; & E_Y &= 25 \text{ kips} \downarrow \\
\Sigma M_E &= 0; & M_E &= 125 \text{ kip-ft} \circlearrowleft
\end{aligned}$$

Resolving E_X and E_Y into normal and tangential components:

$$\begin{aligned}
V_E &= -25 \cos \Theta + 50 \sin \Theta \\
&= -25\left(\frac{2}{\sqrt{5}}\right) + 50\left(\frac{1}{\sqrt{5}}\right) \\
V_E &= 0 \\
N_E &= 25 \sin \Theta + 50 \cos \Theta \\
&= 25\left(\frac{1}{\sqrt{5}}\right) + 50\left(\frac{2}{\sqrt{5}}\right) \\
N_E &= 55.8 \text{ kips} \\
M_E &= 125 \text{ kip-ft}
\end{aligned}$$

acting as shown in Figure 5.8f. The negative sign obtained for the slope at E indicates that the tangent to the parabolic curve slopes down and to the right.

3. The shear, axial force, and moment at D were obtained by passing a cutting plane just to the left of point D. Imagine the cutting plane passing through the arch section just to the right of point D. We have, considering that portion of the arch to the left of the cut,

$$D_X = 50 \text{ kips} \leftarrow \qquad D_Y = 25 \text{ kips} \uparrow \qquad M_D = 375 \text{ kip-ft} \curvearrowright$$

The presence of the 100-kip concentrated applied load on the isolated arch segment gives rise to a change in the vertical force acting at the cut section. Since we are cutting the arch an infinitesimal distance to the left and right

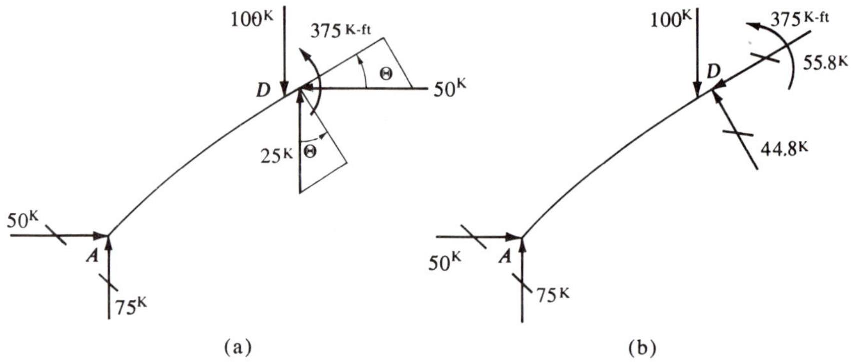

Figure 5.9

of point D, the slope of the arch is essentially equal at both cuts and, thus, both the shear and axial force will change as we pass to sections just to the left and right of a concentrated load. Recall that a concentrated load produces an abrupt change in shear for the case of straight beams, and now, for curved members, i.e., arches, we find that a concentrated load causes an abrupt change to occur in both the axial force and shear. Referring to Figure 5.9a we have:

$$\begin{aligned}
V_D &= 25 \cos \Theta + 50 \sin \Theta \\
&= 25\left(\frac{2}{\sqrt{5}}\right) + 50\left(\frac{1}{\sqrt{5}}\right) \\
V_D &= 44.6 \text{ kips} \\
N_D &= -25 \sin \Theta + 50 \cos \Theta \\
&= -25\left(\frac{1}{\sqrt{5}}\right) + 50\left(\frac{2}{\sqrt{5}}\right) \\
N_D &= 55.8 \text{ kips} \\
M_D &= 375 \text{ kip-ft} \quad \text{(no change)}
\end{aligned}$$

acting as indicated in Figure 5.9b.

Example 5.4

For the three-hinged, semicircular arch shown in Figure 5.10a, determine the shear, axial force, and moment at a section at D.

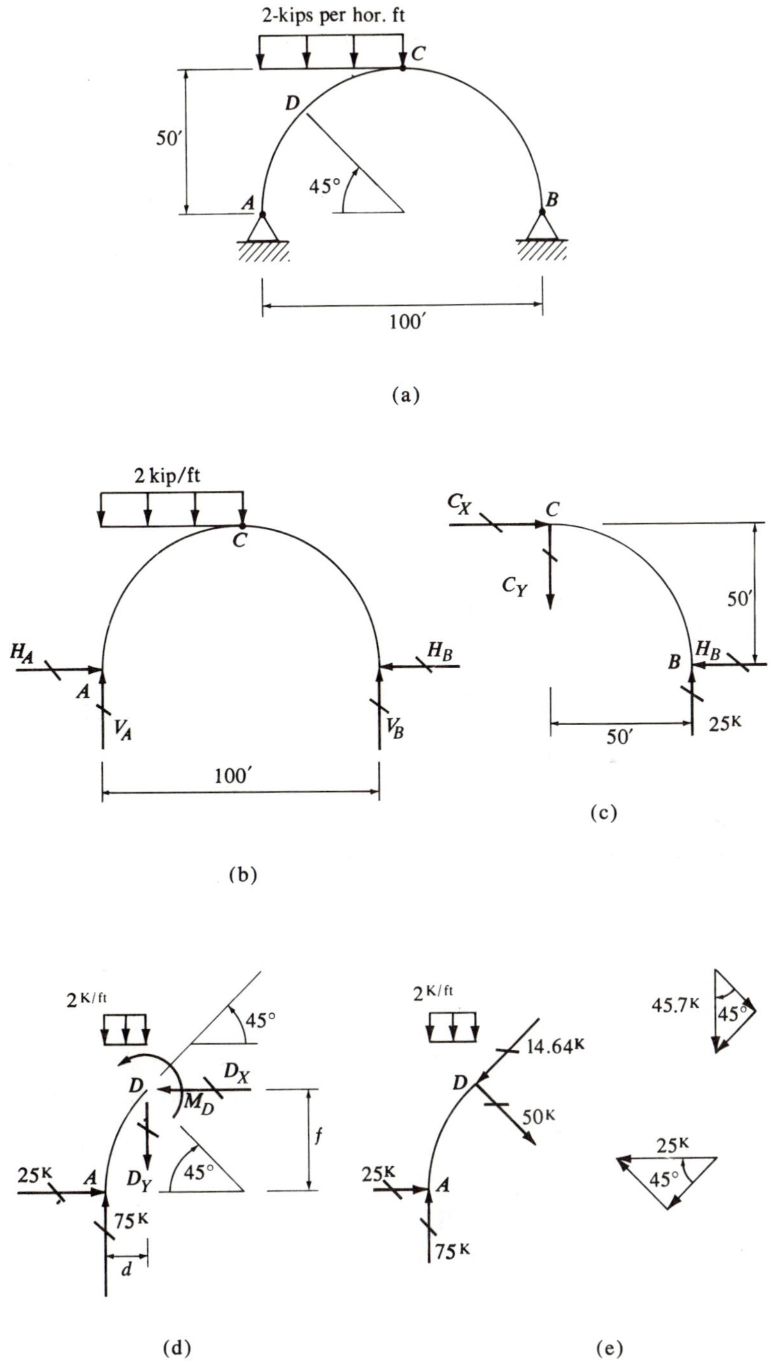
2-kips per hor. ft
C
D
50′
45°
A
B
100′
(a)
2 kip/ft
C
H_A
A
V_A
H_B
V_B
100′
(b)
C_X
C
C_Y
50′
B
H_B
25^K
50′
(c)
2^K/ft
45°
D
D_X
M_D
f
25^K
A
D_Y
45°
75^K
d
(d)
2^K/ft
14.64^K
D
50^K
25^K
A
75^K
45.7^K
45°
25^K
45°
(e)

Figure 5.10

Solution:

1. The solution will proceed as follows: (a) determine the reactions for the arch; (b) cut the arch at point D and determine the vertical force, the horizontal force, and bending moment acting at the cut section; (c) calculate the slope of the arch at D and resolve the vertical and horizontal forces acting at D into components normal and parallel to the cross section at D, thus determining the axial force and shear respectively at D.
2. Referring to the FBD of the arch shown in Figure 5.10b we have:

$$\Sigma M_B = 0 \quad (\curvearrowright +)$$

$$V_A(100) - (2)(50)(75) = 0$$

$$\underline{V_A = +75.0 = 75 \text{ kips} \uparrow}$$

$$\Sigma F_Y = 0 \uparrow +$$

$$75 - 2(50) + V_B = 0$$

$$\underline{V_B = +25 = 25 \text{ kips} \uparrow}$$

The horizontal reaction at B can be obtained by utilizing the hinge at C. Using Figure 5.10c:

$$\Sigma M_C = 0 \quad (\curvearrowright +)$$

$$50H_B - 25(50) = 0$$

$$\underline{H_B = +25 = 25 \text{ kips} \leftarrow}$$

Referring again to the FBD shown in Figure 5.10b:

$$\Sigma F_X = 0 \rightarrow +$$

$$H_A - 25 = 0$$

$$\underline{H_A = 25 \text{ kips} \rightarrow}$$

3. We can pass a cutting plane through the arch at point D and isolate that portion of the arch to the left of the cut. The vertical and horizontal force and the moment acting at D can now be established. Referring to the FBD shown in Figure 5.10d:

$$\Sigma F_Y = 0 \uparrow +$$

$$75 - 2d - D_Y = 0$$

$$d = 50 - 50 \cos 45°; \; d = 14.65 \text{ ft}$$

$$D_Y = 75 - 2(14.65)$$

$$D_Y = 45.7 \text{ kips} \downarrow \quad \text{(on segment } AD\text{)}$$

$$\Sigma F_X = 0 \rightarrow +$$

$$25 - D_X = 0$$

$$D_X = 25 \text{ kips} \leftarrow$$

$$\Sigma M_D = 0 \quad (\curvearrowright +)$$

$$75d - 25f - 2d(d/2) - M_D = 0$$
$$f = 50 \sin 45° = 35.3 \text{ ft}; \ d = 14.65 \text{ ft}$$
$$M_D = 75(14.65) - 25(35.3) - 2\left(\frac{14.65}{2}\right)^2$$
$$M_D \simeq 0$$

4. The tangent to the circular arch at point *D* makes a 45° angle with the horizontal as shown in Figure 5.10d (the angle of inclination of the tangent at *D* is numerically equal to the angle subtended by arc *DC*). With the slope known at *D*, the shear and axial force can be determined by referring to the FBD shown in Figure 5.10e. Thus:

$$V_D = D_Y \cos 45° - D_X \sin 45°$$
$$= 45.7(0.707) - 25(0.707)$$
$$\underline{V_D = 14.64 \text{ kips}}$$

$$N_D = D_Y \sin 45° + D_X \cos 45°$$
$$= 45.7(0.707) + 25(0.707)$$
$$\underline{N_D = 50.0 \text{ kips}}$$

$$\underline{M_D = 0}$$

acting as indicated in Figure 5.10e.

5.3 Composite Structures

It has been stated in Chapter Three that most structures can be classified as beams, rigid frames, arches, or trusses. Beams are subjected primarily to both shear and bending stresses. Rigid frame members and arches are subjected to combined shear, bending, and axial stresses. Truss members, on the other hand, are assumed to be subjected only to axial stress. In a vast majority of instances, bending is predominate in both beams and rigid frame members. Therefore, beams and frames are primarily subjected to bending stresses, while truss members are primarily subjected to axial stresses. There are structures, however, in which some members are subjected to axial stresses only and others to bending stresses only. Such structures are frequently referred to as **composite structures** or **beam trusses.**

In analyzing composite structures, it is of prime concern to recognize which members are **two-force members** and which are **three-force members.** A two-force member can be defined as one which is pin-connected at both ends and is not subjected to any loading between the pinned ends. The only forces acting on a FBD of a two-force member are the forces at the ends. Thus, the member is subjected to the action of two equal tensile or compressive forces at the member ends. The member is then subjected only to axial stress. Each member of a pin-connected truss meets the definition of a two-force member. Any member which does not satisfy the criteria for a two-force member is a three-force member. It is important to identify the two-force members initially when analyzing a composite structure. Since a two-force member is subjected only to axial forces, the forces acting at the pinned ends must have a direction along the longitudinal axis of the member. This knowledge of the direction of the force acting at the member end is often required as a supplement to the equations of equilibrium, i.e., it provides a condition or an additional equation. Failure to recognize a two-force member can often render a problem insolvable. The following illustrative examples are intended to clarify the analysis procedure for composite structures.

Example 5.5

Analyze the structure shown in Figure 5.11a.

Solution:

1. A quick check on the number of unknown reactions encountered for the structure in Figure 5.11a reveals four: D_Y, D_X, A_X, and A_Y. Member *DE*, however, is recognized as a two-force member, i.e., pin-connected and no loading between the ends. A FBD of member *DE*, shown in Figure 5.11b, further reveals that the resultant of D_X and D_Y must have a direction along the longitudinal axis of *DE*. We have the following relationships:

$$D_X = \tfrac{12}{13}T \qquad D_Y = \tfrac{5}{13}T$$

Since both D_X and D_Y can be expressed in terms of T, the number of reaction unknowns now stands at three (A_X, A_Y, and T). The three equations of equilibrium are sufficient for the determination of the reactions.

2. Referring to the FBD shown in Figure 5.11c we have:

$$\Sigma M_D = 0 \quad \circlearrowright +$$

$$-A_X(\tfrac{19}{3}) + (2)(8)(4) + (10)(14) = 0$$

$$\underline{A_X = +32.2 = 32.2 \text{ kips} \rightarrow}$$

$$\Sigma F_X = 0 \rightarrow +$$

$$32.2 - \tfrac{12}{13}T = 0$$

$$\underline{T = +34.9 = 34.9 \text{ kips}} \text{ (as shown in Figure 5.11b)}$$

$$\Sigma F_Y = 0 \uparrow +$$

$$(\tfrac{5}{13})(34.9) + A_Y - (2)(8) - 10 = 0$$

$$\underline{A_Y = +12.59 = 12.59 \text{ kips} \uparrow}$$

3. The steel rod *DE* is subjected to an axial tensile force of 34.9 kips. For members *AC* and *EB*, *V* and *M* diagrams will be required. A FBD of member *AC*

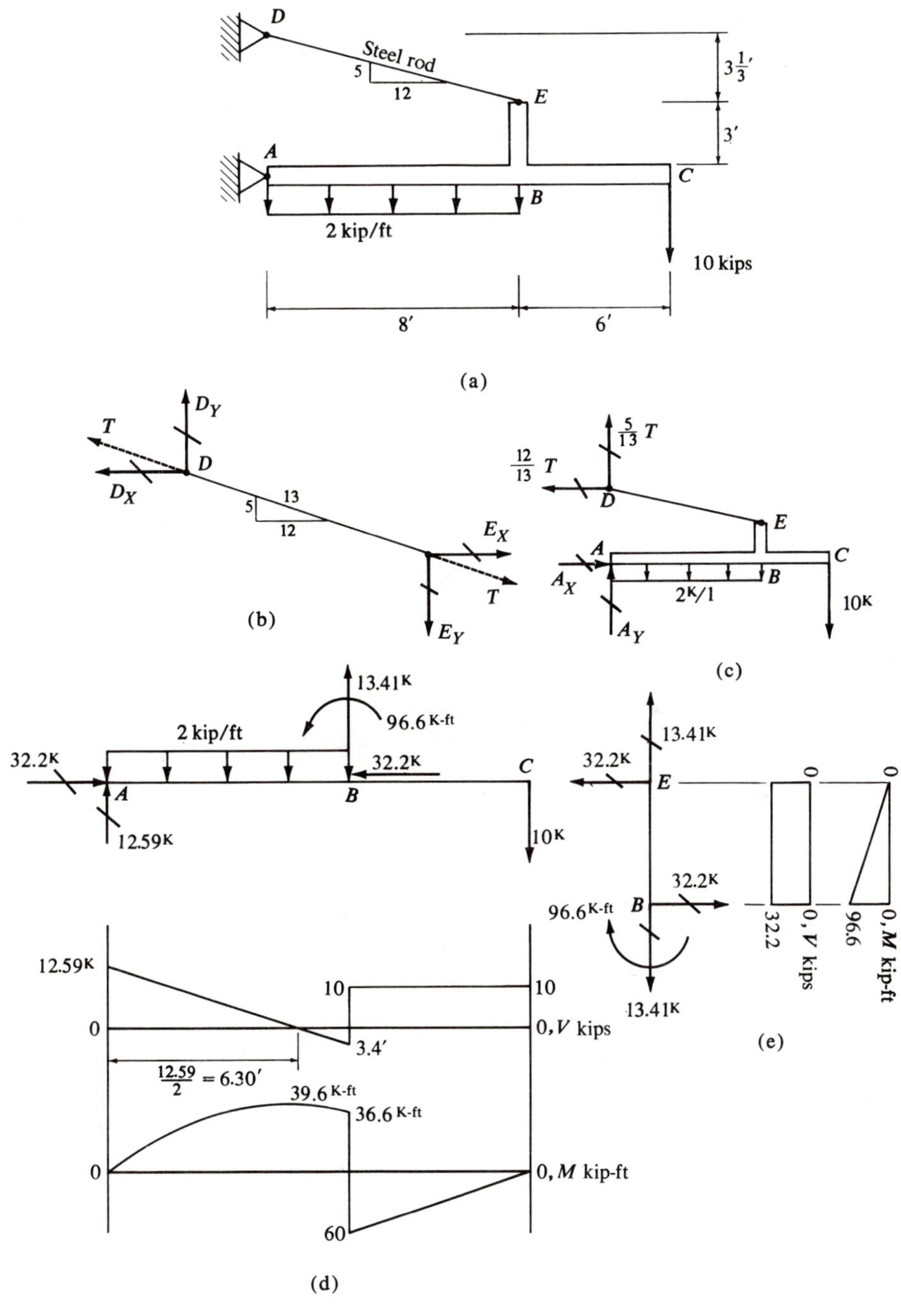
D
Steel rod
5
12
E
$3\frac{1}{3}'$
3'
A
C
B
2 kip/ft
10 kips
8'
6'
(a)
D_Y
T
D
D_X
13
5
12
E_X
T
(b)
E_Y
$\frac{5}{13}$ T
$\frac{12}{13}$ T
D
E
A
C
A_X
B
2K/1
10K
A_Y
(c)
13.41K
96.6 K-ft
2 kip/ft
32.2K
32.2K
C
A
B
12.59K
10K
13.41K
32.2K
E
0
0
32.2K
96.6K-ft
B
32.2
0, V kips
96.6
0, M kip-ft
13.41K
(e)
12.59K
10
10
0
0, V kips
3.4'
$\frac{12.59}{2}$ = 6.30'
39.6 K-ft
36.6 K-ft
0
0, M kip-ft
60
(d)

Figure 5.11

is shown in Figure 5.11d and a FBD of member *EB* is shown in 5.11e. Notice that the action of *EB* on *AC* is shown at *B* on the FBD of member *AC*. Using conventional techniques, the *V* and *M* diagrams for members *AC* and *EB* were plotted and are shown in Figures 5.11d and 5.11e respectively.

4. If member *DE* had not been recognized as a two-force member, only the three equations of equilibrium would have been available to solve for the four reaction unknowns, A_X, A_Y, D_X, and D_Y. The structure may have been classified erroneously as a statically indeterminate structure.

Example 5.6

Analyze the composite structure shown in Figure 5.12a.

Solution:

1. An examination of the supports indicates that there are only three unknown reaction components. Utilizing the FBD of the entire structure shown in Figure 5.12b, the external reactions can be obtained as follows:

$$\Sigma M_A = 0 \quad (\curvearrowright +)$$

$$6(2) - 8C_Y + 2(6) = 0$$

$$\underline{C_Y = +3.0 = 3.0 \text{ kips} \uparrow}$$

$$\Sigma F_Y = 0 \uparrow +$$

$$A_Y - 6 - 2 + 3 = 0$$

$$\underline{A_Y = +5.0 = 5.0 \text{ kips} \uparrow}$$

$$\Sigma F_X = 0 \rightarrow +$$

$$\underline{A_X = 0}$$

2. In order to completely analyze this structure, we must determine all the forces and moments acting on each member of the structure. This can best be accomplished by drawing a FBD of each member as shown in Figures 5.12c to 5.12e inclusive. Since each member is pinned at each end, there exists only the possibility of a horizontal and vertical force acting at each pinned end. Note that at the *B* end of members *AB* and *CB* we have utilized the fact that actions and reactions are always equal and opposite, i.e., the magnitude of the forces acting at *B* on *AB* are equal to those acting at *B* on *CB* but directions must be reversed. Member *AB* is a two-force member, i.e., it is pin-connected at its ends and no loads act between the ends. Members *AC* and *CB*, however, do not meet this criterion. In Figure 5.12d, the resultants of V_3 and H_3 and V_4 and H_4 are equal and have lines of action directed along *AB* but in opposite directions. If any one of the end reaction components is determined for *AB*, the axial force exerted on the member can, therefore, be easily determined.

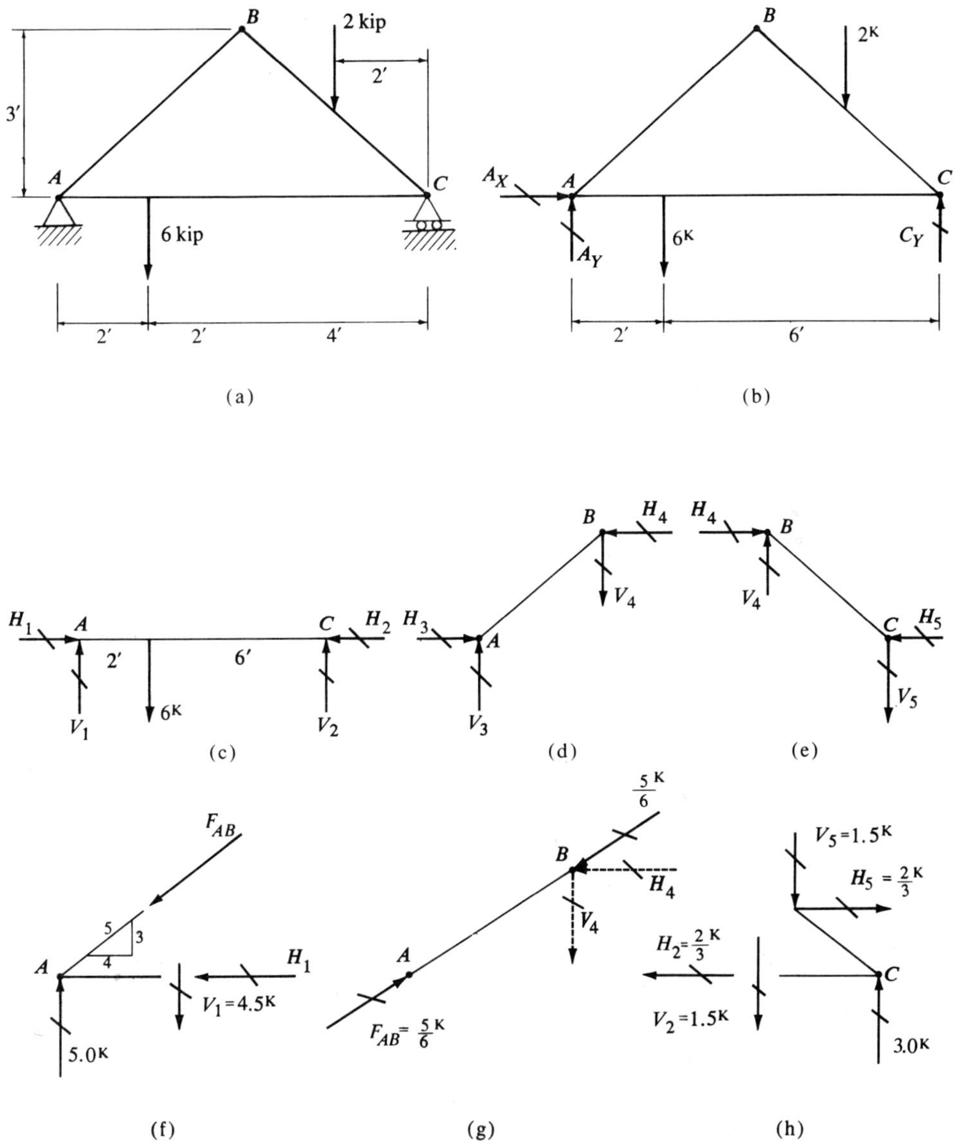

Figure 5.12

3. Referring to the FBDs shown in Figures 5.12c to 5.12e, the order of solution is as follows:

Figure 5.12c:

$$\Sigma M_A = 0 \quad (+\curvearrowright)$$

$$6(2) - 8\,V_2 = 0$$
$$\underline{V_2 = +1.5 = 1.5 \text{ kips} \uparrow} \text{ (as assumed)}$$

$$\Sigma F_Y = 0 \uparrow +$$
$$V_1 + 1.5 - 6 = 0$$
$$\underline{V_1 = +4.5 = 4.5 \text{ kips} \uparrow} \text{ (as assumed)}$$

$$\Sigma F_X = 0 \rightarrow +$$
$$H_1 - H_2 = 0$$
$$H_1 = H_2$$

In order to determine H_1, we may utilize a FBD of the joint at A as shown in Figure 5.12f. Since it has been established that AB is a two-force member, the axial force acting on it is shown as F_{AB} in Figure 5.12f. Notice also that V_1 and H_1 are shown to act equal but opposite to that shown in Figure 5.12c. **It is important to note that A_Y and C_Y are external reactions** and hence do not appear on the FBDs of the individual members of the structure. Referring to Figure 5.12f:

$$\Sigma F_Y = 0 \uparrow +$$
$$5.0 - 4.5 - (\tfrac{3}{5})F_{AB} = 0$$
$$\underline{F_{AB} = +\tfrac{5}{6} = \tfrac{5}{6} \text{ kips compression}}$$

$$\Sigma F_X = 0 \rightarrow +$$
$$-H_1 - \tfrac{4}{5}(\tfrac{5}{6}) = 0$$
$$\underline{H_1 = -\tfrac{2}{3} = \tfrac{2}{3} \text{ kips} \rightarrow \text{at joint } A}$$

Referring to Figure 5.12c again:

$$H_1 = H_2$$
$$H_1 = \tfrac{2}{3} \text{ kips} \rightarrow \text{at joint } A$$
$$\therefore H_1 = \tfrac{2}{3} \text{ kips} \leftarrow \text{on member } AC$$
$$\underline{H_2 = \tfrac{2}{3} \text{ kips} \rightarrow \text{on member } AC}$$

Since the force acting on member AB has been determined as (5/6) kips compression, we have from Figure 5.12g:

$$H_4 = (\tfrac{4}{5})(\tfrac{5}{6}) = \tfrac{2}{3} \text{ kips} \leftarrow \text{on } AB$$
$$V_4 = (\tfrac{3}{5})(\tfrac{5}{6}) = \tfrac{1}{2} \text{ kips} \downarrow \text{ on } AB$$

Referring to Figure 5.12e:

$$H_4 = \tfrac{2}{3} \text{ kips} \rightarrow; \qquad V_4 = \tfrac{1}{2} \text{ kips} \uparrow$$

$$\Sigma F_Y = 0 \uparrow +$$
$$\tfrac{1}{2} - 2 - V_5 = 0$$
$$\underline{V_5 = -\tfrac{3}{2} = \tfrac{3}{2} \text{ kips} \uparrow}$$

$$\Sigma F_X = 0 \rightarrow +$$
$$\tfrac{2}{3} - H_5 = 0$$
$$H_5 = +\tfrac{2}{3} = \tfrac{2}{3} \text{ kips} \leftarrow$$

All the forces acting on each member are now known. As a self-checking procedure, let us isolate joint C as a FBD as shown in Figure 5.12h. Thus, we have as a check:

$$\Sigma F_Y = 0 \uparrow +$$
$$3 - 1.5 - 1.5 = 0$$
$$0 = 0 \quad \text{Check}$$

$$\Sigma F_X = 0 \rightarrow +$$
$$-\tfrac{2}{3} + \tfrac{2}{3} = 0$$
$$0 = 0 \quad \text{Check}$$

4. The procedure to follow for such composite structures is as follows:
   (a) Determine the external reactions if possible.
   (b) Isolate each member of the structure as a free body.
   (c) Count the unknowns acting on each FBD and the number of equations available for solution of these unknowns. For example, in the problem just solved we have:

Unknowns

Figure 5.12c	4	(V_1, V_2, H_1, H_2)
Figure 5.12d	4	(V_3, V_4, H_3, H_4)
Figure 5.12e	2	(V_5, H_5)
Total	10	

Equations

Figure 5.12c	3	$(\Sigma F_Y = 0, \Sigma F_X = 0, \Sigma M = 0)$
Figure 5.12d	3	$(\Sigma F_Y = 0, \Sigma F_X = 0, \Sigma M = 0)$
Figure 5.12e	3	$(\Sigma F_Y = 0, \Sigma F_X = 0, \Sigma M = 0)$
Total	9	

Conditions

Figure 5.12d	1	(Member AB a two-force member) $V_3 = (\tfrac{3}{5})F_{AB}$ $H_3 = (\tfrac{4}{5})F_{AB}$ $\therefore V_3 = (\tfrac{3}{4})H_3$
Total	1	

In this case, the number of unknowns and available equations are equal, rendering the structure statically determinate.

   (d) The forces acting on each member can be determined using the FBDs of the members and the joints along with the equations of equilibrium and special conditions.

(e) The structure is completely analyzed when the axial force acting on each two-force member is known and V and M diagrams are drawn for each three-force member. Using conventional techniques the V and M diagrams for members AC and CB can be easily obtained.

Problems

5–1. Draw the shear and moment diagrams for the rigid frames shown in Figure P 5–1.

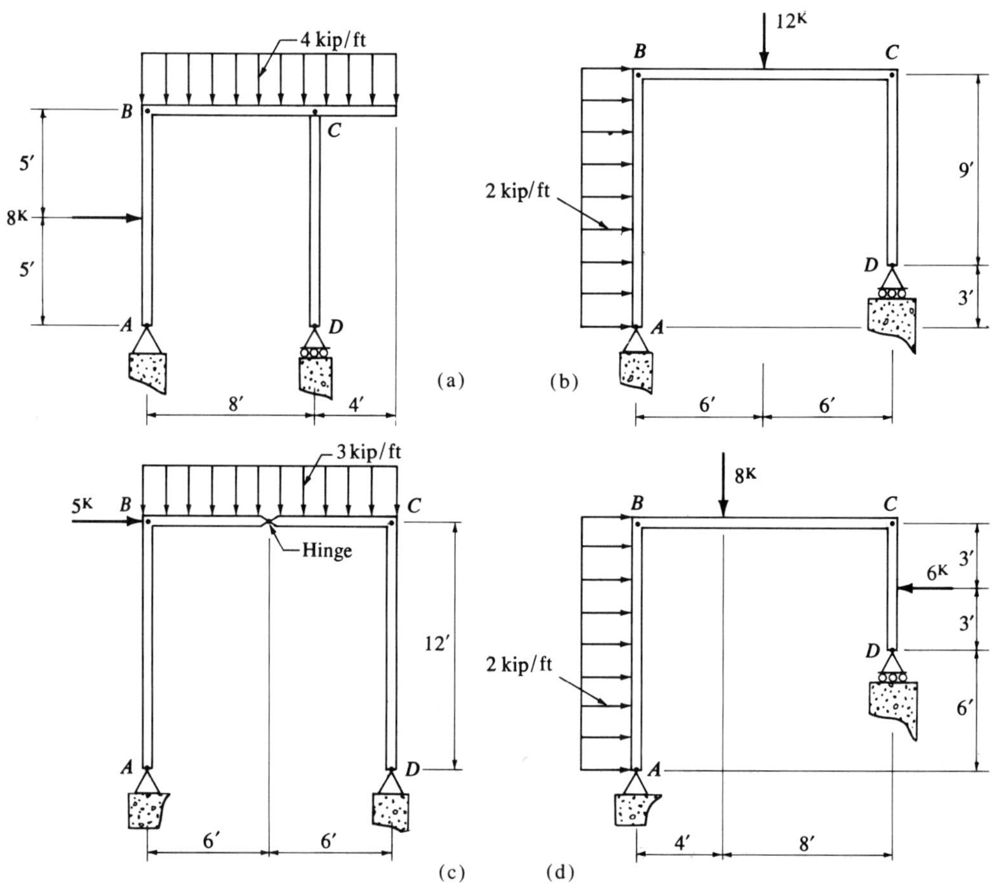

Figure P 5–1

5–2. Draw the shear and moment diagrams for the frames shown in Figure P 5–2.

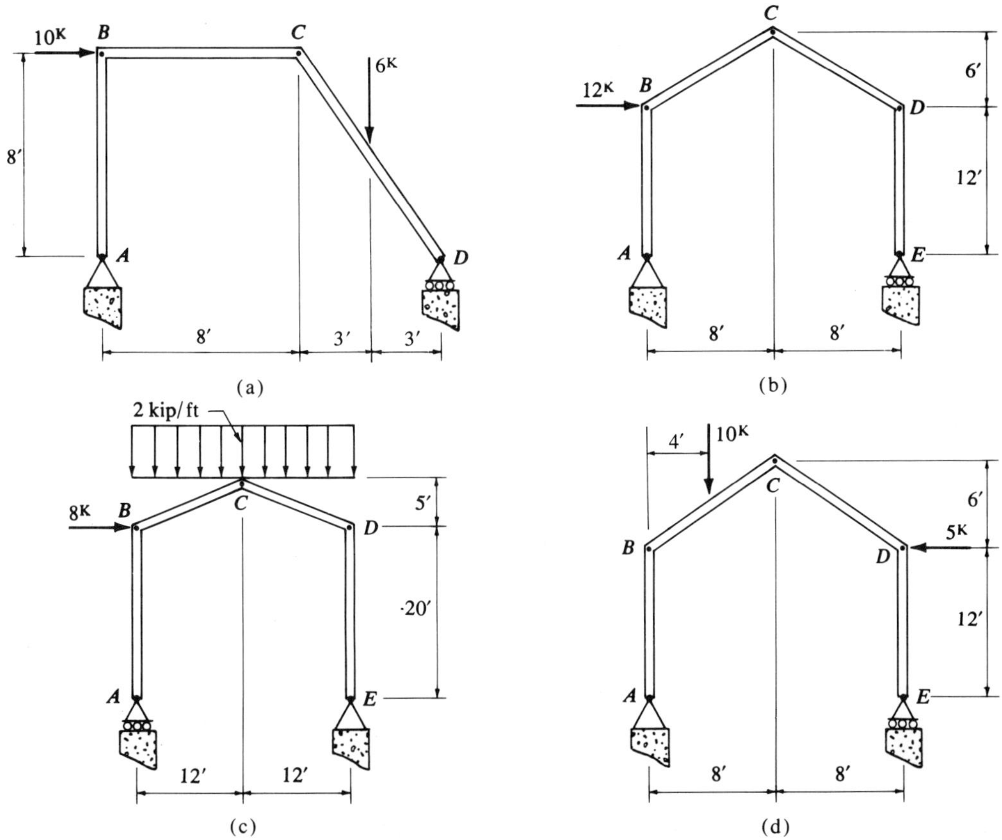

Figure P 5–2

5–3. Draw the shear and moment diagrams for the frame shown in Figure P 5–3. The frames are spaced 10 ft center to center. $\gamma_w = 62.4$ lb per cubic ft.

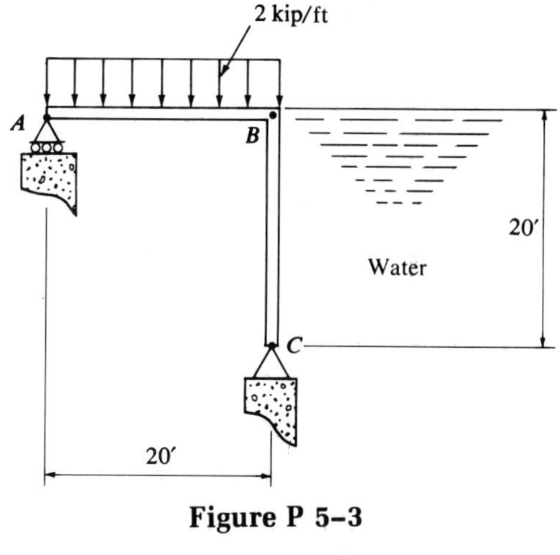

Figure P 5–3

5–4. Draw the shear and moment diagrams for the frame shown in Figure P 5–4.

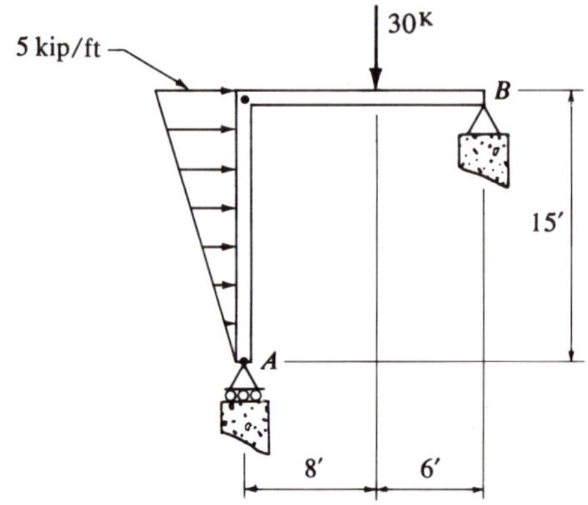

Figure P 5–4

5–5 to 5–8. The axes of the three-hinged arches shown are parabolic. Determine the axial force, shear, and moment at point D of each arch shown.

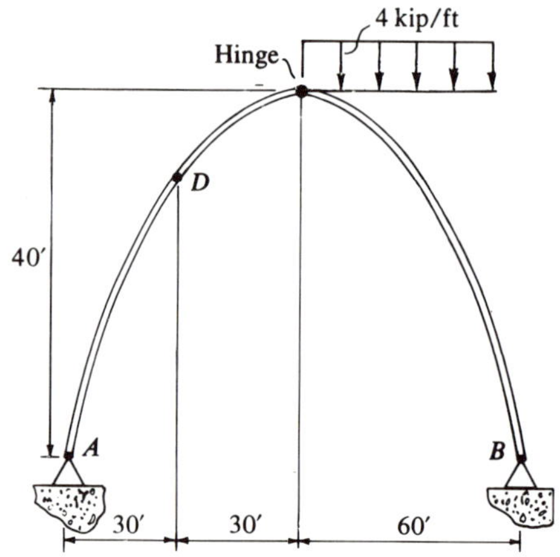

Figure P 5–5

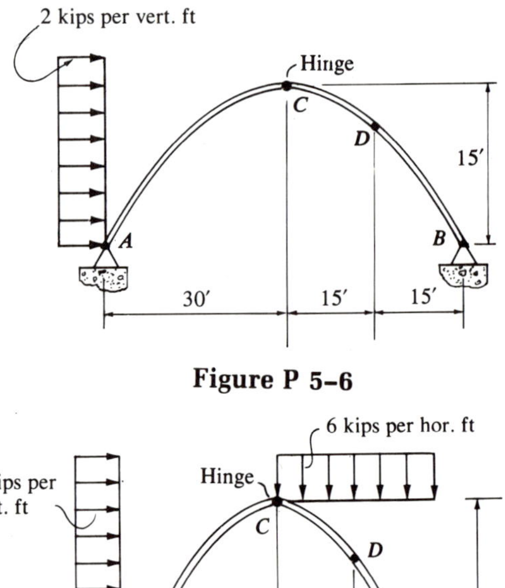

Figure P 5–6

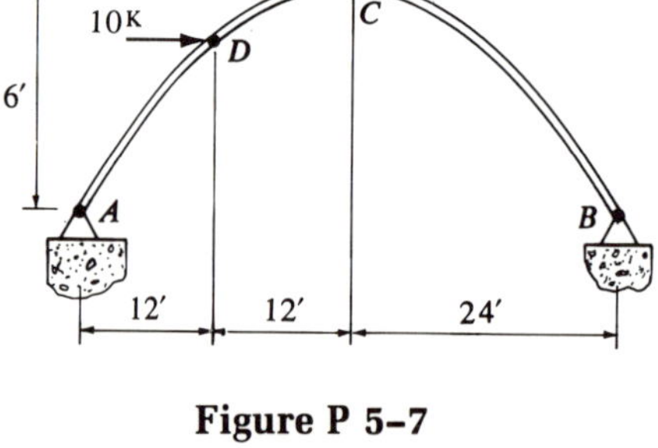

Figure P 5–7

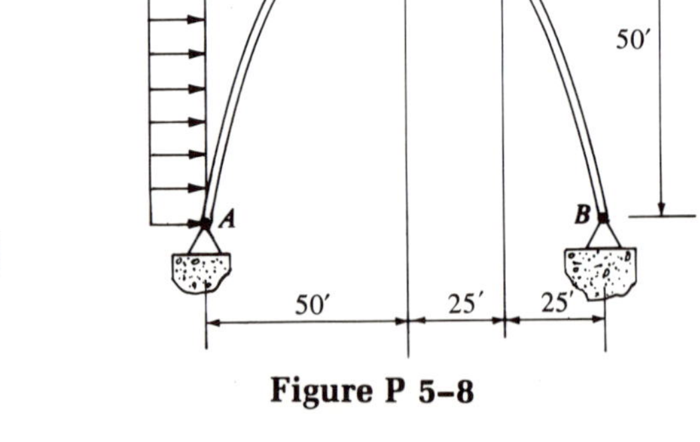

Figure P 5–8

5–9 to 5–11. The axes of the three-hinged arches shown are semicircular. For the loadings shown, find the axial force, shear, and moment at point D.

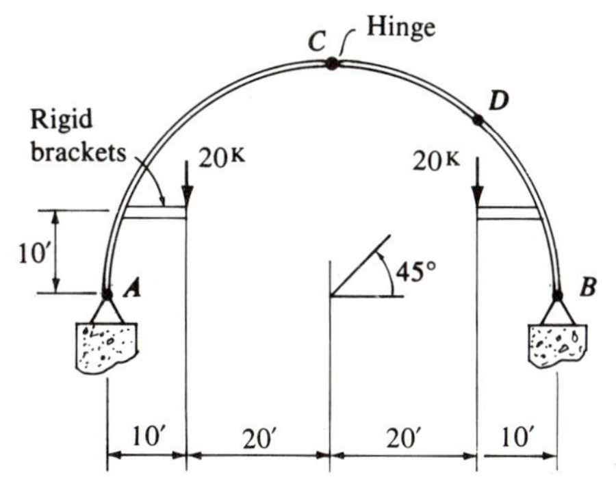

Figure P 5–9

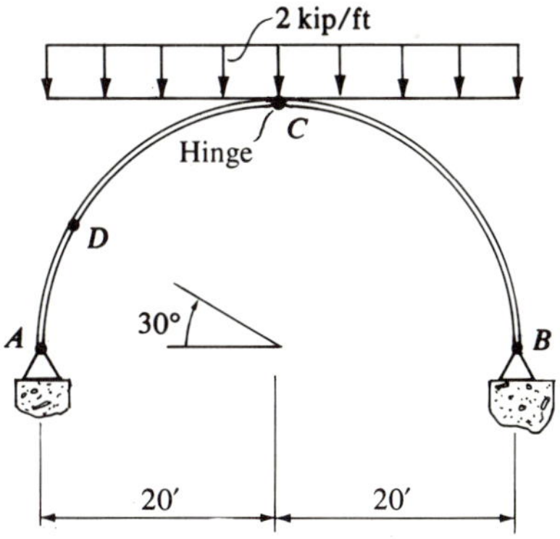

Figure P 5–10

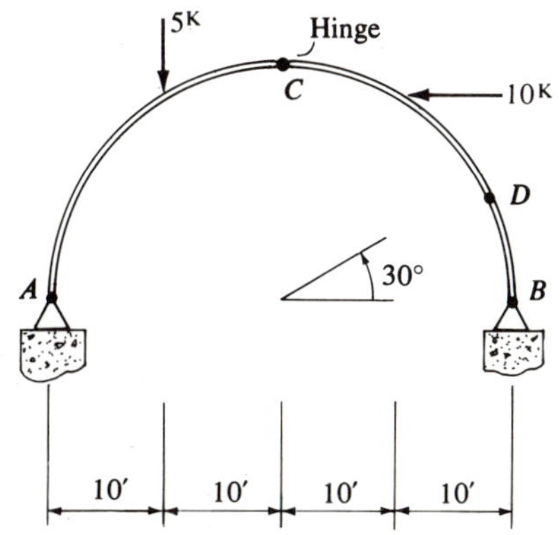

Figure P 5–11

5–12 to 5–15. Completely analyze the composite structure shown.

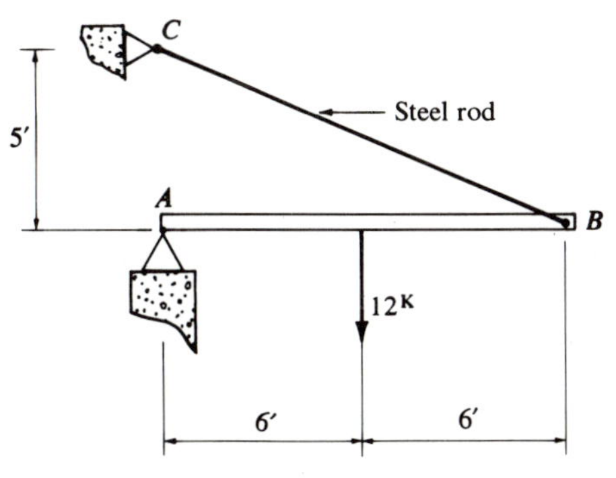

Figure P 5–12

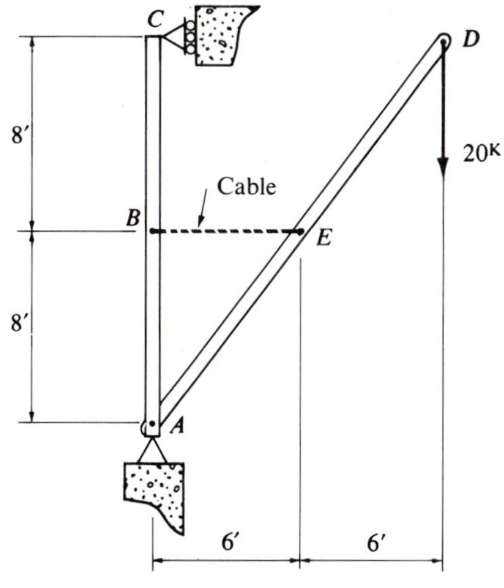

Figure P 5–13

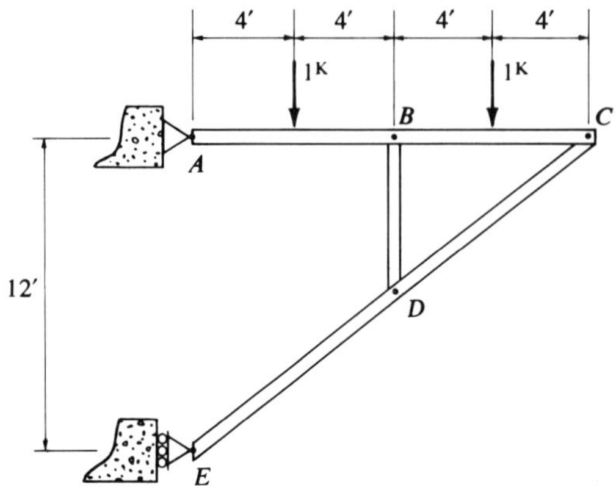

Figure P 5–14

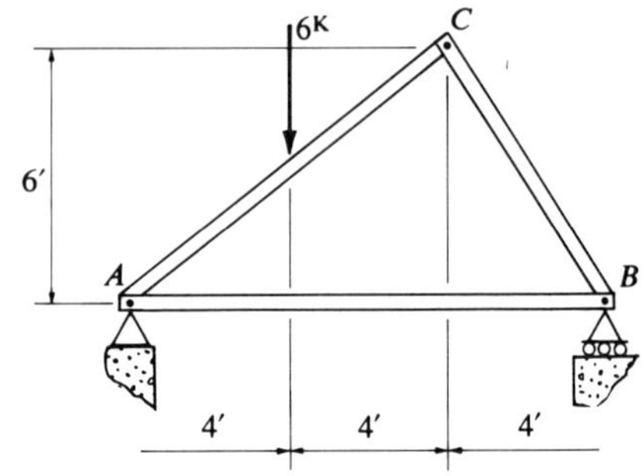

Figure P 5–15

CHAPTER SIX

Influence Lines

6.1 Introduction

Chapters Two to Five have dealt with the various basic methods used to determine reactions, shear, bending moment, and bar forces for simple, statically determinate structures. These determinations, however, were made considering only stationary and constant magnitude loads acting on the structure. We know that structures are subjected to loads that move, or vary in position. Different load positions will have different effects on the structure. There is, however, one position of the loads for each function considered, such as the reaction at a support, that will produce the most severe effect. In other words, there is one position of the loads that will produce the maximum reaction at a support. The same can be said for shear, bending moment, and bar forces. The structural analyst must be able to determine the positions of the movable loads that produce the most severe effects in the structure.

As in Chapter One, a distinction must be made between the loads acting on the structure. Recall that immovable loads were called **dead loads** and loads that vary in position were designated as **live loads.** These designations will be used throughout this chapter.

6.2 Definitions

Before we can determine the positions of the live loads on a structure to produce the most severe effects, we must know how each function, such as a reaction, shear, etc., varies as these loads move along the span of the structure. Let us consider the simply supported beam shown in Figure 6.1a. In order to

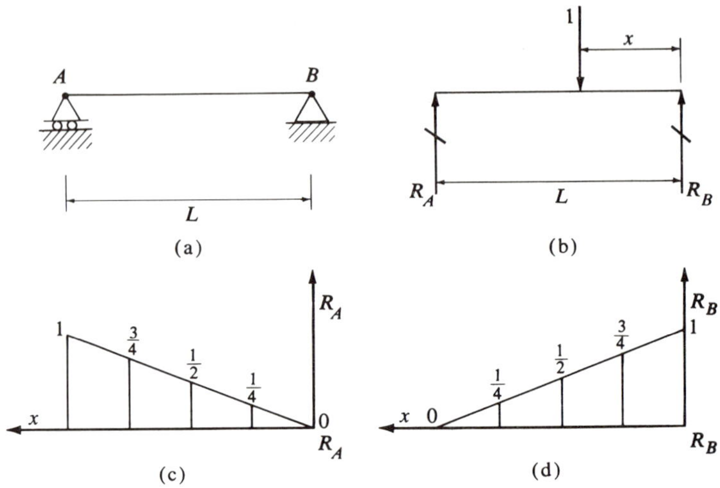

Figure 6.1

design the support at A, the value of the reaction at its maximum value must be known. If the beam AB is subjected to live loading, the position of such loading to produce the maximum reaction at A must be known. If we knew how the reaction at A varied as loads moved along the span, it would be relatively easy to then position the live loads to produce the maximum reaction at A. The easiest way to see how the reaction at A varies with different loading positions is to compute R_A for various load positions as a load travels from B to A or, for that matter, from A to B. In order to simplify matters even further, we shall compute R_A for various positions of a **unit load.** A unit load is a load that has a magnitude of unity, i.e., one. The units on the unit load may be pounds, kips, grams, etc., depending upon whatever units the analyst will be working in. Computing R_A for various positions of a unit load as it travels along the span will indicate how R_A varies for different load positions. Furthermore, the position of the unit load for a maximum value for R_A will be determined along with the magnitude for R_A of its maximum value. For any load other than unity, say P kips, the load position for maximum R_A will remain the same as for the unit load and the magnitude will be P times as great as for the unit load. The use of the unit load is, therefore, applicable to any other load system. Referring again to Figure 6.1, let us determine how R_A varies as a unit load moves across the span L. Figure 6.1b shows a unit load placed arbitrarily on the span. Solving for R_A, we obtain:

Figure 6.1b:

$$\Sigma M_B = 0 \quad (+\circlearrowright)$$

$$R_A(L) - 1(x) = 0$$

$$R_A = \frac{x}{L}$$

Substituting in various values for x, the following are obtained:

$$\text{Unit Load @ } x = L/4; \quad R_A = \frac{1}{4} \uparrow$$

$$\text{Unit Load @ } x = L/2; \quad R_A = \frac{1}{2}$$

$$\text{Unit Load @ } x = 3L/4; \quad R_A = \frac{3}{4}$$

$$\text{Unit Load @ } x = L; \quad R_A = 1$$

Since a positive magnitude is obtained for each value of x, R_A is directed upward as assumed in Figure 6.1b for each position of the unit load. The maximum value for R_A is 1 and occurs when the unit load is at A. If a live load of 5 kips were to be carried by the beam AB shown in Figure 6.1a, R_A would be maximum when the 5-kip load is positioned at A and the corresponding maximum magnitude for R_A is 5(1) or 5 kips since the effect for a 5-kip load would be 5 times as great as for a 1-kip load. It should be noted that the unit load takes on units of kips since the applied live loads are in kips. It is more convenient to show the variation of R_A graphically as the unit load moves along the span. Such a plot is shown in Figure 6.1c for R_A. Any point on the x-axis represents the position of the unit load along the span and the corresponding ordinate represents the value of R_A for that position of the unit load. Such a plot is commonly referred to as an **influence line.** More generally, an influence line is defined as a curve, the ordinates of which at any point equal the value of some function due to the presence of a unit load acting at that point. Thus, from Figure 6.1c it can be seen that with the unit load at midspan, R_A is equal to 1/2.

An influence line for the vertical reaction at B may also be drawn. Since an influence line for R_A has already been drawn, it can be utilized for the construction of the influence line for R_B as follows:

Figure 6.1:

$$\Sigma F_Y = 0 \uparrow +$$
$$1 - R_A - R_B = 0$$
$$R_B = 1 - R_A$$

$$\text{Unit Load @ } x = L/4; \quad R_B = 1 - \frac{1}{4} = \frac{3}{4}$$

$$\text{Unit Load @ } x = L/2; \quad R_B = 1 - \frac{1}{2} = \frac{1}{2}$$

$$\text{Unit Load @ } x = 3L/4; \quad R_B = 1 - \frac{3}{4} = \frac{1}{4}$$

$$\text{Unit Load @ } x = L; \quad R_B = 1 - 1 = 0$$

Coordinates for the influence line for R_B may also be obtained by summing moments about point A for various positions of the unit load. Thus:

Figure 6.1b:

$$\Sigma M_A = 0 \quad (+\curvearrowright)$$

$$1(L - x) - (L)R_B = 0$$

$$R_B = 1 - x$$

$$\text{Unit Load @ } x = L/4; \quad R_B = \frac{3}{4}\uparrow$$

$$\text{Unit Load @ } x = L/2; \quad R_B = \frac{1}{2}$$

$$\text{Unit Load @ } x = 3L/4; \quad R_B = \frac{1}{4}$$

$$\text{Unit Load @ } x = L; \quad R_B = 0$$

Again, R_B is always directed upward as assumed in Figure 6.1b since all positive results were obtained.

Referring to Figure 6.1c, the following conclusions can be drawn: (1) as the unit load travels from B to A, R_A increases linearly, reaching a maximum magnitude when the unit load is at A; (2) since all the values for R_A are plotted above the base line, R_A is always directed upward for any position of the unit load; (3) the influence line is a straight line, and, hence, only two points were needed for construction (more generally, the influence lines for statically determinate structures consist of straight-line segments and only a few select points are needed to accomplish the construction); (4) to obtain the reaction at A for any load with a magnitude other than unity, simply multiply the appropriate influence line coordinate by the magnitude of the applied live load. To obtain the maximum effect, place the live load at the point of having the maximum influence line ordinate.

Influence lines are valuable tools in the analysis of structures since the majority of structures are subjected to live loading. As stated previously, influence lines are not limited to reactions but can be constructed for shear and bending moment at any point. The following examples will further illustrate the construction of influence lines.

Example 6.1

Draw the influence lines for the vertical reaction at C, the moment at C, and the shear at B for the cantilever beam shown in Figure 6.2.

Solution:

1. Before proceeding with the solution, a sign convention must be established for the construction of influence lines. Upward reactions, positive shear, and positive bending moments, i.e., moments that tend to bend the member concave upward, will all be plotted above the baseline.

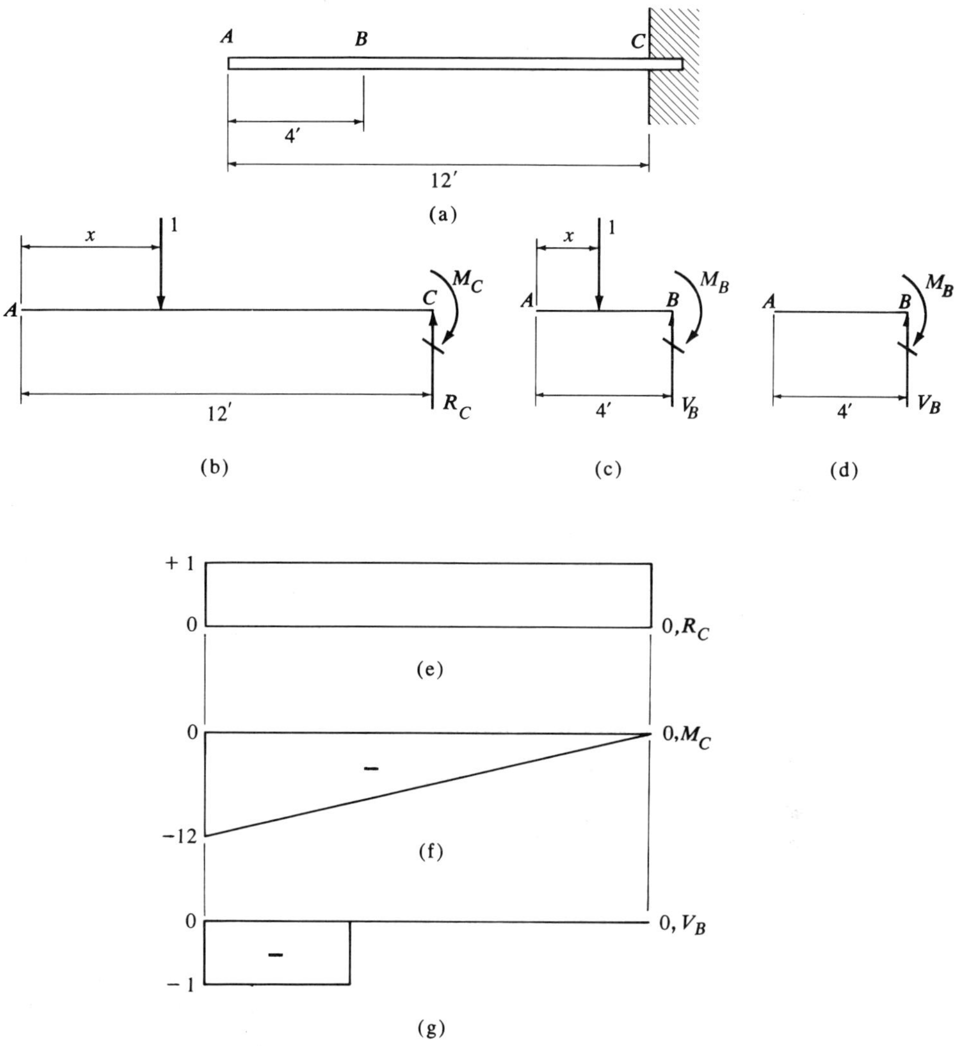

Figure 6.2

2. To obtain the influence line for the vertical reaction at C, the FBD shown in Figure 6.2b is utilized. The unit load is placed at an arbitrary distance of x-ft from A so that a general expression for R_C may be determined. Notice from the FBD that the only vertical forces acting on the beam are the downward directed unit load and the assumed upward reaction R_C. To maintain equilibrium in the vertical direction, R_C must always be directed upward and equal in magnitude to the unit load for any position of the unit load along the span. The influence line for R_C is shown in Figure 6.2e.
3. The influence line for the moment at C may also be obtained by utilizing the FBD shown in Figure 6.2b. Summing moments about point C will yield a general expression for M_C as follows:

Figure 6.2b:

$$\Sigma M_C = 0 \; (+\curvearrowright)$$

$$M_C - 1(12 - x) = 0$$
$$M_C = 12 - x$$

Unit Load @ $x = 0'$; $M_C = 12$
Unit Load @ $x = 6'$; $M_C = 6$
Unit Load @ $x = 12'$; $M_C = 0$

Since a positive answer for M_C was obtained each time, M_C always acts in the direction shown in Figure 6.2b (clockwise) for any position of the unit load. A moment in the clockwise direction at C would tend to bend the member concave downward, and, hence, is negative moment at C. The corresponding influence line for M_C using the coordinates just obtained is shown in Figure 6.2f.

4. To obtain the influence line for the shear at B, we again draw a FBD that will yield a general expression for the shear at B. Cutting the beam at point B and drawing a FBD of that portion of the span to the left of the cut is shown in Figure 6.2c for an arbitrary position of the unit load. Solving for the shear at B we obtain the following:

Figure 6.2c:

$$\Sigma F_Y = 0 \uparrow +$$
$$V_B - 1 = 0$$
$$V_B = 1$$

The shear force is directed upward for any position of the unit load on segment AB shown in Figure 6.2c. This means that an upward shear force at B is required to keep the segment AB from moving downward under the action of the applied unit load. By our previously adopted sign convention, this constitutes negative shear. If the unit load moves to the right of point B, the FBD shown in Figure 6.2d replaces that FBD shown in Figure 6.2c. The shear at B is nonexistent for the unit load anywhere on segment BC. The corresponding influence line for V_B is shown in Figure 6.2g.

Example 6.2

Draw the influence lines for the shear at B and the moment at C for the beam shown in Figure 6.3.

Solution:

1. To obtain an influence line for the shear at B, we must first obtain a general expression for the shear at B, V_B, for any position of the unit load. This involves cutting the member at B to release the internal shear force. If we

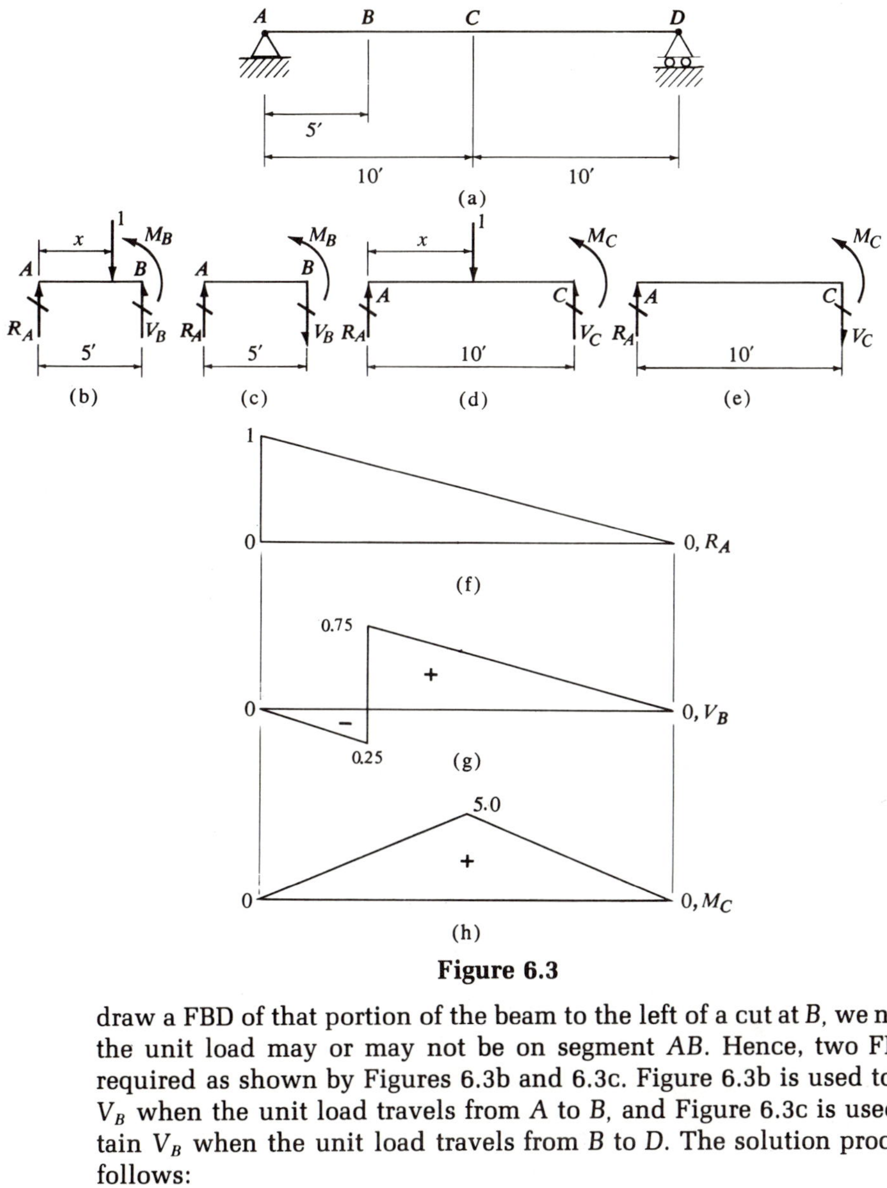

Figure 6.3

draw a FBD of that portion of the beam to the left of a cut at B, we note that the unit load may or may not be on segment AB. Hence, two FBDs are required as shown by Figures 6.3b and 6.3c. Figure 6.3b is used to obtain V_B when the unit load travels from A to B, and Figure 6.3c is used to obtain V_B when the unit load travels from B to D. The solution proceeds as follows:

Figure 6.3b:

$$\Sigma F_Y = 0 \uparrow +$$
$$R_A + V_B - 1 = 0$$
$$\underline{V_B = 1 - R_A} \quad (x \leq 5')^*$$

---

* This term in parentheses is read, **for all values of x less than and equal to 5 feet,** and means that the equation $V_B = 1 - R_A$ is only applicable for the previously stated values of x. Since x is the distance from point A (see Figure 6.3b) to the unit load, the expressed equation for V_B is applicable for any position of the unit load up to and including 5 feet from A.

Figure 6.3c:

$$\Sigma F_Y = 0 \uparrow +$$
$$R_A - V_B = 0$$
$$\underline{V_B = R_A} \quad (x > 5')$$

In order to determine V_B we must know R_A. Recall that R_A also varies for various positions of the unit load. The best way to determine R_A for various positions of the unit load is to draw an influence line for R_A. It is often necessary to construct influence lines for functions, such as R_A in this case, that are not required but serve only as an aid in the solution of the desired functions such as V_B. The influence line for R_A is shown in Figure 6.3f. Its construction is straightforward. If the unit load is placed at point A in Figure 6.3a, R_A must be +1 (the positive signing indicating an upward directed force); and if placed at D, R_A must be zero. Recalling that influence lines for functions of statically determinate structures are straight-line segments, these two points provide the necessary data for the influence line for R_A. Using the influence line for R_A and the two equations for V_B obtained previously, we may proceed as follows:

$$V_B = 1 - R_A \quad (x \leq 5')$$
$$@\ x = 0'; \quad R_A = +1 \text{ (from Figure 6.3f)}, \quad V_B = 0$$
$$@\ x = 2.5'; \quad R_A = 0.875, \quad V_B = 1 - 0.875 = 0.125$$

The positive sign obtained for V_B indicates that the shear force at B is directed upward and indicates negative shear (an upward force is required at B to prevent the segment AB from moving downward, i.e., negative shear). Continuing:

$$@\ x = 5'; \quad R_A = 0.75, \quad V_B = 1 - 0.75 = 0.25$$

Now using $V_B = R_A \quad (x > 5')$

$$@\ x = 5' + \Delta x; \quad R_A \simeq 0.75, \quad V_B \simeq 0.75$$

Notice that the unit load was placed an infinitesimal distance to the right of point B. This was accomplished because the shear at B changes abruptly from negative to positive shear as the unit load passes over point B. The ordinate for V_B with the unit load at B was to be −0.25 and when placed at a distance of Δx to the right of point B we obtain +0.75. The positive sign obtained is due to a downward shear force required at section B (see Figure 6.3c) for equilibrium of segment AB. Continuing:

$$@\ x = 12.5'; \quad R_A = 0.375, \quad V_B = 0.375$$
$$@\ x = 20'; \quad R_A = 0, \quad V_B = 0$$

Using those ordinates, the influence line for V_B is plotted and shown in Figure 6.3g. This shape is typical for simply supported spans.

2. To obtain the influence line for M_C, we again cut the beam AD at point C to release the internal resisting moment M_C at point C. The corresponding FBDs for the left portion of the cut are shown in Figures 6.3d and 6.3e. Again, two FBDs are required: Figure 6.3d when the unit load is on segment AC, and Figure 6.3e when the unit load is on segment CD. In both FBDs the moment at C, M_C, is assumed to be positive, i.e., causing the member to bend concave upward. The solution proceeds as follows:

Figure 6.3d:

$$\Sigma M_C = 0 \quad (\curvearrowright +)$$

$$(10)R_A - 1(10 - x) - M_C = 0$$

$$\underline{M_C = 10R_A + x - 10} \quad (x \leq 10')$$

Figure 6.3e:

$$\Sigma M_C = 0 \quad (\curvearrowright +)$$

$$10R_A - M_C = 0$$

$$\underline{M_C = 10R_A} \quad (x > 10')$$

Substituting in various values of x, we may obtain the corresponding ordinates for the influence line for M_C:

$$M_C = 10R_A + x - 10 \quad (x \leq 10')$$

@ $x = 0'$; $R_A = +1$ (from Figure 6.3f), $M_C = 0$
@ $x = 5'$; $R_A = 0.75$, $M_C = 7.5 + 5 - 10 = +2.5$
@ $x = 10'$; $R_A = 0.50$, $M_C = 5.0 + 10 - 10 = +5.0$

$$M_C = 10R_A \quad (x > 10')$$

@ $x = 10' + \Delta x$; $R_A = 0.50$, $M_C = +5.0$
@ $x = 15'$; $R_A = 0.25$, $M_C = +2.5$
@ $x = 20'$; $R_A = 0.0$, $M_C = 0.0$

Plotting these values, the influence line for M_C is obtained as shown in Figure 6.3h.

3. The shapes of the influence lines shown in Figures 6.3g and 6.3h are typical for shear and moment at any section of a simply supported beam. Hence, for shear only two ordinates are required and for moment only one ordinate is required for construction of the influence lines. This simplifies the calculations necessary for the construction of the influence lines.

Example 6.3

Draw the influence lines for the reactions at B and D and the moment at B and C for the beam shown in Figure 6.4.

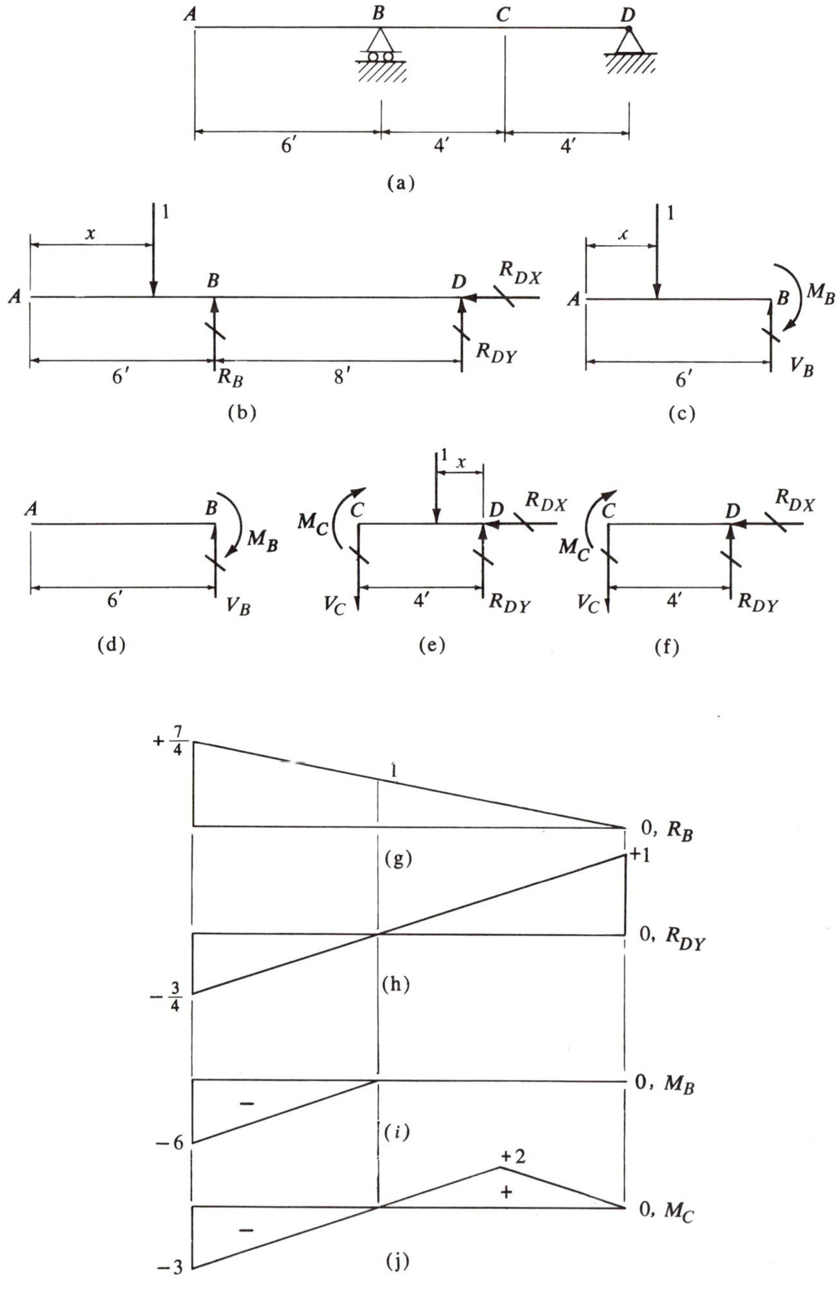

Figure 6.4

Solution:

1. To obtain the influence lines for the reactions at B and D, the FBD of the entire beam AD shown in Figure 6.4b will be used. General expressions for R_B and R_{DY} will be obtained for an arbitrary position of the unit load as follows:

   Figure 6.4b:

$$\Sigma M_D = 0 \quad (\curvearrowright +)$$

$$8R_B - 1(14 - x) = 0$$

$$\underline{R_B = (\tfrac{1}{8})(14 - x)}$$

$$@\ x = 0'; \quad R_B = \tfrac{7}{4}$$
$$@\ x = 6'; \quad R_B = (\tfrac{1}{8})(8) = 1$$
$$@\ x = 14'; \quad R_B = (\tfrac{1}{8})(0) = 0$$

   Since only straight-line segments are involved, the above three points are more than sufficient for construction of the influence line for R_B. The positive result obtained each time confirms the upward sense assumed for R_B in Figure 6.4b. Continuing:

   Figure 6.4b:

$$\Sigma F_Y = 0 \uparrow +$$
$$R_B + R_{DY} - 1 = 0$$
$$\underline{R_{DY} = 1 - R_B}$$

$$@\ x = 0'; \quad R_B = \tfrac{7}{4}, \quad R_{DY} = 1 - \tfrac{7}{4} = -\tfrac{3}{4} = \tfrac{3}{4} \downarrow$$
$$@\ x = 6'; \quad R_B = 1, \quad R_{DY} = 1 - 1 = 0$$
$$@\ x = 14'; \quad R_B = 0, \quad R_{DY} = 1 - 0 = +1 = 1 \uparrow$$

   Note that the values for R_B were required for various positions of the unit load as an aid in determination of values of R_{DY}.

   The corresponding influence lines for R_B and R_{DY} are shown in Figures 6.4g and 6.4h respectively.

2. Cutting the beam AD at a point just to the left of point B releases the internal moment at B. Again, two FBDs are required for a general solution for M_B: Figure 6.4c when the unit load is on segment AB and Figure 6.4d when the unit load is on segment BD. The solution proceeds as follows:

   Figure 6.4c:

$$\Sigma M_B = 0 \quad (\curvearrowright +)$$

$$M_B - 1(6 - x) = 0$$

$$\underline{M_B = 6 - x} \quad (x \leq 6')$$

Figure 6.4d:

$$\Sigma M_B = 0 \quad (\curvearrowright +)$$

$$\underline{M_B = 0} \quad (x > 6')$$

Notice that M_B was assumed in both FBDs to be producing negative bending, i.e., causing the member to bend concave downward. Hence, a positive answer for M_B will indicate negative moment. Continuing:

$$\begin{aligned}
& M_B = 6 - x \quad (x \leq 6') \\
@\ x = 0; \quad & M_B = 6 \\
@\ x = 6; \quad & M_B = 0 \\
& M_B = 0 \quad (x > 6') \\
& M_B = 0 \quad \text{all values of } x > 6'
\end{aligned}$$

The corresponding influence line for M_B is shown in Figure 6.4i.

3. The FBDs necessary to find M_C for all positions of the unit load are shown in Figures 6.4e and 6.4f. It should be noted that FBDs were drawn for that portion of the structure to the right of a cut at C. This was done simply because it is easier, i.e., less structure is involved to the right of a cut at C. Notice also that x is now measured from D. The sense of M_C is assumed in both FBDs to cause positive bending at C. The solution proceeds as follows:

Figure 6.4e:

$$\Sigma M_C = 0 \quad (\curvearrowright +)$$

$$M_C + 1(4 - x) - 4R_{DY} = 0$$

$$M_C = 4R_{DY} + x - 4 \quad (x \leq 4')$$

$$\begin{aligned}
@\ x = 0'; \quad & R_{DY} = 1, \quad & M_C = 4 - 0 - 4 = 0 \\
@\ x = 4'; \quad & R_{DY} = \tfrac{1}{2}, \quad & M_C = (4)(\tfrac{1}{2}) + 4 - 4 = 2
\end{aligned}$$

Figure 6.4f:

$$\Sigma M_C = 0 \quad (\curvearrowright +)$$

$$M_C - 4R_{DY} = 0$$

$$M_C = 4R_{DY} \quad (x > 4')$$

$$\begin{aligned}
@\ x = 8'; \quad & R_{DY} = 0, \quad & M_C = 0 \\
@\ x = 14'; \quad & R_{DY} = -\tfrac{3}{4}, \quad & M_C = 4(-\tfrac{3}{4}) = -3 = 3 \ \circlearrowleft
\end{aligned}$$

The values for R_{DY} were obtained from the influence line shown in Figure 6.4h. The influence line for M_C is shown in Figure 6.4j.

6.3 Applications of Influence Lines

The discussion and examples presented in this chapter up to this point have dealt with the construction of influence lines for various functions. The application of the influence line, however, needs considerable explanation. Recall that the ordinate of an influence line for a particular function at any point is the magnitude of that particular function when a unit load is placed at that point. The point of a maximum ordinate, therefore, yields the position of the unit load for a maximum response of the plotted function. If we know the position of a unit load to produce a maximum response, then a load of magnitude P will also produce a maximum response at that location with a magnitude P times as large as that magnitude produced by the unit load. Hence, an influence line can readily tell us where to position a single concentrated load to produce a maximum response. The application can be extended to include distributed loads as well. Consider the segment of a loaded beam and the corresponding influence line for a particular function S shown in Figure 6.5.

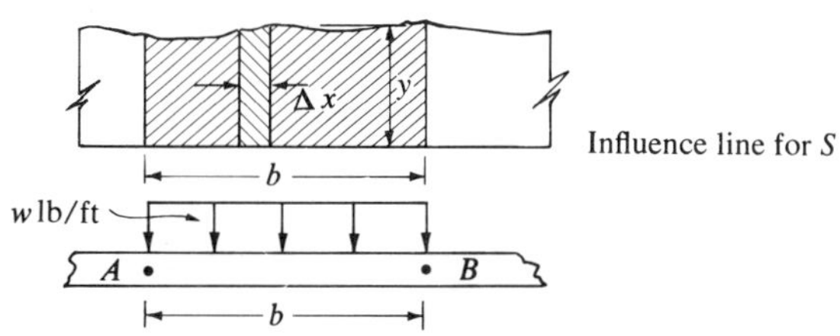

Figure 6.5

The portion of the uniformly distributed load acting on the distance Δx may be treated as a concentrated load of magnitude $w\Delta x$. It has been stated previously that the value of the function S due to this concentrated load may be taken as $w(\Delta x)y$ where y is the ordinate of the influence line at the point of application of the concentrated load. Hence, we have $\Delta S = w(\Delta x)y$. Since there are an infinite number of these concentrated loads acting between points A and B (there are an infinite number because Δx is an infinitesimally small distance), the total value of S due to the total load between A and B is obtained by summing all the partial effects ΔS resulting from the infinite number of concentrated loads acting between A and B. Thus:

$$\Delta S = w(\Delta x)y$$

$$\therefore S = \sum_{A}^{B} w(\Delta x)y$$

Since w is a constant value we have:

$$S = w \sum_{A}^{B} (\Delta x)y$$

where $\sum_{A}^{B} (\Delta x)y$ is the area under the influence line between A and B. For uniformly distributed loads, the following rule can be stated: **the value of a function due to a uniformly distributed live load is equal to the product of the intensity of the live loading and the area under the influence line, for the function being considered, for that portion of the structure which is loaded.** The following examples will demonstrate the application of influence lines.

Example 6.4

Determine the maximum reaction at A, the maximum shear at C, and the maximum moment at B for the beam shown in Figure 6.6 due to the passage of a moving uniform load of 1-kip per foot without limit in length and a roving concentrated load of 10 kips.

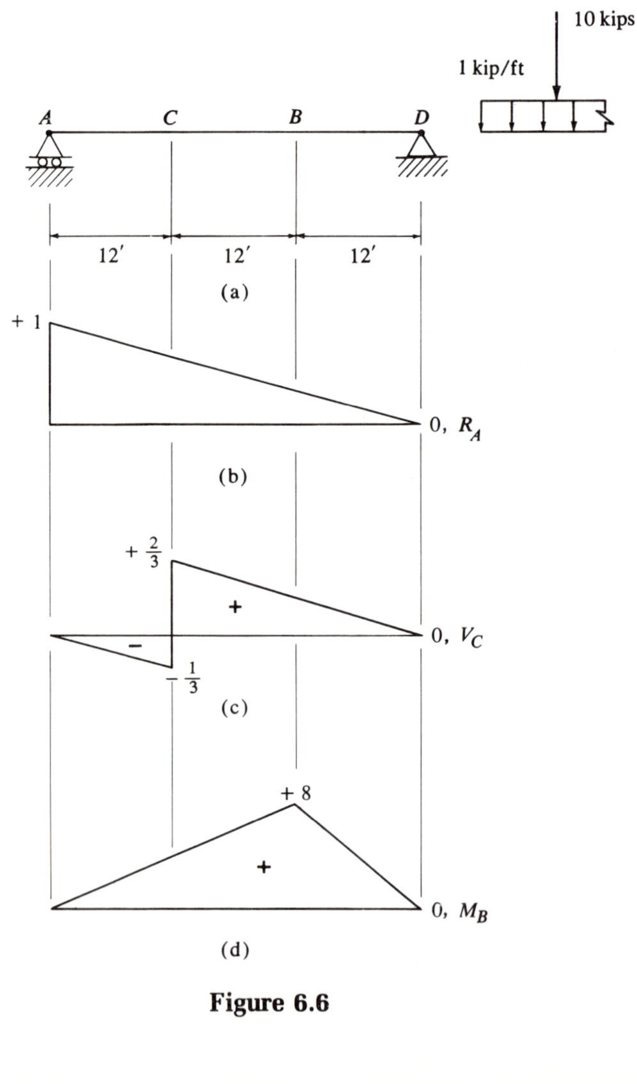

Figure 6.6

Solution:

1. Before proceeding with the solution, a short explanation concerning the live loading encountered in this example will be made. The live loading for a highway bridge consists of many various types of vehicles. A critical loading for each lane, however, could consist of series of closely spaced heavy trucks. For design purposes it is permissible to substitute an equivalent loading for the closely spaced trucks consisting of a uniform load per foot of lane plus an additional roving single concentrated load. The magnitudes of these equivalent loads depend upon the bridge type, use, and anticipated traffic. The American Association of State Highway Officials (AASHO) has specified various equivalent loadings depending upon the bridge requirements. Hence, this example has direct application to the analysis of highway bridges.
2. Using the methods outlined in the previous sections, the influence lines for R_A, V_C, and M_B shown in Figures 6.6b, 6.6c, and 6.6d were constructed. Considering the influence line for R_A, we may note the following characteristics: (1) R_A is at a maximum when a single concentrated load is placed at point A; (2) since all the influence line ordinates for R_A are positive, R_A is directed upward for any position of a downward applied load. For the given loading condition, therefore, we must place the roving 10-kip load at point A as noted by the first observed characteristic. The second observed characteristic dictates that we load the entire span AD with the 1-kip per foot uniform load since a load at each point produces an upward directed result in R_A. Numerically, we obtain the following:

$$\text{Maximum } R_A = 10(1) + (1.0)(\tfrac{1}{2})(1)(36)$$
$$= 10 + 18$$
$$\underline{\text{Maximum } R_A = 28^{K} \uparrow}$$

3. Considering the influence line for V_C shown in Figure 6.6c, we note that the maximum ordinate occurs just to the right of point C and pinpoints the location of the roving 10-kip concentrated load for a maximum effect in the shear at C. Placing the 1-kip per foot uniform load, however, presents a more difficult problem. A load placed anywhere on segment AC produces negative shear at C and a load anywhere on segment CD produces positive shear at C. If we load the entire span with the uniform load, that portion acting on segment AC will produce a negative effect on the shear at C while that portion acting on segment CD will produce a positive effect on the shear at C. We may conclude, therefore, that a maximum effect on the shear at C will not occur if we load the entire span, but only that segment producing negative shear at C or that segment producing only positive shear at C. The segment to be loaded is that segment having the greater area under its corresponding portion of the influence line. In this case, the larger area corresponds to segment CD. In addition, we must check to see that this selected segment produces the same sign on the shear at C as does the roving concentrated load. In this case, both produce positive

shear and for maximum V_C we must place the roving 10-kip load just to the right of point C and we must load segment CD with the 1-kip per foot uniform load. Numerically, we obtain:

$$\text{Maximum } V_C = \text{Single Load} \times (\text{Max. orainate}) + \text{load intensity} \times (\text{Influence area})$$
$$\text{Maximum } V_C = 10(\tfrac{2}{3}) + (1)(\tfrac{1}{2})(\tfrac{2}{3})(24)$$
$$= 6.66 + 8.0$$
$$\underline{\text{Maximum } V_C = +14.66}$$

4. Turning our attention now to the influence line for M_B shown in Figure 6.6d, we observe that the moment at B is positive for any position of the unit load and the maximum ordinate occurs at B. Therefore, we must place the roving 10-kip load at point B and load the entire span with the 1-kip per foot uniform load. Thus:

$$\text{Maximum } M_B = 10(8) + (1)(\tfrac{1}{2})(8)(36)$$
$$= 80 + 144$$
$$\underline{\text{Maximum } M_B = 224 \text{ kip-ft}}$$

Example 6.5

The simply supported beam shown in Figure 6.7 carries moving loads of 5 kips, 10 kips, and 10 kips spaced 5 ft apart. Calculate the maximum left reaction and the maximum shear and the moment at a point 15 ft from the left support.

Solution:

1. The first step in the solution is to construct the influence lines for R_A, V_B, and M_B. These are presented in Figures 6.7b, 6.7c, and 6.7d. It will be assumed that the wheel loads may travel in either direction across the span.
2. Since this problem deals with concentrated loads only, we must seek positions of these loads at or near maximum influence ordinates in order to produce the maximum effects. For maximum R_A, the shape of the influence line dictates that we place the loads as near to point A as physically possible (recall that the loads are a fixed distance apart). Figure 6.7b shows the wheel group placed as close to point A as possible. Naturally, the heaviest wheel is placed at point A, the point having the maximum influence ordinate. Using similar triangles, we may calculate the influence ordinates for each load position shown as follows:

$$R_A = 10(1) + 10(\tfrac{40}{45}) + 5(\tfrac{35}{45})$$
$$R_A = 22.8^K$$

Attempts to position the loads in any other fashion will all yield results for R_A less than 22.8 kips. Hence:

$$\underline{\text{Maximum } R_A = 22.8^K \uparrow}$$

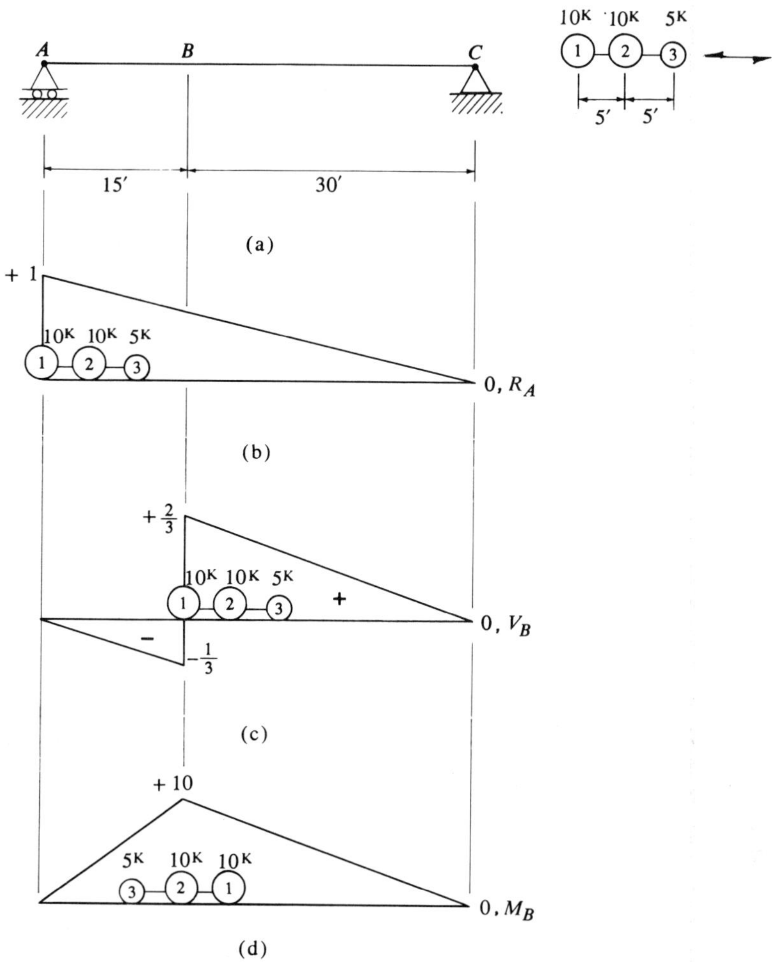

Figure 6.7

3. The reasoning used to position the loads for maximum R_A can be extended to obtain a maximum V_B. Examination of the influence line for V_B shown in Figure 6.7c reveals that we must place the wheels as close to point C as possible, whether we approach point C from the left or from the right. For the wheel position shown and using similar triangles, we obtain:

$$V_B = 10(\tfrac{2}{3}) + 10(\tfrac{25}{30})(\tfrac{2}{3}) + 5(\tfrac{20}{30})\tfrac{2}{3}$$
$$V_B = 14.44^{K}$$

Repositioning the loads on either side at point C or approaching point C from either direction will not yield a magnitude for V_B greater than 14.44 kips. Hence:

$$\underline{\text{Maximum } V_B = 14.44^{K}}$$

4. Considering the influence line for M_B shown in Figure 6.7d, we note that the maximum ordinates for M_B occur in the vicinity of point B with the maximum ordinate at point B. We want, therefore, to position the heaviest wheel loads at and near point B. For the wheel positions shown we obtain the following for M_B:

$$M_B = 10(10) + 10(\tfrac{25}{30})(10) + 5(\tfrac{10}{15})(10)$$
$$M_B = 216.7 \text{ kip-ft}$$

Trying other wheel positions will not yield any greater value for M_B. Again, therefore:

$$\underline{\text{Maximum } M_B = 216.7 \text{ kip-ft}}$$

6.4 Criteria for Maximum Effect on Beams

In the previous example problem the analyst had to determine the position of a series of fixed spaced, concentrated loads that would produce the maximum result in several functions. Summarizing, the method utilized the shape of the corresponding influence line and a trial-and-error method of placing the series of live loads to produce the maximum effect. Since only three live loads were involved, the procedure was neither laborious nor complex. Imagine, however, a very long series of concentrated loads moving across the span of a beam. Let us further assume that the loads are varied in magnitude and that the length of the series is larger than the beam's span. For a given function, the placement of such a series to produce a maximum effect would become very tedious and time-consuming. Not only would the analyst be required to find numerous influence ordinates at each load position, but the problem would be made even more complex by the fact that upon each movement of the series, loads could leave and/or enter the span length. Hence, a criterion has been developed for each typical influence line shape for maximum effect by such live loadings. An explanation of this criterion is beyond the scope of this text and the reader is referred to any standard text on structural analysis if such an explanation is required. This criterion is based upon finding the maximum moment and shear at a specific point due to a long series of concentrated loads. For design purposes, however, the analyst is concerned with the maximum shear and maximum moment that can occur at any section in the member due to the live loading, i.e., **the absolute maximum live shear** and **the absolute maximum live moment.** Consider the series of moving concentrated loads shown in Figure 6.8. The shear and moment diagrams for the beam AB under the action of the series of live loads at one position are shown in Figure 6.7b and c. From these diagrams, the following conclusions may be made: (1) the absolute maximum shear in a beam resulting from a series of moving concentrated loads will occur at a section immediately adjacent to one of the reactions; and (2) the absolute maximum live moment must occur directly

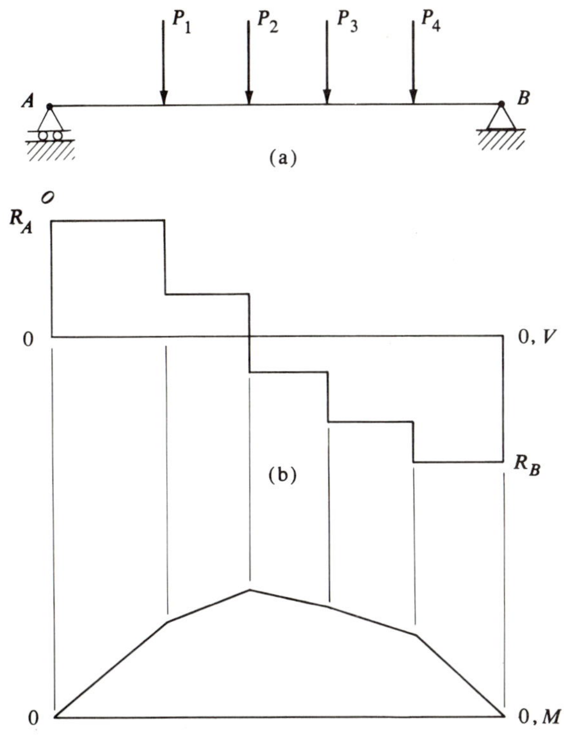

Figure 6.8

beneath one of the loads. To find the absolute maximum live moment we must ask the following questions: (a) under which load does this maximum moment occur? and (b) what is the position of this load? To answer these questions, we must turn our attention to the simply supported beam and the corresponding wheel loading shown in Figure 6.9. The following assumptions are also made:

(a) The wheels are at a fixed distance apart.
(b) Since the maximum moment must occur under one of the loads, it will be assumed that the maximum moment occurs under wheel no. 2.
(c) Wheel no. 2, henceforth referred to as the critical wheel, is x-feet to the left of the beam's centerline.
(d) W_T is the resultant of the wheel loading and its position is known.
(e) The distance from the critical wheel to the resultant is d-feet.

Since we have assumed that the absolute maximum moment occurs under wheel no. 2, we must determine the distance x that will make the moment under that wheel a maximum. An expression for the moment under wheel no. 2 must, therefore, be formed. Applying the equations of equilibrium to the beam shown in Figure 6.9 we obtain the following:

$$\Sigma M_B = 0 \quad (+\curvearrowright)$$

$$R_A(L) - W_T(L/2 + x - d) = 0$$

$$R_A = (W_T)(1/L)(L/2 + x - d)$$

$\Sigma M_2 = 0 \quad (+\curvearrowright)$; where M_2 is the moment under wheel no. 2.

$$M_2 = R_A(L/2 - x) - W_1(a)$$

$$= \frac{W_T}{L}\left(\frac{L}{2} + x - d\right)\left(\frac{L}{2} - x\right) - W_1(a)$$

$$M_2 = \frac{W_T}{L}\left[\left(\frac{L^2}{4} + \frac{Ld}{2}\right) + (xd - x^2)\right] - W_1a$$

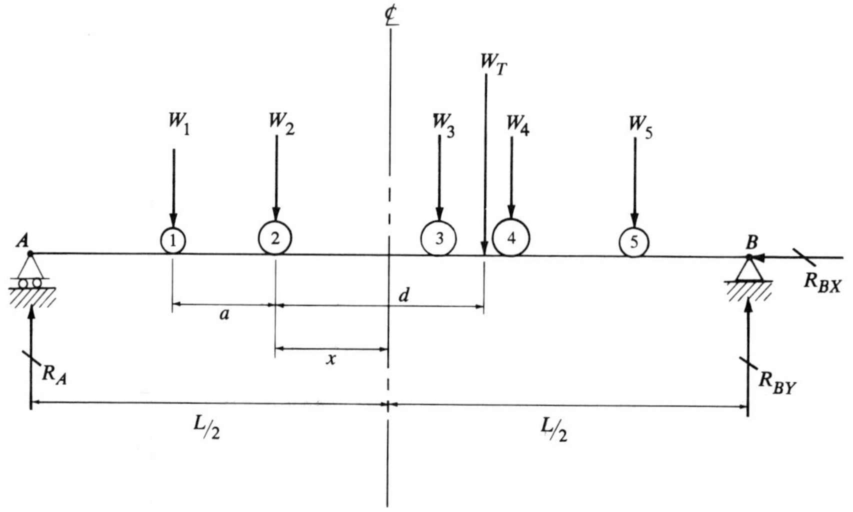

Figure 6.9

For M_2 to be a maximum, the terms contained in the bracket must have a maximum positive value. A further examination of the terms within the bracket indicates that both L and d are constant values; therefore, the terms within the bracket will be maximum when the expression $(xd - x^2)$ has a maximum positive value. Setting $f(x) = (xd - x^2)$ and plotting $f(x)$ for various values of x will reveal that $f(x) = (xd - x^2)$ has a maximum positive value when $x = d/2$. From this determination, we may conclude that **the absolute maximum moment directly beneath one of a series of concentrated moving loads acting on a simply supported beam occurs when the centerline of the span is exactly halfway between that particular load and the resultant of all the loads on the span.**

Generally speaking, the absolute maximum moment will occur under the load that is nearest to the line of action of the resultant. Otherwise, several other loads adjacent to this load may be tried.

Example 6.6

Find the absolute maximum bending moment for the beam and loading shown in Figure 6.10a.

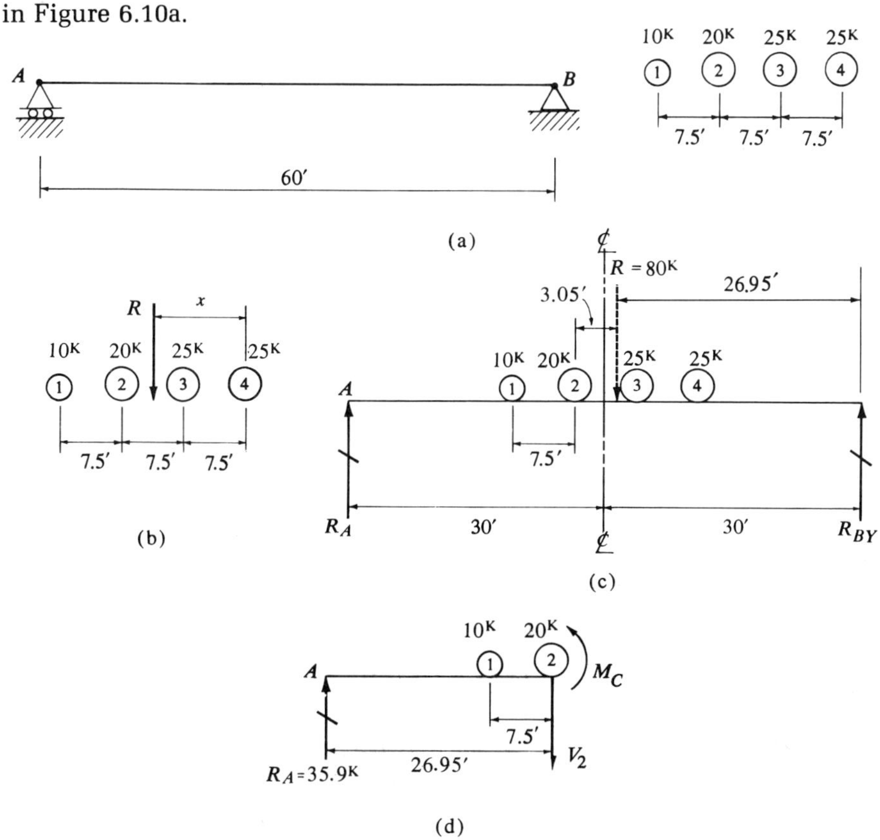

Figure 6.10

Solution:

1. The first step in problems of this nature is to determine the resultant of the series of concentrated live loads and its location relative to the wheel loading group. Referring to Figure 6.10b, we may proceed as follows:

$$R = \Sigma W \text{ (for a parallel force system)}$$
$$= 10 + 20 + 25 + 25$$
$$R = 80^{K}$$

Using the condition that the moment about any point due to the component forces must be equal to the moment produced by the resultant force about the same point, we may determine the location of the resultant by equating moments about wheel no. 4 as follows:

Figure 6.10b:

$$\Sigma M_4 = \Sigma M_4$$
$$Rx = (10)(22.5) + (20)(15) + 25(7.5)$$
$$80x = 712.5; \qquad x = 8.90'$$

2. The resultant of the wheel group lies between wheel no. 2 and wheel no. 3. Wheel no. 2, therefore, will be tried as the critical wheel. The wheel group will be placed on the span in such a fashion that the centerline of the span bisects the distance between wheel no. 2 and the resultant. This placement is shown in Figure 6.10c. The vertical reaction at A is determined for this loading configuration as follows:

   Figure 6.10c: $\Sigma M_B = 0$ (+ clockwise)

   $$R_A(60) - 80(26.95) = 0$$
   $$R_A = +35.93^K = 35.93^K \uparrow$$

   It should be noted that the resultant R was used rather than its components in the computations used to establish R_A. This was done simply to expedite the computations. This can only be done when dealing with external equilibrium.

3. In order to determine the moment beneath wheel no. 2, the beam is cut at wheel no. 2 and a FBD of that portion to the left of the cut is drawn as shown in Figure 6.10d. M_2, the moment beneath wheel no. 2, is found as follows:

   Figure 6.10d: $\Sigma M_2 = 0$ (+ clockwise)

   $$35.9(26.95) - 10(7.5) - M_2 = 0$$
   $$M_2 = 894.3 \text{ kip-ft}$$

4. Attempts to position the wheels in a configuration to yield a moment anywhere in the span greater than 894 kip-ft are fruitless. The absolute maximum moment due to the given wheel loading is, therefore, 894 kip-ft.

6.5 Influence Line for Simple Trusses

Since trusses, like beams, are also subjected to live loading, it is desirable to know how the bar forces vary as a unit load passes over the span. No new techniques or methods are involved. The construction of an influence line for the bar force in a simple truss member involves finding the most convenient general expression for determining the bar force as a unit load traverses the span. After the influence line has been constructed for the bar force in a truss member, the application of the influence line with respect to actual moving loads is identical to that used for beams. The method of joints or the method of sections may be used to construct the influence line for a bar force in a truss.

Example 6.7

Draw the influence lines for the bar forces in members FH, CJ, and BC of the truss shown in Figure 6.11a.

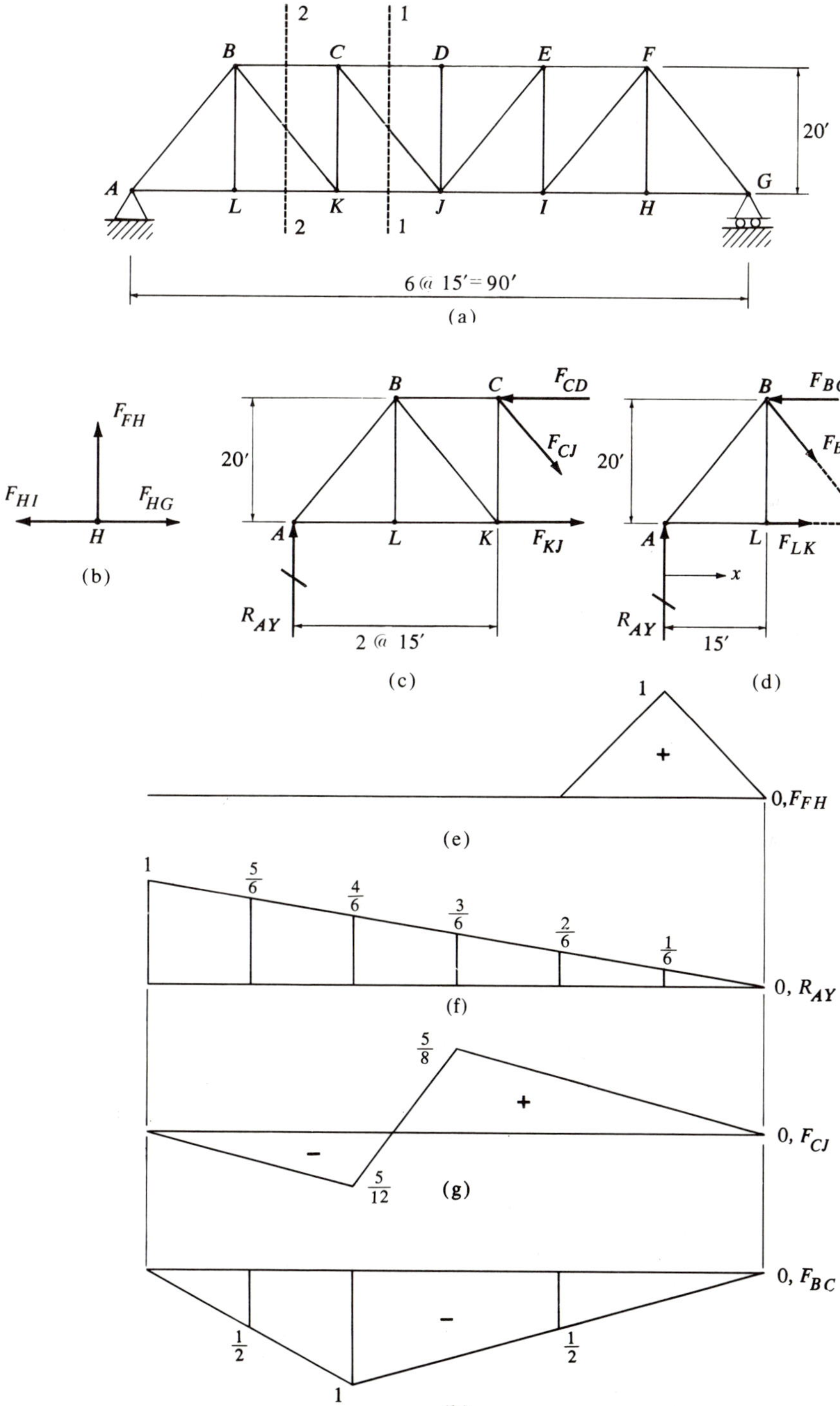

2
1
B
C
D
E
F
20′
A
G
L
K
J
I
H
2
1
6 @ 15′ = 90′
(a)
F_{FH}
F_{HI}
F_{HG}
H
(b)
B
C
F_{CD}
20′
F_{CJ}
A
L
K
F_{KJ}
R_{AY}
2 @ 15′
(c)
B
F_{BC}
F_{BK}
20′
A
L
F_{LK}
K
x
R_{AY}
15′
(d)
1
+
0, F_{FH}
(e)
1
$\frac{5}{6}$
$\frac{4}{6}$
$\frac{3}{6}$
$\frac{2}{6}$
$\frac{1}{6}$
0, R_{AY}
(f)
$\frac{5}{8}$
+
0, F_{CJ}
−
$\frac{5}{12}$
(g)
0, F_{BC}
−
$\frac{1}{2}$
$\frac{1}{2}$
1
(h)

Figure 6.11

Solution:

1. The first step in the construction of an influence line for a truss member is to compute the most convenient general expression for the bar force in that truss member. Considering the bar force in member *FH* first, we see from Figure 6.11a that member *FH* is the only member acting at joint *H* that has the ability to resist a vertical force applied at the joint *H*. A general expression for *F* may be obtained by applying the method of joints at *H*. A FBD of joint *H* is shown in Figure 6.11b. By summing the forces in the vertical direction at *H*, we see that F_{FH} can only have a value when the unit load is placed at joint *H*. If the unit load is directed downward, F_{FH} must be directed upward for equilibrium, or member *FH* is in tension. In Figure 6.11e, an ordinate of +1 (tension will be plotted above the baseline for all truss member influence lines) is plotted at *H*. Since we have stated that F_{FH} only takes on a value for a unit load at *H*, the ordinates at all the remaining panel points are zero. As the unit load travels from G to *H*, we recall that the load between panel points is transmitted to the truss panel points by the floor system. It is assumed, therefore, that the force in *FH* increases linearly from 0 to +1 as the unit load travels from G to *H*. This corresponds to a unit load traveling across the **loaded chord,** i.e., the chord containing the panel points at which the live load is applied. A similar reasoning is used to determine the variation between *H* and *I*.
2. Cutting section 1–1 in Figure 6.11a and the corresponding FBD shown in Figure 6.11c provide a means of obtaining a general expression as follows:

Figure 6.11c:

$$\Sigma F_Y = 0 \uparrow +$$
$$R_{AY} - (\tfrac{4}{5})F_{CJ} = 0$$
$$\underline{F_{CJ} = (\tfrac{5}{4})R_{AY}}, \text{ unit load to right of panel point K.}$$

or

$$R_{AY} - 1 - (\tfrac{4}{5})F_{CJ} = 0$$
$$\underline{F_{CJ} = (\tfrac{5}{4})(R_{AY} - 1)}, \text{ unit load at or to left of panel point K.}$$

Again, we see that another variable, R_{AY}, has entered into the general expressions for F_{CJ}. Since values for R_{AY} will be required for positions of the unit load, an influence line will be required for R_{AY}. An influence for R_{AY} is shown in Figure 6.11f. The construction was accomplished just as for any vertical reaction. Continuing with the solution:

1^K @ G; $R_{AY} = 0$, $F_{CJ} = 0$
1^K @ H; $R_{AY} = \frac{1}{6}$, $F_{CJ} = (\frac{5}{4})(\frac{1}{6}) = +\frac{5}{24} = \frac{5}{24}$ t
1^K @ I; $R_{AY} = \frac{2}{6}$, $F_{CJ} = (\frac{5}{4})(\frac{2}{6}) = \frac{5}{12}$ t
1^K @ J; $R_{AY} = \frac{3}{6}$, $F_{CJ} = (\frac{5}{4})(\frac{3}{6}) = \frac{5}{8}$ t
1^K @ K; $R_{AY} = \frac{4}{6}$, $F_{CJ} = (\frac{5}{4})(\frac{4}{6} - 1) = -\frac{5}{12} = \frac{5}{12}$ c
1^K @ L; $R_{AY} = \frac{5}{6}$, $F_{CJ} = (\frac{5}{4})(\frac{5}{6} - 1) = -\frac{5}{24} = \frac{5}{24}$ c
1^K @ A; $R_{AY} = 1$, $F_{CJ} = (\frac{5}{4})(1 - 1) = 0$

The influence line for F_{CJ} is shown in Figure 6.11g. Recall that the vertical component of the diagonal *CJ* carries the shear in panel *KJ*. We may note the similarity between the shape of the influence line for F_{CJ} and that for shear at a point in a beam.

3. Using cutting section 2–2 in Figure 6.11a and the corresponding FBD shown in Figure 6.11d, a general expression for F_{BC} may be obtained as follows:

$$\Sigma M_K = 0 \quad (+\curvearrowright)$$

$$(30)R_{AY} - (20)F_{BC} = 0$$

$$F_{BC} = (\tfrac{3}{2})R_{AY}, \text{ unit load to right of panel point } L$$

or

$$(30)R_{AY} - (1)x - (20)F_{BC} = 0$$

$$F_{BC} = \tfrac{3}{2}R_{AY} - \left(\frac{x}{20}\right)(1), \text{ unit load at or to left of panel point } L.$$

Where x is the distance in feet from *A* to the unit load,

$$\begin{array}{lll}
1^{K} @ G; & R_{AY} = 0, & F_{BC} = 0 \\
1^{K} @ I; & R_{AY} = \tfrac{2}{6}, & F_{BC} = (\tfrac{3}{2})(\tfrac{2}{6}) = +(\tfrac{1}{2}) = \tfrac{1}{2}\ c \\
1^{K} @ K; & R_{AY} = \tfrac{4}{6}, & F_{BC} = (\tfrac{3}{2})(\tfrac{4}{6}) = 1\ c \\
1^{K} @ L; & R_{AY} = \tfrac{5}{6}, & F_{BC} = (\tfrac{3}{2})(\tfrac{5}{6}) - (\tfrac{15}{20})(1) = \tfrac{1}{2}\ c \\
1^{K} @ A; & R_{AY} = 1, & F_{BC} = (\tfrac{3}{2})(1) - (\tfrac{30}{20})(1) = 0
\end{array}$$

The influence line for F_{BC} is shown in Figure 6.11h. Recall again that the top and bottom chords form a couple to resist bending moment about the cut panel. We note the similarity between the shape of the influence line for the bar force in top chord *BC* and that for moment at a point in a beam.

4. Member *FH* is known as a **secondary member,** i.e., a member stressed for only certain positions of the unit load. Member *CJ* and *BC* are known as **primary members** since their influence lines extend over the entire span, which indicates that they are stressed for all positions of the unit load. Member *CJ*, however, is subjected to a **stress reversal,** i.e., member *CJ* may be in compression or tension depending upon where the loads are applied. It is extremely important to know which members are subject to reversals since design procedures for compression and tension are quite different.

Example 6.8

Determine the maximum bar forces in members *ML* and *CM* due to the moving uniform load and the roving concentrated load shown in Figure 6.12.

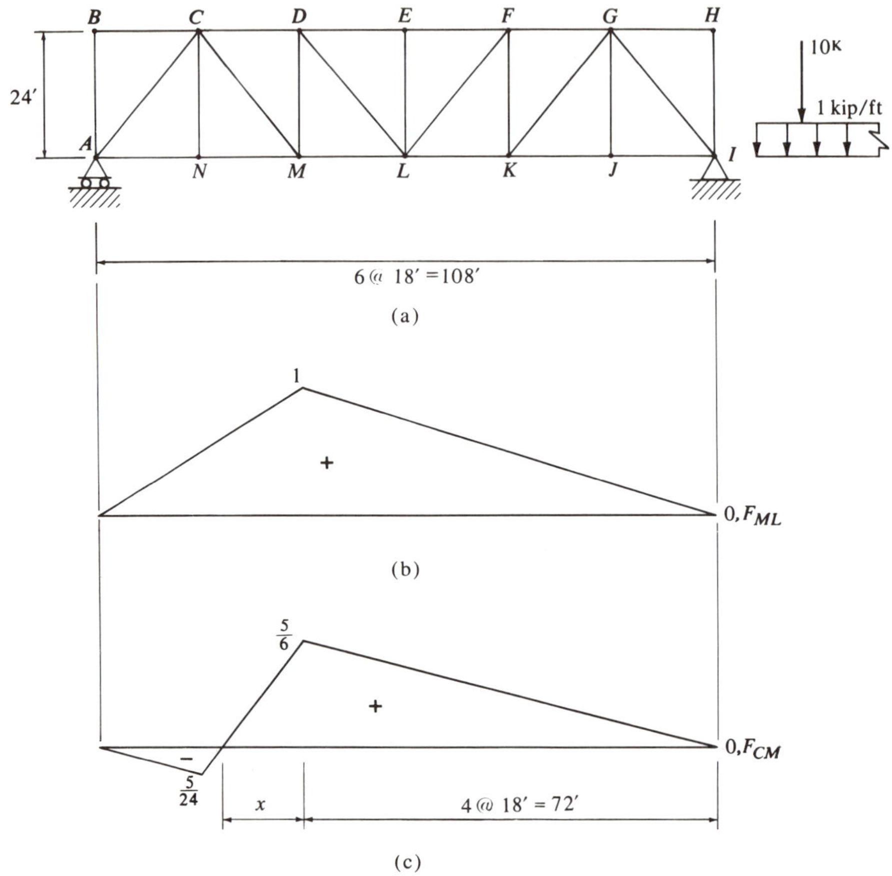

Figure 6.12

Solution:

1. Problems of this nature, i.e., finding maximum forces due to moving loads, are best handled by constructing influence lines. The influence lines for F_{ML} and F_{CM} are shown in Figures 6.12b and 6.12c. From the previous example, the general shape of influence lines for such diagonals and chord members were discussed; they may be constructed with a few select points.
2. Consider member *ML* first. We observe that the maximum influence ordinate occurs at panel point *M*. This pinpoints the position of the roving 10-kip concentrated load for a maximum effect. Since the bar force in *ML* is tension for all positions of the unit load, the moving uniform load will produce the most critical effect when it is placed over the entire span. Hence:

$$\text{Maximum } F_{ML} = 10(1) + (1)(\tfrac{1}{2})(108)(1)$$
$$= 10 + 54$$
$$\underline{\text{Maximum } F_{ML} = 64^{K}\ t}$$

3. From the influence line for F_{CM}, it can be seen that member CM is subject to a stress reversal. The larger influence area, however, is that which produces tension in member CM. We should place the roving 10-kip concentrated load at panel point M and the moving uniform load over that portion of the span beneath the positive influence area. The length of the base of the positive area can be determined using similar triangles as follows:

$$\frac{(5/6 + 5/24)}{18} = \frac{(5/6)}{x}$$

$$x = \frac{(5)(24)(18)}{(25)(6)}$$

$$x = 14.40 \text{ ft}$$

Proceeding with the determination of the maximum force in member CM we obtain the following:

$$\text{Maximum } F_{CM} = 10\left(\tfrac{5}{6}\right) + (1)\left(\tfrac{1}{2}\right)\left(\tfrac{5}{6}\right)(72 + 14.40)$$
$$= 8.33 + 36.0$$
$$\underline{\text{Maximum } F_{CM} = 44.3^{K}\ t}$$

Problems

6–1. Draw the influence lines for all the reactions for the structures shown in Figure P 6–1.

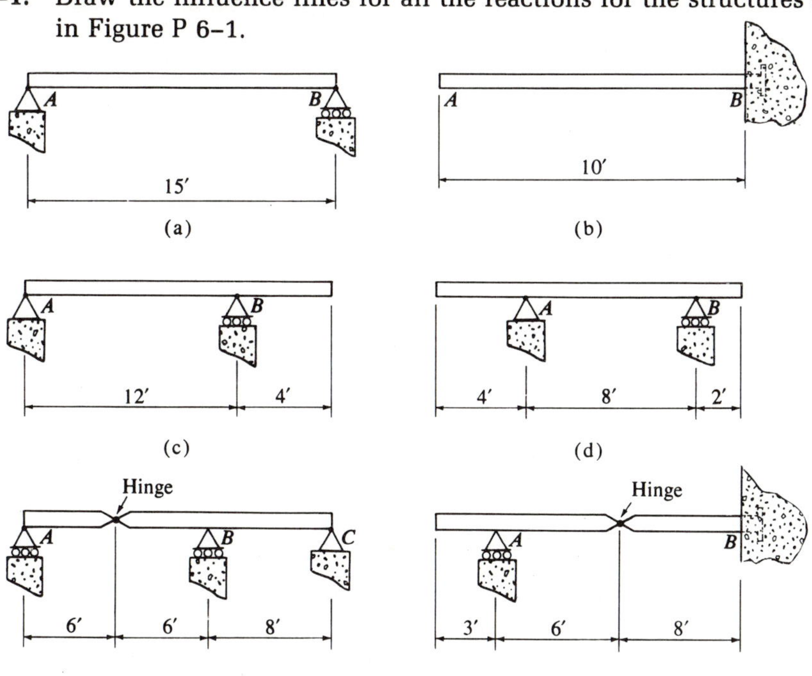

Figure P 6–1

6–2. For the structure shown in Figure P 6–2, draw the influence lines for the reactions at A, C, and D for a unit load traveling from B to D.

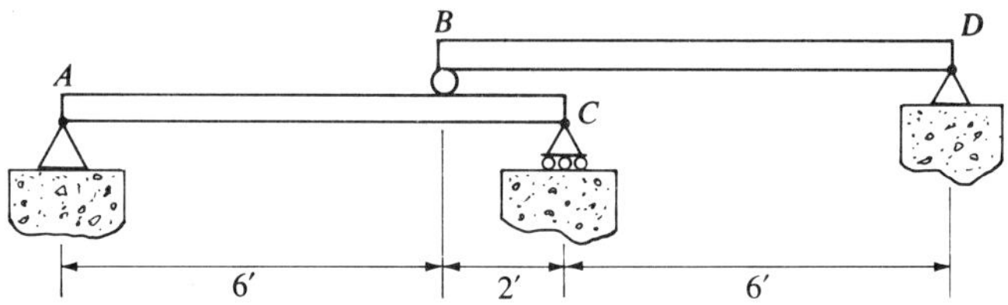

Figure P 6–2

6–3. Draw the influence lines for the reactions at B and D and for the shear and moment at the section at C for the beam shown in Figure P 6–3.

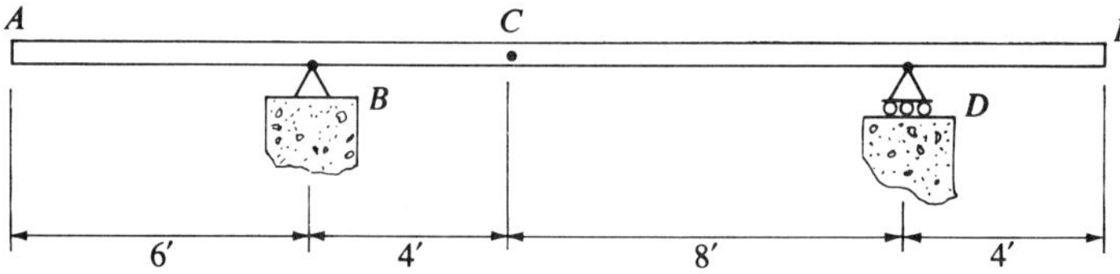

Figure P 6–3

6–4. For the beam shown in Figure P 6–4, determine the maximum reaction at A, the maximum moment at B, and the maximum shear just to the right of point B due to a roving 10-kip load and a moving uniform load of 4 kips per ft.

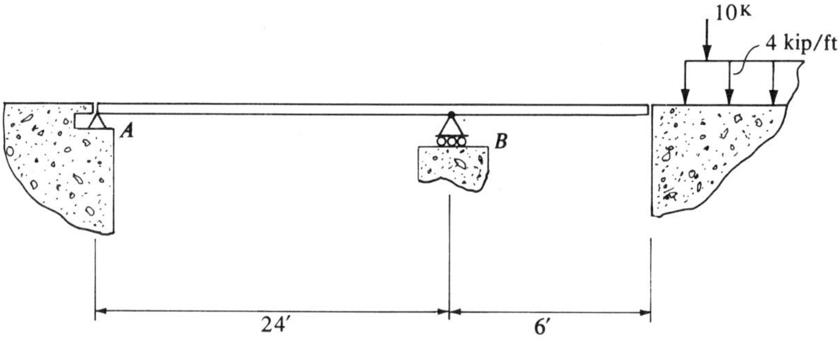

Figure P 6–4

6–5. Draw the influence line for the vertical and horizontal reactions at point D for the frame shown in Figure P 6–5. Assume the unit load travels from A to C.

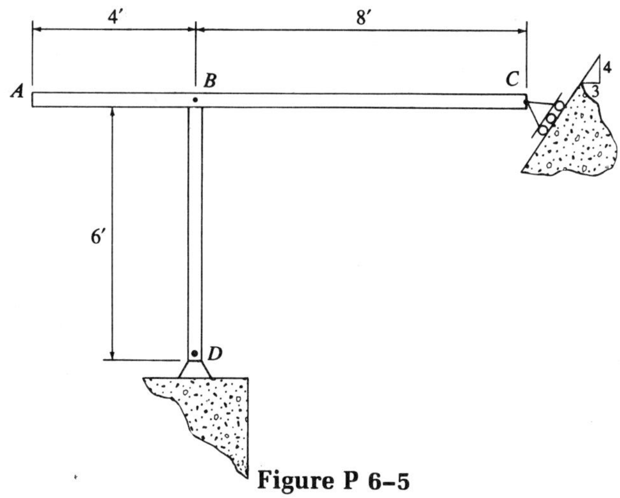

Figure P 6–5

6–6. Draw the influence line for the reactions at both supports *B* and *D*, the moment at *C*, and the shear at a section just to the right of point *B*. The unit load travels from *A* to *C*.

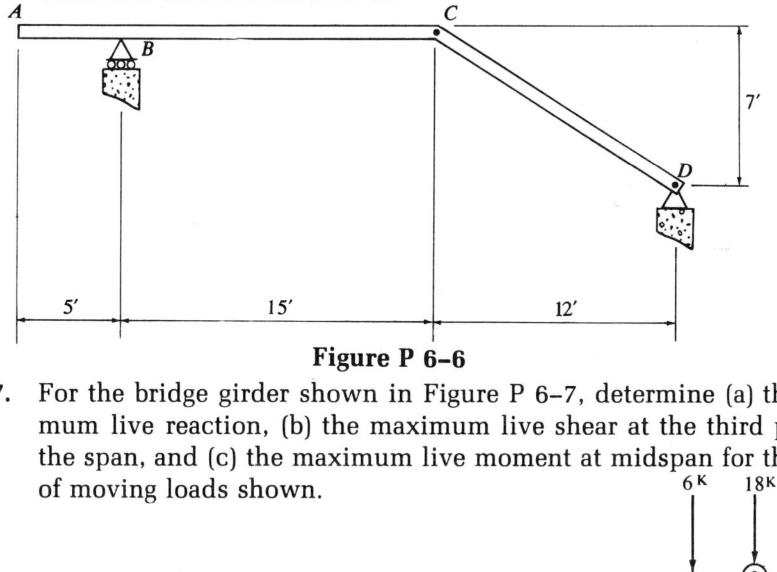

Figure P 6–6

6–7. For the bridge girder shown in Figure P 6–7, determine (a) the maximum live reaction, (b) the maximum live shear at the third point of the span, and (c) the maximum live moment at midspan for the series of moving loads shown.

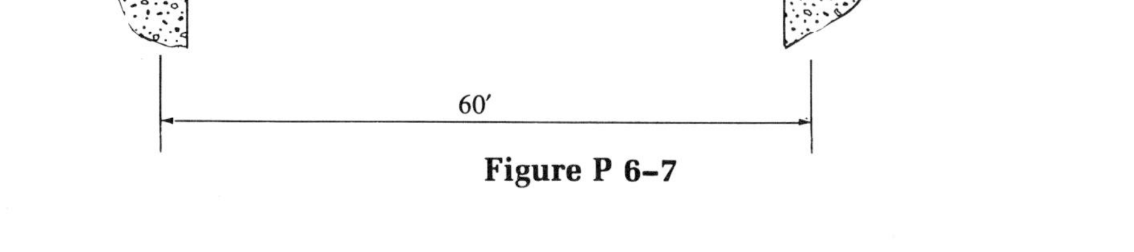

Figure P 6–7

6–8. Determine the absolute maximum live moment in a simply supported beam of 60 ft due to the series of moving loads shown in Figure P 6–8.

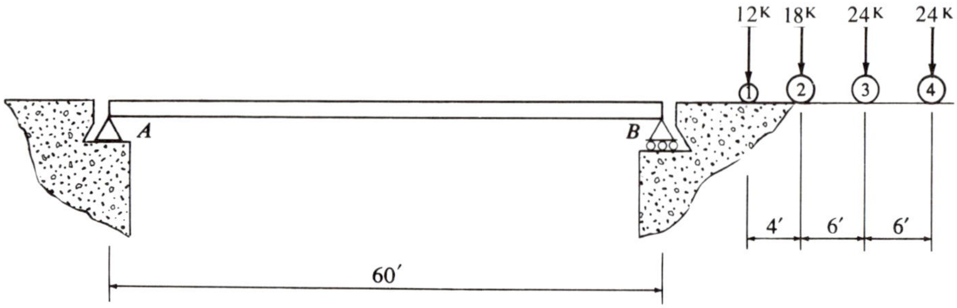

Figure P 6–8

6–9. Determine the absolute maximum shear and moment in the 64-ft crane grider shown in Figure P 6–9. Assume beam and rail weigh approximately 300 lb per foot of span.

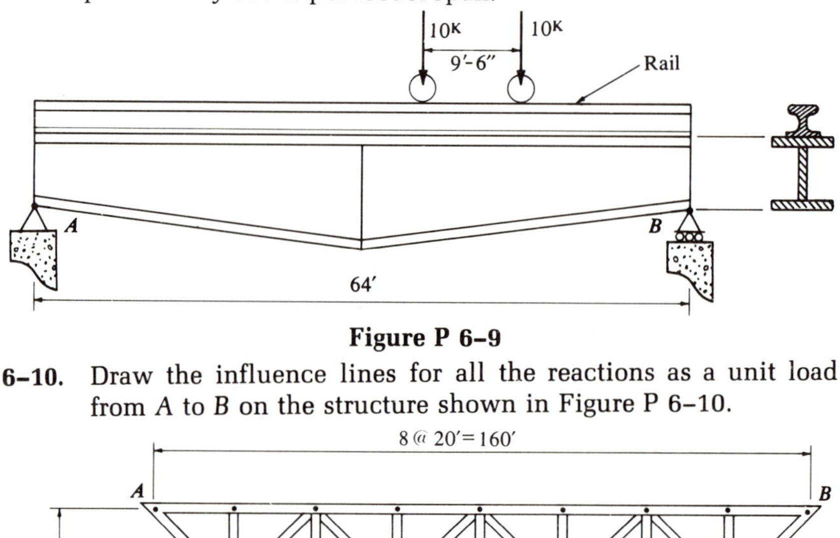

Figure P 6–9

6–10. Draw the influence lines for all the reactions as a unit load travels from A to B on the structure shown in Figure P 6–10.

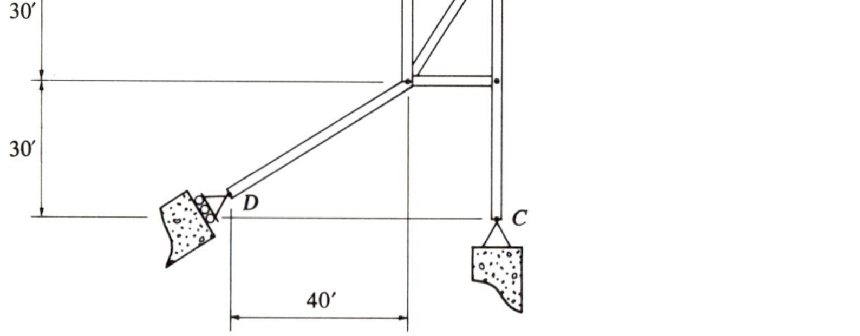

Figure P 6–10

6–11. Referring to Figure P 6–11, determine the maximum reaction at A and the maximum moment at A due to the series of moving loads shown.

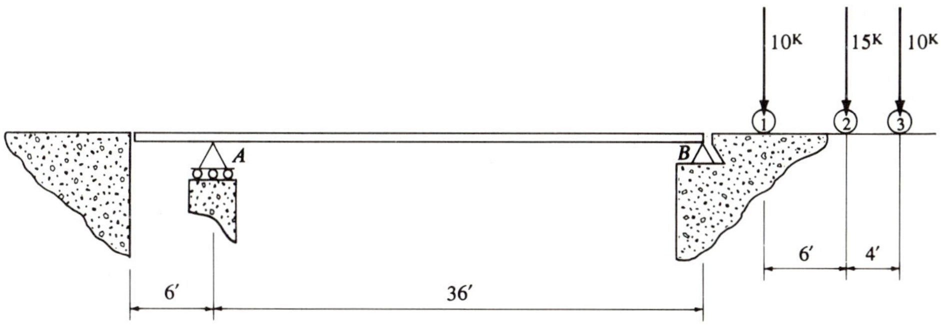

Figure P 6–11

6–12. Referring to the truss shown in Figure P 6–12, draw the influence lines for the bar forces in members KI, KJ, BC, KH, and HI.

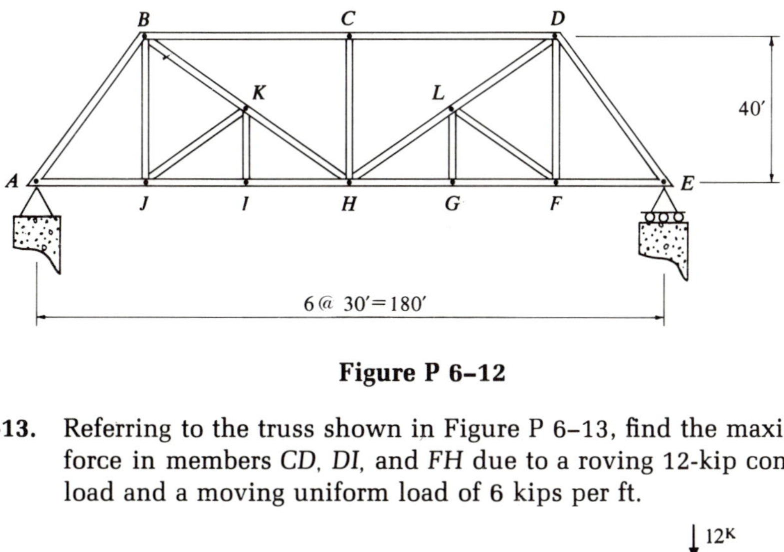

Figure P 6–12

6–13. Referring to the truss shown in Figure P 6–13, find the maximum live force in members CD, DI, and FH due to a roving 12-kip concentrated load and a moving uniform load of 6 kips per ft.

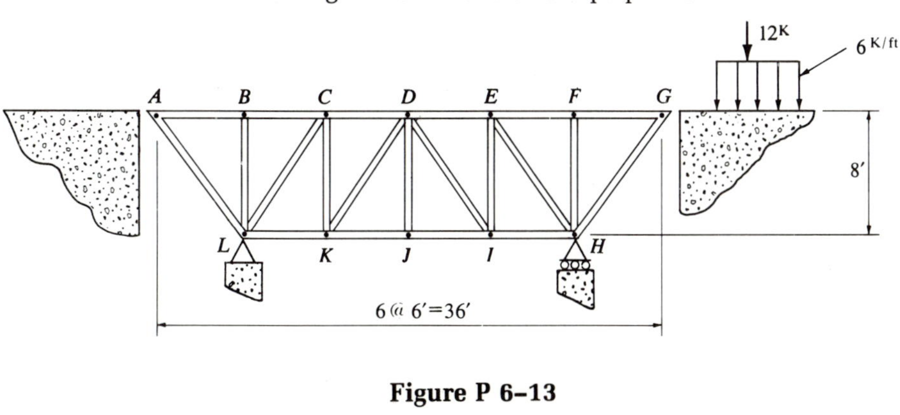

Figure P 6–13

6–14. For the truss shown in Figure P 6–14, draw the influence lines for the bar forces in members *DC* and *DG* as a unit load travels from *E* to *I*.

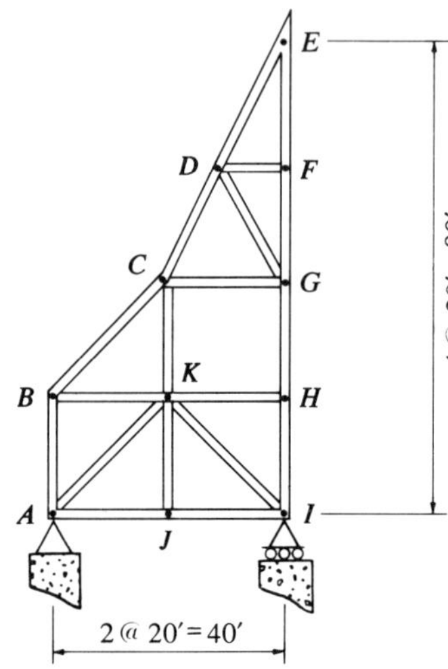

Figure P 6–14

6–15. Parts are to be carried by the monorail system shown in Figure P 6–15. The loads are 500 lb at each trolley wheel. Determine the maximum reaction at *A*, the maximum force in member *CJ*, and the maximum force in member *KJ*. The supporting truss weighs 300 lb per ft and the monorail, with hangers, weighs 80 lb per ft.

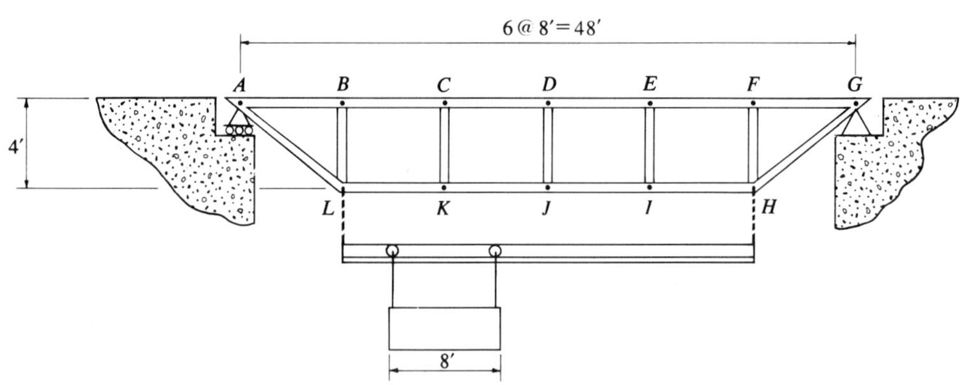

Figure P 6–15

6–16. Two identical trusses are used to support the conveyor and live load. The dead load of the conveyor, rollers, and truss is 50 lb per ft for one truss. The live load is 250 lb per foot of conveyor, or 125 lb per foot of truss. Find the maximum force in members *CD* and *DH*.

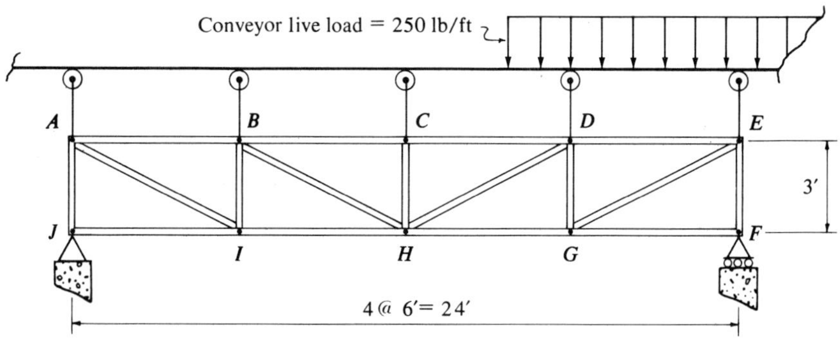

Figure P 6–16

6–17. A truss is erected to support five transmission lines as shown in Figure P 6–17. The maximum possible tension in each line is 1000 lb. Line 3(P_3) is connected initially and the remaining lines will be connected at a later time as needed, i.e., additional lines can be mounted in any possible combination and in no prescribed order. Determine the maximum bar force that can be expected in members *BL* and *CD*.

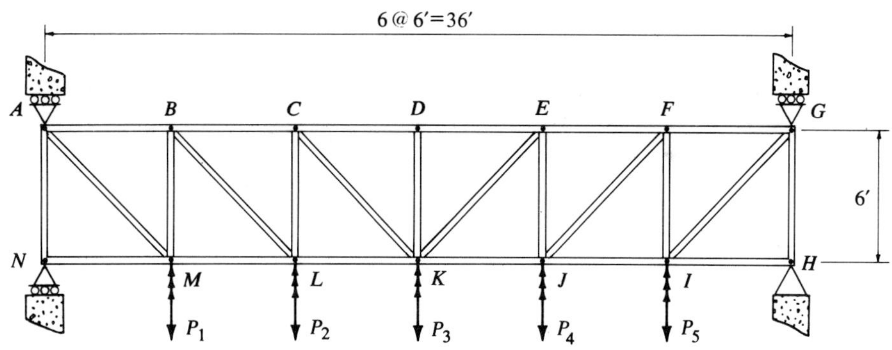

Figure P 6–17

CHAPTER SEVEN

Deflections of Beams

7.1 Introduction

The structures dealt with thus far in the text have been assumed to be rigid structures, i.e., the structures do not deform under loading. It is known, however, that these so-called rigid structures do deform. Since these deformations are small, they may be neglected when analyzing the structure. Since the deformations, or deflections, are relatively small they do not significantly affect the equilibrium conditions of the problem, but they may become troublesome in some instances. Excessive deflections in beams can cause substantial cracking in plaster ceilings and plate glass windows. Even very minute deformations in machine parts can lead to abrasion and friction under working conditions. It would seem, therefore, that it would be beneficial to the analyst to investigate the magnitude and direction of displacements resulting from a loaded structure. In fact, most design codes specify allowable limits on maximum deflections. The American Concrete Institute (ACI) recommends a maximum allowable deflection of 1/240 of the clear span while the American Institute of Steel Construction (AISC) specifies 1/360 of the span as a maximum allowable limit for deflection. In addition to the damage that may result for excessive structural displacements, imagine the psychological effect large visible deflections would have on an observing public. The deflected shape of a structure is also useful in the analysis of statically indeterminate structures.

In some instances, a structure is fabricated with initial deflections in the direction opposite to that expected in the field under the action of the applied loading. This is commonly referred to as **camber.** By cambering a structure it is hoped that the deflections caused as a result of the applied loading will just eliminate the intentional deflections fabricated into the structure. The final deflected shape of the structure is, therefore, the same as that structure analyzed, which was a rigid structure, i.e., little or no deformation.

There are numerous methods available which will enable the analyst to determine the deflections along the span of a beam. Some of these methods are the moment–area method, the conjugate beam method, the integration method, and various energy methods. The moment–area method is a semigraphical method which is easily understood, frequently used, and does not require a knowledge of calculus. For these reasons, it is discussed in the present chapter.

7.2 Development of the Moment–Area Method

After application of the applied loading, the longitudinal centroidal axis becomes a curve which is called the **elastic curve.** Consider the beam shown in Figure 7.1a, which is bent only with equal couples. A small segment of the beam of length L is removed and the corresponding FBD is shown in Figure 7.1b in its extremely exaggerated deformed configuration. In regions of constant bending moment, the elastic curve is an arc of a circle of radius R as in-

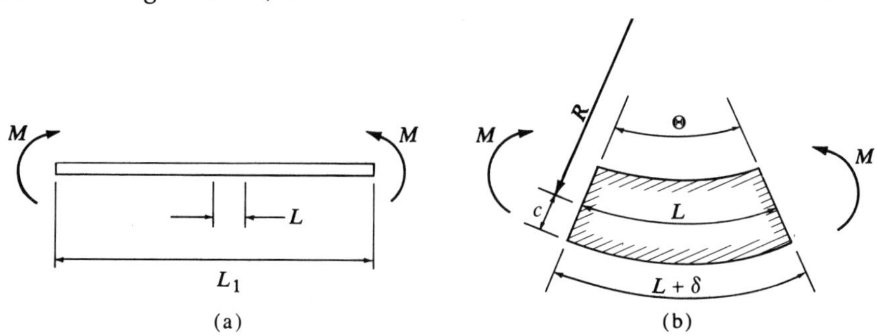

Figure 7.1

dicated in Figure 7.1b. Assuming that deformation (elongation and compression) of the beam's fibers is proportional to the distance from the neutral surface (the longitudinal centroidal axis in this instance) which is unchanged in length, we may proceed as follows:

$$R(\Theta) = \text{arc length, where } \Theta \text{ is in radians}$$

hence

$$\Theta = L/R$$

and

$$\Theta = \frac{L + \delta}{R + c}$$

or equating equal values,

$$\frac{L}{R} = \frac{L + \delta}{R + c}$$

from which $$\frac{c}{R}=\frac{\delta}{L}$$

Recall that $$\frac{\delta}{L}=\text{strain}=\epsilon$$

and $$\epsilon=\frac{\sigma}{E}$$

where σ is the stress at the point of the strain and E is the modulus of elasticity of the beam's material. From any Strength of Materials text,

$$\sigma=\frac{Mc}{I}$$

where M is the bending moment, c is the distance from the neutral surface to the outermost fiber, and I is the moment of inertia of the cross-sectional area. Hence:

$$\frac{c}{R}=\frac{\delta}{L}=\epsilon=\frac{\sigma}{E}=\frac{Mc}{EI}$$

or $$\frac{1}{R}=\frac{M}{EI}$$

and $$R=\frac{EI}{M} \tag{7–1}$$

Equation (7–1) was derived for a beam under the action of a constant bending moment. The bending moment is not usually constant; hence, the elastic curve is not circular. This means that the **radius of curvature** R does not remain constant (recall that a circular curve has a constant radius). This does not mean, however, that Eq. (7–1) is not valid for all loading cases. If L in Figure 7.1b becomes smaller and smaller and approaches zero, the radius of curvature does approach a constant value and Eq. (7–1) is valid for all types of loading.

The moment–area method utilizes both the moment diagram and the properties of the elastic curve of the beam. Figure 7.2 shows a simply supported beam with a zone of constant moment, the corresponding moment diagram, and a greatly exaggerated elastic curve. Before proceeding any further, the elastic curve shown in Figure 7.2c along with its corresponding labels should be closely scrutinized. From the elastic curve:

$$\tan\Delta\Theta=\frac{\Delta y}{\Delta x}=\Delta\Theta \text{ (for small angles)}$$

$$\Delta L=R\Delta\Theta; \qquad \Delta\Theta=\frac{\Delta L}{R}$$

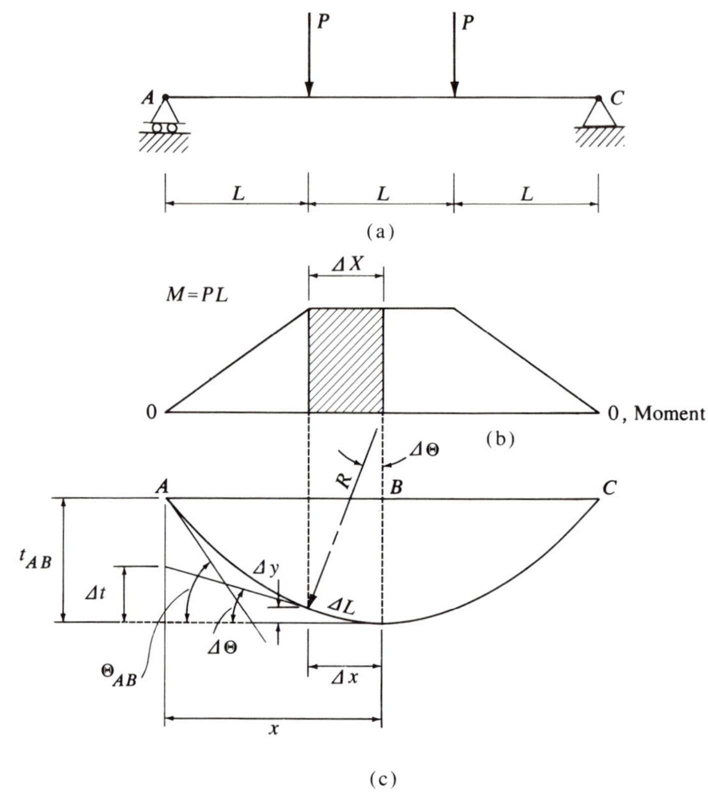

Figure 7.2

For small angles $$\Delta x \simeq \Delta L$$

Hence $$\Delta\Theta = \frac{\Delta x}{R}$$

From Eq. (7–1) $$\frac{1}{R} = \frac{M}{EI}$$

and $$\Delta\Theta = (\Delta x)\frac{M}{EI}$$

From Figure 7.2c it can be seen that $\Delta\Theta$ is the angle between tangents drawn to the elastic curve at a distance Δx apart. The angle Θ between tangents drawn to the elastic curve at any two points x distance apart, such as points *A* and *B* in Figure 7.2c, may be obtained by summing all the $\Delta\Theta$s for each increment Δx between *A* and *B* or:

$$\Theta_{AB} = \sum_{x\,@\,A}^{x\,@\,B} \frac{(\Delta x)M}{EI}$$

Since *EI* is constant, we have:

$$\Theta_{AB} = \frac{1}{EI} \sum_{x @ A}^{x @ B} M\Delta x, \quad \Theta \text{ in radians}$$

From Figure 7.2b, $M\Delta x$ is an area under the moment diagram. From this derivation, the **first moment–area theorem** may be stated as follows:

The angle Θ_{AB} between the tangents drawn at two points A and B on the elastic curve of a beam	**=**	**The area under the moment diagram between points A and B, divided by EI**

Referring again to the elastic curve shown in Figure 7.2c, we may proceed as follows:

$$\Delta t = (x)(\Delta\Theta)$$

$$\Delta t = (x)\frac{M(\Delta x)}{EI}$$

Again summing all the Δts for each increment Δx between *A* and *B*, we obtain the following:

$$t_{AB} = \sum_{x @ A}^{x @ B} \frac{xM\Delta x}{EI}$$

Since *EI* is constant, we have:

$$t_{AB} = \frac{1}{EI} \sum_{x @ A}^{x @ B} xM\Delta x$$

From Figure 7.2b, it can be seen that the product $xM\Delta x$ is the first moment of the area $M\Delta x$ about point *A*. From this derivation, the **second moment–area theorem** may be stated as follows:

The vertical distance of any point A on the elastic curve of a beam from a tangent drawn to the curve at any point B	**=**	**The first moment about point A of all the area under the moment diagram between points A and B, divided by EI**

As with most analytical methods, a sign convention must be established. Referring to Figure 7.3, the following sign convention may be established:

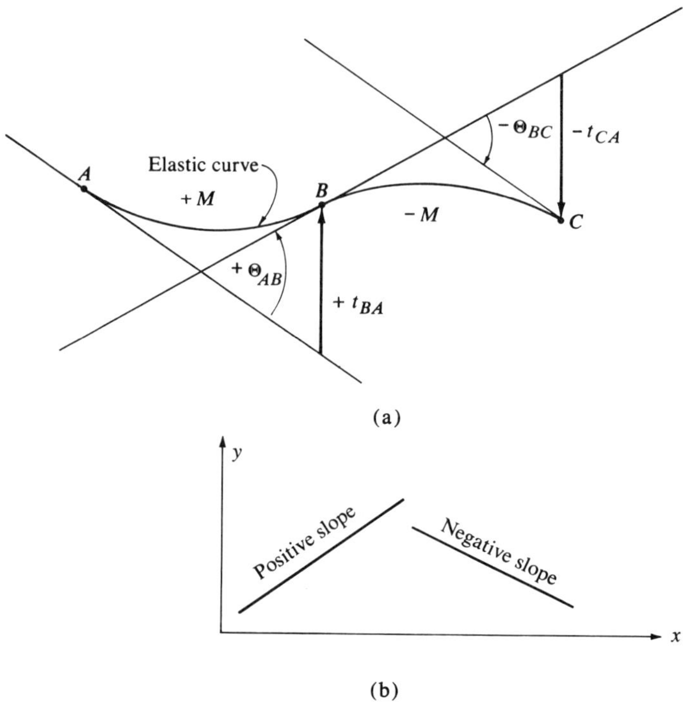

Figure 7.3

1. The vertical distance of any point on the elastic curve from a tangent drawn to the curve at another point is commonly referred to as a **tangential deviation.** A tangential deviation is positive when the point on the elastic curve is above the tangent as t_{BA} in Figure 7.3a and negative when the point is below the tangent as t_{CB} in Figure 7.3a.
2. The angle between two intersecting tangents drawn to the elastic curve can be referred to as a **change in slope.** For example, the change in slope between a tangent drawn at point A in Figure 7.3a and a tangent drawn at point B is Θ_{AB}. The change in slope is positive if the initial slope, such as at point A in Figure 7.3a, changes from a negative slope (see Figure 7.3b) to a less negative slope or a positive slope such as at point B. The change in slope is negative if the initial slope, such as at point B, changes from a positive slope (see Figure 7.3b) to a less positive slope or a negative slope such as at point C.
3. Since a moment diagram is necessary to implement the moment–area method, the sign convention established in Chapter Three for the construction of shear and moment diagrams will be followed throughout this chapter.

It has been the author's experience that students very often have initial difficulty applying the moment–area method and/or make numerous mistakes simply because they do not apply the established moment–area theorems exactly as they are stated. It is highly recommended, therefore, that the two moment–area theorems be memorized verbatim before proceeding any further in this chapter. The definitions of tangential deviation and change of slope should also be clearly understood.

A brief mention of the double subscripts used throughout the previous derivations should also be made at this time. The first subscript used when dealing with tangential deviation symbols refers to the point on the elastic curve while the second subscript refers to the point at which the tangent is drawn. Also, the first subscript indicates the point to be used as a moment center when applying the second moment–area theorem. The order of subscripts used when referring to a change in slope is somewhat less important. Care must be exercised, however, as a misinterpretation of the meaning of the subscripts can lead to an incorrect sign for a change in slope.

7.3 Application of the Moment–Area Method

The moment–area method provides the analyst a means of obtaining tangential deviations and angles between intersecting tangents to the elastic curve. The analyst, however, is interested in determining the deflection, i.e., the vertical displacement of the neutral surface from its unloaded position and in some instances the amount of rotation incurred at a particular point. Fortunately, the moment–area method can be applied in such a manner so as to obtain displacements and rotations. It should be clearly understood, however, that tangential deviation and change in slope do not mean deflection and slope respectively.

Consider the simply supported beam along with its elastic curve shown in Figure 7.4. Suppose that the deflection beneath the concentrated load is desired. The problem simply becomes one of finding a way to manipulate the tangential deviations in such a fashion so as to obtain the desired deflection. The initial step in applying the moment–area method is to sketch the elastic curve. Secondly, we must search for a way of obtaining the required displacement by using only tangential deviations. Referring to Figure 7.4, we see that the deflection at B may be obtained as follows:

$$\Delta = \overline{BF} - t_{BA}$$

We may obtain t_{BA} simply by applying the second moment–area theorem. The determination of $\overline{BF}$, however, presents a different problem, which can be handled by using similar triangles as follows:

$$\frac{\overline{CD}}{\overline{AC}} = \frac{\overline{BF}}{\overline{AB}}$$

or

$$\frac{t_{CA}}{L} = \frac{\overline{BF}}{a}$$

from which $\overline{BF} = (a/L)t_{CA}$

and $\underline{\Delta = (a/L)t_{CA} - t_{BA}}$

Again, t_{CA} may be determined by an application of the second moment–area theorem. Suppose that it is further required that Θ_A and Θ_C be determined. The first moment–area theorem provides only a means of determining a change in slope from one tangent to the elastic curve to another tangent. From Figure 7.4, however:

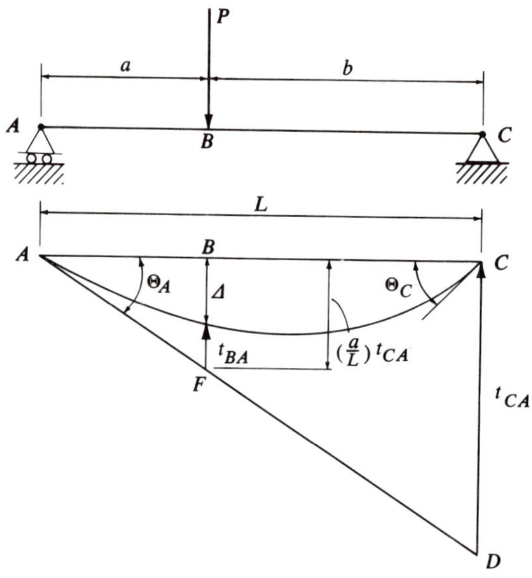

Figure 7.4

$$\tan \Theta_A = \frac{\overline{CD}}{\overline{AC}} = \frac{t_{CA}}{L}$$

for small angles $\underline{\Theta_A = \frac{t_{CA}}{L}}$, Θ_A in radians

and $\underline{\Theta_C = \Theta_A + \Theta_{AC}}$

The problem can be handled just as easily by solving for Θ_C initially. Or:

$$\underline{\Theta_C = \frac{t_{AC}}{L}}, \quad (t_{AC} \neq t_{CA})$$

$$\underline{\Theta_A = \Theta_C + \Theta_{CA}}, \quad (\Theta_{CA} = \Theta_{AC})$$

This problem follows a more or less basic procedure that is used when applying the moment–area method. Although the problem may change, the procedure remains essentially the same.

It is extremely important that an accurate sketch of the elastic curve be made initially. The analyst may look to the moment diagram as a means of determining how the beam is being bent, i.e., concave upward as indicated by positive moment or concave downward as indicated by negative moment. Using the elastic curve, the analyst may then sketch in the appropriate distances and tangential deviations necessary for the determination of the required deflection or slope. From the author's experience, it is advantageous to solve the problem with symbols and then substitute in the appropriate numerical values.

7.4 Moment Diagram by Parts

It should be evident from the moment–area theorems that the proper construction of the moment diagram is essential for the application of these theorems. In Chapter Three moment diagrams were plotted using the left end of the beam as the origin. The resulting moment diagram, plotted from left to right along the span, provided the analyst with a visual description of how the moment varied along the span as well as the points of maximum positive and negative moment. Such a moment diagram is referred to as a **composite moment diagram** and it is essential to the structural designer. In the application of the moment–area method for obtaining deflections, we recall that the area under the moment diagram and the first moment of that area are required. The areas encountered using a composite moment diagram, in part or in full, are sometimes very cumbersome to work with, i.e., the area itself and its respective centroid may be rather difficult to obtain. It is, therefore, sometimes easier and more practical to draw a series of diagrams showing the moment due to each load and/or reaction on a separate sketch. The areas encountered in this manner under each individual moment diagram are very common shapes whose properties are well known and easily obtainable. These individual moment diagrams are commonly referred to as a **moment diagram by parts.**

The **principle of superposition** enables us to plot a moment diagram by parts. The principle may be briefly stated as follows: **When a linear relationship exists between stresses or deflections and the applied loads, the resultant effect of several loads acting on a structure simultaneously may be represented as the sum of the contributions from each of the loads applied individually.** Based upon this principle, a moment diagram by parts consists of a moment diagram drawn from each load to an arbitrarily selected point. In all actuality, what is being accomplished is that we are graphically summing moments about a point and plotting that moment which tends to bend the member in the positive direction (concave upward) above the base line. Any point may be selected as the reference point although in most instances it is preferable to select the reference point at either end of the beam.

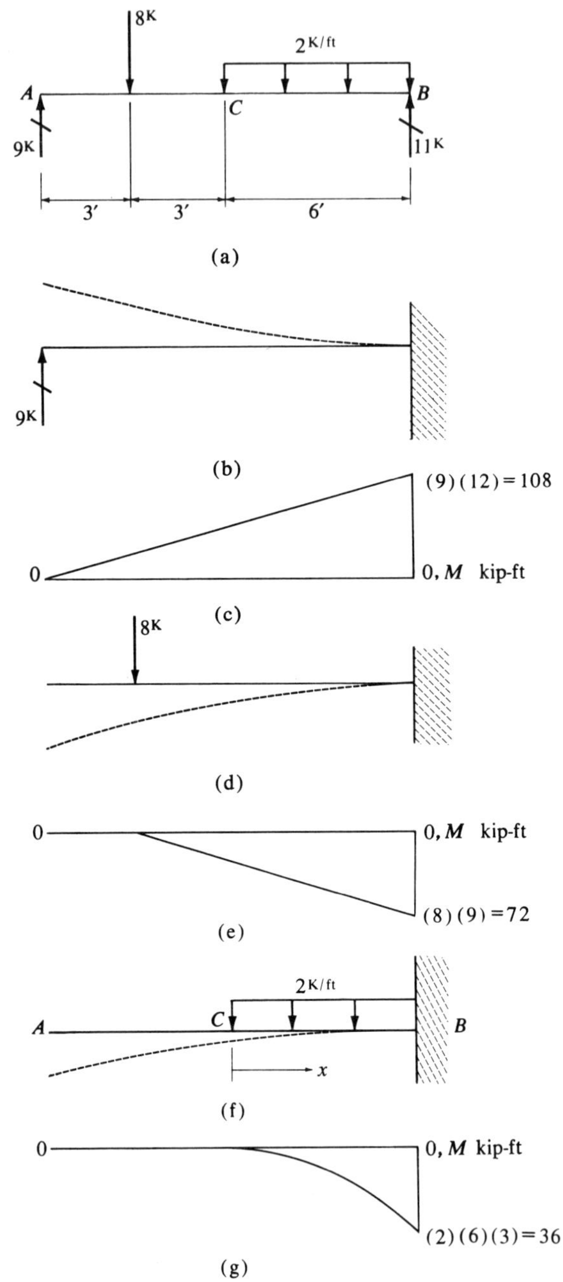

Figure 7.5

Consider the beam shown in Figure 7.5a and suppose that we wish to draw the moment diagram by parts about point B, i.e., use point B as the reference point. To draw the moment diagram by parts, we must answer the following about each load: (1) What is the magnitude of the moment produced about the reference point, or moment center? (2) How does the moment vary between the load and the moment center?, i.e., does it vary linearly, parabolically, cubically, etc.? and (3) Does the member bend positively or negatively under action of the load? To answer the first question, we simply find the moment of each load about the moment center in the conventional manner. We answer number two by looking at a general expression for the moment produced by each load. For example, a concentrated load produces a moment which varies as the distance from the load. A uniformly distributed load produces moment which varies as the square of the distance from the beginning of the loading ordinates (as long as the moment center is taken beneath or at the end of the uniform load). The resulting moment plots are linear (straight line) and parabolic $[f = (x^2)]$ respectively. The last question is answered by "imagining" the selected reference point, or moment center, to be fixed, i.e., unable to rotate as at a fixed support. A rough sketch, therefore, made mentally or physically will depict whether the member bends concave upward, i.e., positive moment, or concave downward, i.e., negative moment. Referring back to the beam shown in Figure 7.5a, the procedure for drawing a moment diagram by parts is as follows:

1. Remove all the loads, reactions, and couples from the beam and select the reference point, say point *B* in this case.
2. Replace the acting forces, or couples, one at a time such as the reaction at *A* shown in Figure 7.5b.
3. This 9-kip reaction produces a moment of magnitude (9)(12) = 108 kip-ft about point *B*. Since we are dealing with a concentrated load, the moment varies linearly as we move from the point of application, i.e., the moment plot is a straight line. If we imagine point *B* to be fixed against rotation, the member will deflect as shown by the dashed curve shown in Figure 7.5b. This constitutes positive bending moment. The resulting moment plot is shown in Figure 7.5c.
4. Applying the same logic to the 8-kip load shown in Figure 7.5d, we obtain the moment plot shown in Figure 7.5e. Actually, it would be more convenient to show both moment plots, Figure 7.5c and 7.5e, in the same diagram, i.e., a linear or triangular plot above and below the base line.
5. The last load to produce moment about point *B* is the uniformly distributed load of 2 kips per foot. If we imagine a cut at a distance x to the right of point *C*, the moment at the cut is $2(x)(x/2)$, or x^2 kip-ft. A plot of the moment is, therefore, a function of the square of the distance, i.e., the moment varies parabolically. The moment at *B* is (2)(6)(6/2), or 36 kip-ft while the moment at *C* due to the uniform load is 0. The resulting plot is shown in Figure 7.5g.
6. The moment at any point due to all the loads can be obtained by summing all the respective ordinates from all the moment plots at that point. For example, the moment at *B* is known to be zero since there is

no applied couple acting at that point. If the ordinates at B from all the partial moment plots are summed, we obtain $M_B = (108 - 72 - 36)$, or $M_B = 0$, which checks our work.

Figure 7.6 shows a moment diagram by parts for the same beam with point A selected as the reference point, or moment center. The moment plots for the two concentrated loads are shown in Figures 7.5b and 7.5c and were obtained

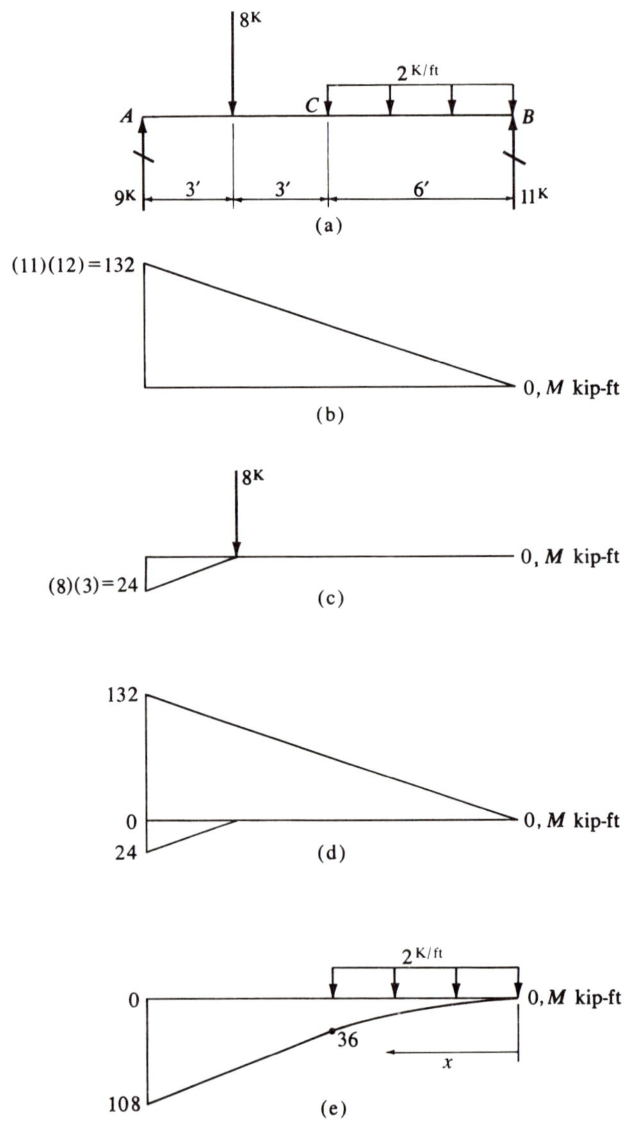

Figure 7.6

using the techniques outlined for the previous plots. It is, however, more convenient to represent these plots in a more or less combined form as shown in Figure 7.6d. The partial moment diagram shown in Figure 7.6e due to the uniform load needs some additional explanation. Referring to Figure 7.6e, the moment at any point beneath the uniform load is 2(x)(x/2), or x^2. This plots parabolically as long as we stay beneath the uniform load $(0 \leqslant x \leqslant 6)$. The moment at any point between C and A due to the uniform load is (2)(6)(x − 3), or 12(x − 3). This plots linearly from C to A. The moments are as follows:

$$M @ B = 0$$
$$M @ C = x^2 = 36 \text{ kip-ft}$$
$$M @ A = 12(x - 3) = 108 \text{ kip-ft}$$

The corresponding plot is shown in Figure 7.6e. The moment at A is known to be zero. Adding the ordinates at A, we obtain $M_A = (132 - 24 - 108) = 0$. Thus, a check is performed.

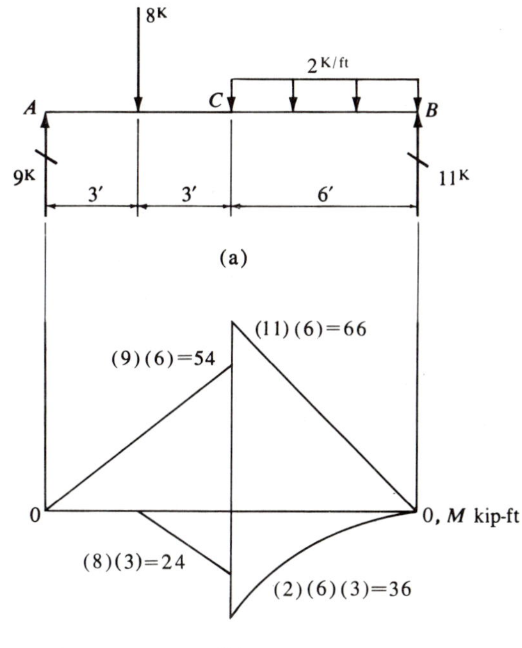

Figure 7.7

As was stated previously, any reference point may be selected when constructing a moment diagram by parts. Selecting point C this time, Figure 7.7 shows the corresponding moment diagram by parts for the same beam and using point C as a moment center. As before, the techniques utilized are as follow:

1. The loads, reactions, and couples are placed on the span one at a time and the moment between the point of application of the load and the selected reference point is plotted.
2. To obtain the sign of the moment, the reference point is imagined to be fixed from rotation. In this case, point C was assumed fixed when considering both segments CB and CA.

Since there are no applied couples, there can only be one moment value at each point. Hence, the moment slightly to the right of C must equal the moment slightly to the left of C, or:

$$M_{CR} = M_{CL}$$
$$66 - 36 = 54 - 24$$
$$30 = 30 \qquad \text{Check}$$

Thus, a check is performed. The net moment at any point on the beam AB is the same whether we use the moment diagrams shown in Figure 7.5, 7.6, or 7.7. The only difference in these moment diagrams is how they are presented. In all cases, the areas encountered under the moment curves are well-known shapes whose properties are well-known. This is a definite advantage when applying the moment–area theorems. If the beam being considered only carries one load, it has been found to be easier to use a composite moment diagram rather than a moment diagram by parts.

7.5 Simply Supported Beams

When dealing with statically determinate beams, the analyst generally encounters only simply supported beams, cantilever beams, or beams with overhangs. Regardless of the loading conditions, the procedures followed when applying the moment–area theorems to obtain deflections for the previously mentioned beams are generally consistent for each type of beam. Hence, each type will be considered separately starting with the simply supported beam.

Example 7.1

Determine the deflection at midspan for the simply supported beam shown in Figure 7.8a. Assume both E and I to be constant along the span.

Solution:

1. The initial steps in the solution consist of drawing the proper FBD, the shear diagram, the moment diagram, and the elastic curve as shown in Figures 7.8b, 7.8c, 7.8d, and 7.8e. To sketch the elastic curve, reference should be made to the moment diagram. In this case the moment is positive at all points, thereby fixing the shape of the elastic curve as concave upward. The moment diagram is symmetrical and, hence, so is the elastic curve.

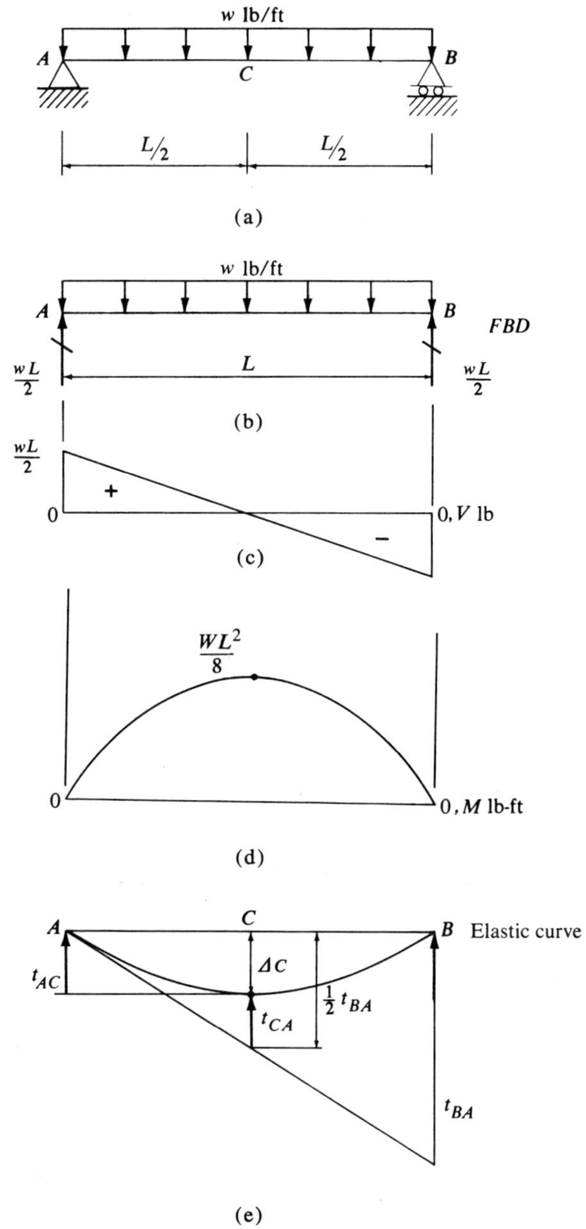

Figure 7.8

2. Using the procedure outlined in Section 7.3, the required deflection is located on the elastic curve and the corresponding tangential deviations necessary for its determination are also located and sketched as shown in Figure 7.8e.

3. Using Figure 7.8e, the following relationships are obtained:

$$\Delta_C = (\tfrac{1}{2})t_{BA} - t_{CA}$$

EIt_{BA} = First moment of area beneath the moment diagram between A and B about point B.

EIt_{CA} = First moment of area beneath the moment diagram between A and C about point C.

The areas encountered are segments of second-degree parabolas whose properties, along with other commonly encountered areas, are shown in Figure 7.9. Continuing:

$$EIt_{BA} = \left(\frac{2}{3}\right)ab\left[\left(\frac{1}{2}\right)b\right]$$

$$= \left(\frac{2}{3}\right)\left(\frac{wL^2}{8}\right)(L)\left(\frac{1}{2}\right)L = \frac{wL^4}{24}$$

$$EIt_{CA} = \left(\frac{2}{3}\right)ab\left[\left(\frac{3}{8}\right)b\right]$$

$$= \left(\frac{2}{3}\right)\left(\frac{wL^2}{8}\right)(L/2)\left(\frac{3}{8}\right)(L/2) = \frac{wL^4}{128}$$

$$\Delta_C = \left[\left(\frac{1}{2}\right)\left(\frac{wL^4}{24}\right) - \frac{wL^4}{128}\right]\frac{1}{EI}$$

$$= \frac{(8-3)}{384}\frac{wL^4}{EI}$$

$$\Delta_C = \frac{5wL^4}{384EI}\downarrow$$

The deflection at C is known to be downward since the distance $(1/2)t_{BA}$ is greater than t_{CA}.

4. It has been shown that the elastic curve for a symmetrically loaded span is symmetrical with respect to the centerline. The following observation can be made: **for a symmetrically loaded, simply supported beam with the product EI constant along the span, the maximum deflection occurs at midspan.** A tangent to the elastic curve at the point of maximum deflection has a zero slope, i.e., the tangent is a horizontal line. Such a tangent is shown at point C in Figure 7.8e. Notice that the tangential deviation t_{AC} is equal in magnitude to Δ_C. The required deflection at midspan in this case could have been obtained by this equality. Thus:

$$\Delta_C = t_{AC} = \left(\frac{1}{EI}\right)\left(\frac{2}{3}\right)\left(\frac{wL^2}{8}\right)\left(\frac{L}{2}\right)\left(\frac{5}{8}\right)\left(\frac{L}{2}\right)$$

$$\Delta_C = \frac{5wL^4}{384EI}\downarrow$$

The deflection @ C is downward since t_{AC} is positive (recall that a positive tangential deviation means that the point on the elastic curve is above the tangent. Hence, point C must be below point A, which does not move).

5. Since there is only one applied load, a composite moment diagram was utilized.

Shape	Area, A	Centroidal Distance, $\bar{x}$
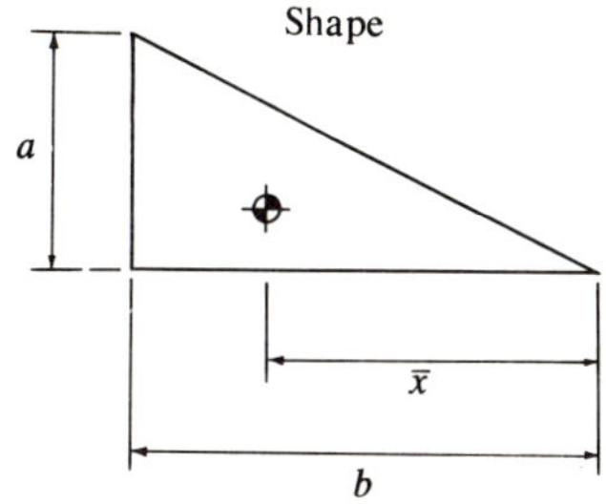	$A = (\frac{1}{2})\,ab$	$\bar{x} = (\frac{2}{3})\,b$
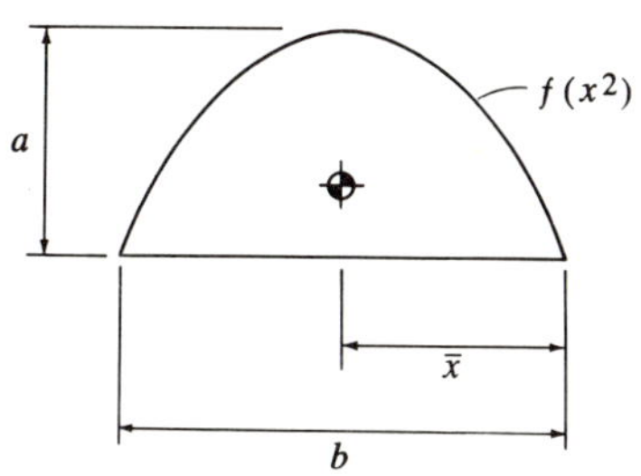	$A = (\frac{2}{3})\,ab$	$\bar{x} = (\frac{1}{2})\,b$
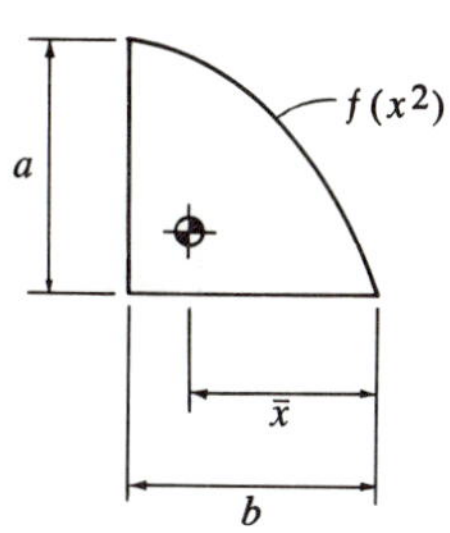	$A = (\frac{2}{3})\,ab$	$\bar{x} = (\frac{5}{8})\,b$
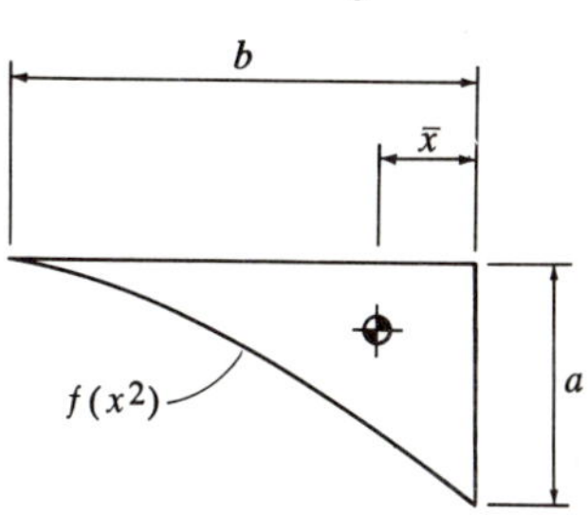	$A = (\frac{1}{3})\,ab$	$\bar{x} = (\frac{1}{4})\,b$
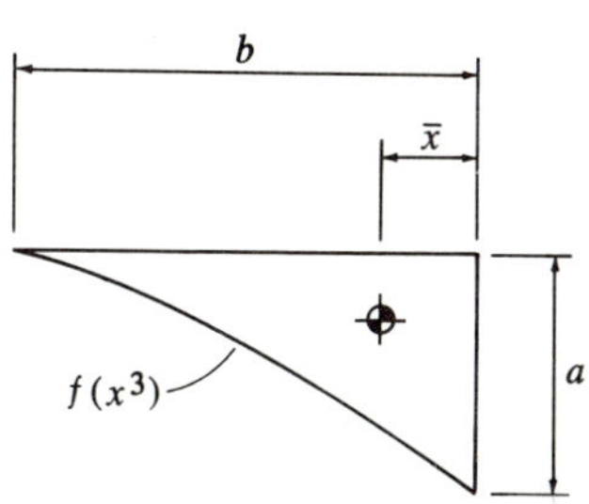	$A = (\frac{1}{4})\,ab$	$\bar{x} = (\frac{1}{5})\,b$

Figure 7.9

Example 7.2

Determine both the deflection and the slope of the elastic curve at the point beneath the concentrated load for the simply supported beam shown in Figure 7.10a. $E = 30 \times 10^6$ psi (steel); $I = 200$ in.^{4}

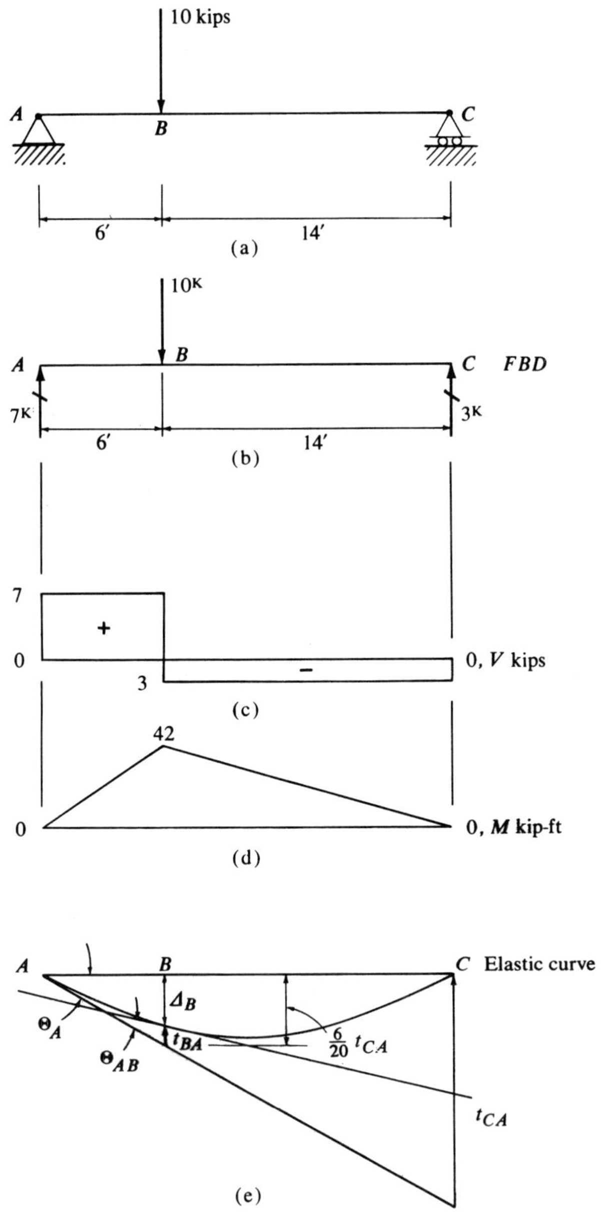

Figure 7.10

Solution:

1. The FBD, shear diagram, moment diagram, and the elastic curve are shown in Figures 7.10b, 7.10c, 7.10d, and 7.10e respectively.
2. From the tangential deviations sketched on Figure 7.10e, the following relationships can be obtained:

$$\Delta_B = (\tfrac{6}{20})t_{CA} - t_{BA}$$

$$\Theta_A = \frac{t_{CA}}{L}$$

$$\Theta_B = \Theta_A + \Theta_{AB}$$

Numerical results may be obtained by substitution of the proper values as follows:

$$EIt_{CA} = \left[\left(\frac{1}{2}\right)(42)(6)\right]\left(14 + \frac{6}{3}\right) + \left[\left(\frac{1}{2}\right)(42)(14)\right]\left(\frac{2}{3}\right)14$$

$$= 2016 + 2744$$

$$EIt_{CA} = 4760 \text{ kip-ft}^3$$

$$EIt_{BA} = \left[\left(\frac{1}{2}\right)(42)(6)\right]\left(\frac{1}{3}\right)6$$

$$= 252 \text{ kip-ft}^3$$

$$\Delta_B = \frac{1}{EI}\left[\left(\frac{6}{20}\right)(4760) - 252\right]$$

$$= \frac{1}{EI}(1176)$$

$$\Delta_B = \frac{(1176)\text{ kip-ft}^3}{30(10)^3\left(\frac{\text{kip}}{\text{in.}^2}\right)(200)\text{ in.}^4} = \frac{196}{(10)^6}\left(\frac{\text{ft}^3}{\text{in.}^2}\right)$$

In order to unify the dimensions, we must multiply by the conversion factor of 1728 in.^{3} per 1 ft^3 or:

$$\Delta_B = \frac{196}{(10)^6}\left(\frac{\text{ft}^3}{\text{in.}^2}\right)\left(\frac{1728\text{ in.}^3}{1\text{ ft}^3}\right)$$

$$\underline{\Delta_B = 0.338\text{ in.} \downarrow}$$

The deflection at B is downward since the distance $(6/14)t_{CA}$ is greater than the distance t_{BA}.

3. The slope of the elastic curve at B may be obtained by adding the change in slope of the elastic curve between A and B to the predetermined slope at A. Thus:

$$\Theta_A = \frac{t_{CA}}{L} \text{ for small angles}$$

$$\Theta_B = \Theta_A + \Theta_{AB}$$

Substituting the proper numerical values,

$$\theta_A = \frac{-4760}{20EI} = \frac{-238}{EI}$$

$EI\theta_{AB}$ = area beneath the moment diagram between points A and B

$EI\theta_{AB} = (\frac{1}{2})(42)(6) = 126$ kip-ft²

$$\theta_B = \frac{-238}{EI} + \frac{126}{EI} = \frac{-112}{EI}$$

θ_A is taken as negative since the slope is directed down and to the right. θ_{AB} is positive since all the moment area between points A and B is positive. Substituting in the values for E and I and the proper conversion factor we have:

$$\theta_B = \frac{-112 \text{ kip-ft}^2}{30(10)^6 \left(\frac{\text{kips}}{\text{in.}^2}\right) (200) \text{ in.}^4} \times \left(\frac{144 \text{ in.}^2}{1 \text{ ft}^2}\right)$$

$$\underline{\theta_B = -0.00000269 \text{ radians}}$$

4. The point of maximum deflection is not known since the loading is not symmetrical. The method used to determine the maximum deflection in such cases will be postponed until a later section.

Example 7.3

Determine the deflection at point C for the simply supported beam shown in Figure 7.11a. EI is constant along the span.

Solution:

1. The FBD and the elastic curve are shown in Figures 7.11b and 7.11c. Since more than one load is applied to the beam, it is more convenient to use a moment diagram by parts. Selecting point A as the reference point, the moment diagram by parts is drawn as shown in Figures 7.11d and 7.11c. Point A was selected as the reference point because experience has shown that it is preferable to choose the reference point at either end of a uniformly distributed load.
2. Using vertical distances sketched on Figure 7.11c we may obtain the following:

$$\Delta_C = \frac{1}{3} t_{AD} - t_{CD}$$

Using the area properties tabulated in Figure 7.9, we may determine the tangential deviations as follows:

$$EIt_{AD} = [(\tfrac{1}{2})(216)(18)](\tfrac{1}{3})(18) - [(\tfrac{1}{2})(144)(12)](\tfrac{1}{3})(12) - [(\tfrac{1}{3})(72)(6)](\tfrac{1}{4})(6)$$

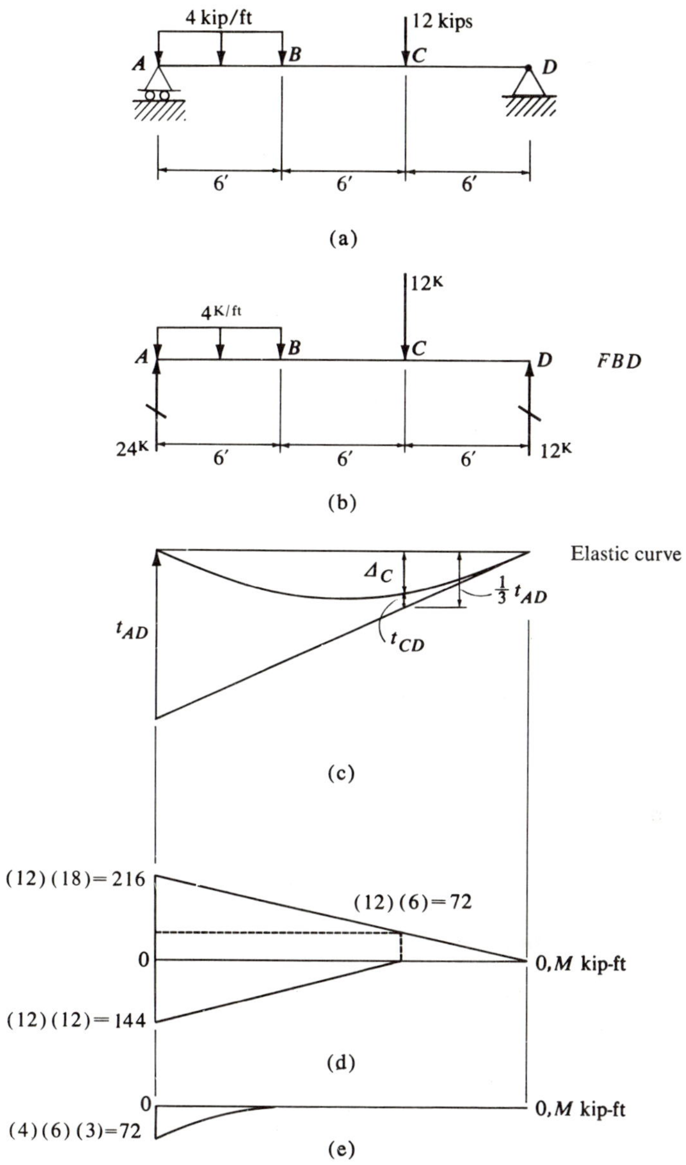

Figure 7.11

The bracketed terms are the areas between points A and D beneath the moment curves and terms outside the brackets are the corresponding lever arms of these areas from point A. Continuing:

$$EIt_{AD} = 7992 \text{ kip-ft}^3$$

In a similar manner:

$$EIt_{CD} = \left[\left(\frac{1}{2}\right)(72)(6)\right]\left(\frac{1}{3}\right)(6)$$
$$= 432 \text{ kip-ft}^3$$
$$\Delta_C = \left(\frac{1}{3}\right)\frac{7992}{EI} - \frac{432}{EI}$$
$$\Delta_C = \frac{2232}{EI} \downarrow$$

The procedure followed in the previous examples deviates only very slightly for most loading conditions on simply supported beams. If these examples are clearly understood, then no difficulty should be encountered for most simply supported beams. The procedure presented is consistent for both symmetrically and unsymmetrically loaded spans, the major difference being that the location of the point of maximum deflection is known to occur at midspan for the symmetrical case.

7.6 Beams With Overhangs

The procedure followed when applying the moment–area theorems to obtain deflections for beams with overhangs is essentially the same as for simply supported beams. In fact, the procedure is exactly the same for the deflections between the supports and the only deviation enters in when the deflection at the free end of the overhang is desired. The following examples illustrate the procedure.

Example 7.4

Determine the deflection of the right end of the overhanging beam shown in Figure 7.12a. $E = 1.8(10)^6$ psi and $I = 60$ in.^{4}

Solution:

1. The FBD and the elastic curve are shown in Figures 7.12b and 7.12c. A moment by parts with point C as the reference point is shown in Figure 7.12c. Point C was selected because only triangular shapes are encountered when tangential deviations shown in Figure 7.12c are solved for. This selection was not intuitive. After the tangential deviations necessary for the determination of the required deflection are sketched on the elastic curve diagram, a little forethought and sometimes some preliminary sketches are necessary before the best reference point is selected.
2. The deflection at the right end may be determined as follows:

$$\Delta_D = t_{DA} - \tfrac{15}{12} t_{CA}$$

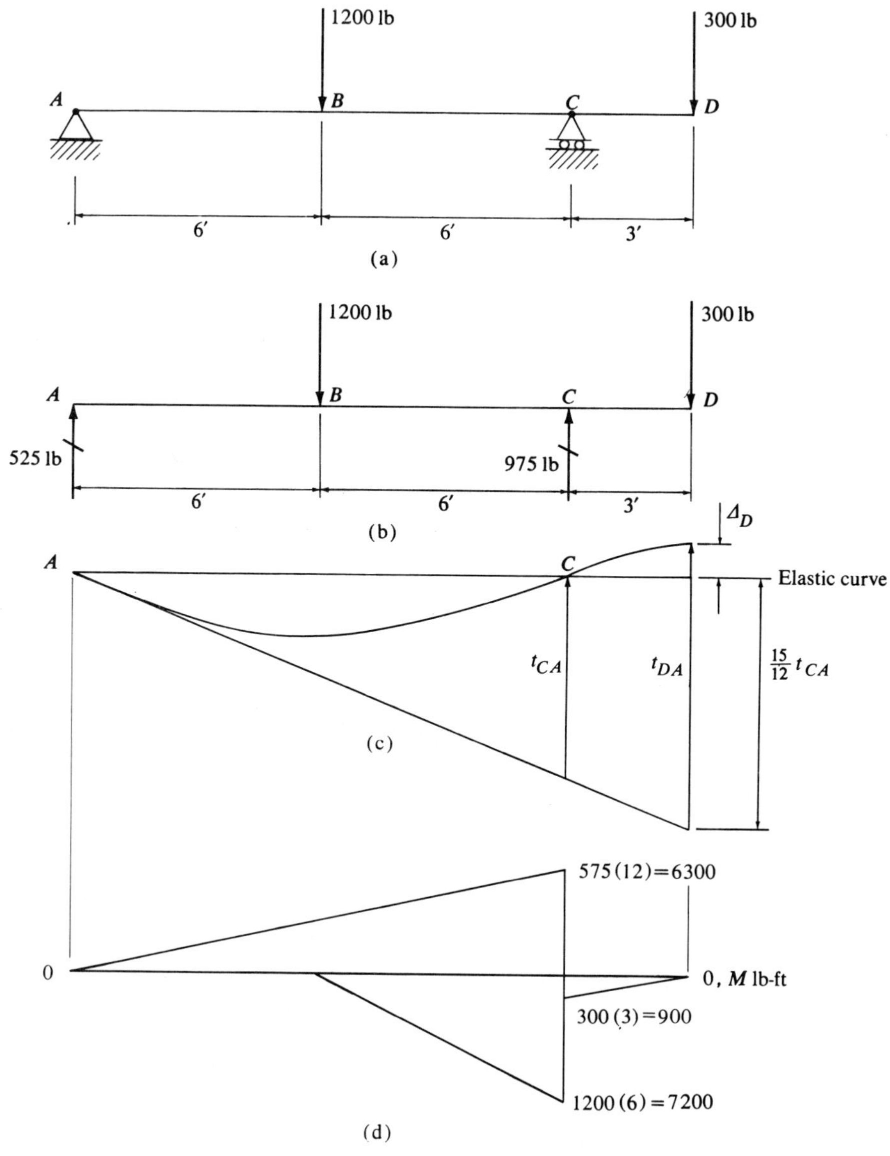

Figure 7.12

The term $(15/12)t_{CA}$ was obtained from similar triangle relationships. If t_{DA} is larger than $15/12\ t_{CA}$, then the deflection is upward since $15/12\ t_{CA}$ is measured to the undeflected position of point D. Continuing:

$$EIt_{CA} = [(\tfrac{1}{2})(6300)(12)](\tfrac{1}{3})(12) - [(\tfrac{1}{2})(7200)(6)](\tfrac{1}{3})(6)$$

Again, the bracketed terms are the areas beneath the moment curves between C and A and the terms outside are the respective lever arms from point C. Continuing:

$$EIt_{CA} = 151{,}200 - 43{,}200$$
$$= 108{,}000 \text{ lb-ft}^3$$
$$EIt_{DA} = \left[\left(\frac{1}{2}\right)(6300)(12)\right]\left(3 + \frac{12}{3}\right) - \left[\left(\frac{1}{2}\right)(7200)(6)\right]\left(3 + \frac{6}{3}\right)$$
$$- \left[\left(\frac{1}{2}\right)(900)(3)\right]\left(\frac{2}{3}\right)(3)$$
$$= 264{,}600 - 108{,}000 - 2700$$
$$EIt_{DA} = 153{,}900 \text{ lb-ft}^3$$
$$\Delta_C = \frac{153{,}900}{EI} - \left(\frac{15}{12}\right)\left(\frac{108{,}000}{EI}\right)$$
$$\Delta_C = \frac{18{,}900}{EI}$$
$$= \frac{18{,}900 \text{ lb-ft}^3}{1.8(10)^6\left(\frac{\text{lb}}{\text{in.}^2}\right)(60) \text{ in.}^4} \times \frac{1728 \text{ in.}^3}{1 \text{ ft}^3}$$
$$\underline{\Delta_C = 0.302 \text{ in.} \uparrow}$$

The deflection is known to be upward since $t_{DA} > (15/12)t_{CA}$ (see Figure 7.12c).

3. It should be noted by observing Figure 7.12c that any deflections that may be required between points A and C would be determined using the procedure used for simply supported spans.

Example 7.5

Determine the deflection of the left end of the overhanging beam shown in Figure 7.13a. EI is constant along the spans.

Solution:

1. The FBD and the sketch of the elastic curve are shown in Figures 7.13b and 7.13c. After studying to see which tangential deviations are required in order to determine the deflection of A, point B is selected as the reference point for the moment diagram by parts. Figures 7.13d and 7.13e show the resulting moment diagrams. Note that all the areas encountered can be easily handled using the properties tabulated in Figure 7.9.
2. Using the tangential deviations labeled in Figure 7.13c, the following expression for the deflection position of point A may be written:

$$\Delta_A = t_{AD} - \tfrac{12}{9}t_{BD}$$

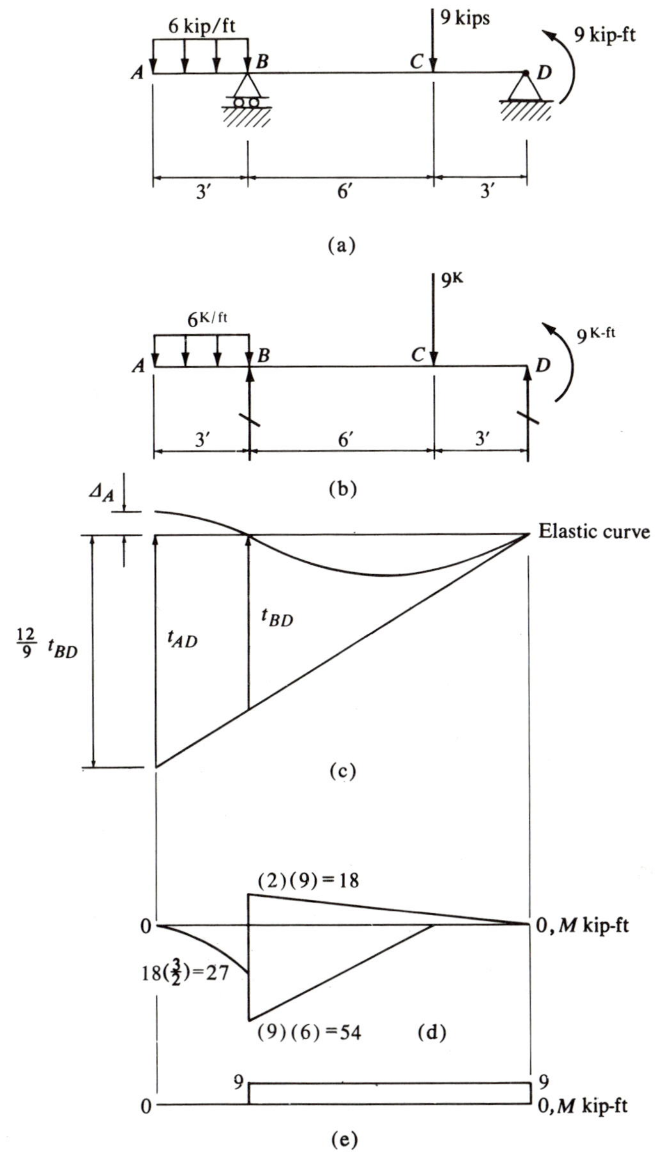

Figure 7.13

If a positive result is obtained for Δ_A, then point A moves upward since $(12/9)t_{BD}$ is measured to the undeflection position of point A. Continuing:

$$EIt_{BD} = \left[\left(\frac{1}{2}\right)(18)(9)\right]\left(\frac{1}{3}\right)(9) - \left[\left(\frac{1}{2}\right)(54)(6)\right]\left(\frac{1}{3}\right)(6) + [(9)(9)]\left(\frac{1}{2}\right)(9)$$

$$= 243 - 324 + 365$$

$$EIt_{BD} = 284 \text{ kip-ft}^3$$

$$EIt_{AD} = \left[\left(\frac{1}{2}\right)(18)(9)\right]\left(3+\frac{9}{3}\right) - \left[\left(\frac{1}{2}\right)(54)(6)\right]\left(3+\frac{6}{3}\right) + [(9)(9)]$$

$$\left(3+\frac{9}{2}\right) - \left[\left(\frac{1}{3}\right)(27)(3)\right]\left(\frac{3}{4}\right)3$$

$$= 486 - 810 + 608 - 61$$

$$EIt_{AD} = 425 \text{ kip-ft}^3$$

$$\Delta_A = \frac{425}{EI} - \left(\frac{12}{9}\right)\left(\frac{284}{EI}\right)$$

$$\Delta_A = +\frac{47.0}{EI} = \frac{47.0}{EI}\ \uparrow$$

7.7 Cantilever Beams

The moment-area method can be applied to cantilever beams with relative ease. This can be easily accomplished because the slope of the elastic curve at one point is known prior to performing any calculations. To help matters even further, the known slope is zero, i.e., the tangent to the curve is a horizontal line. Consider the cantilever beam shown in Figure 7.14a. The fixity at A prevents any rotation of the beam under loading. Therefore, the elastic curve before and after loading remains horizontal at A, i.e., the slope of the elastic curve remains zero. The fixed support at A may be represented by the hinge and roller system shown in Figure 7.14b. Figure 7.14c shows the deflected shape of the beam in an exaggerated manner. Notice that the straight segment between the roller and hinge at A remains horizontal and, therefore, is simply an extension of the tangent to the elastic curve at A. Since the tangent at A is horizontal, it lies along the same line as the longitudinal centroidal axis of the beam before the impending deformation. Hence, a tangential deviation at any point from the tangent at A is equal to the deflection at that point because the horizontal tangent at A and the undeflected elastic curve are collinear. In a similar fashion, the slope of the elastic curve at any point may be obtained by algebraically adding the change in slope between point A and any other point to the slope at A. Since the slope at A is known to be zero, the slope at any other point simply becomes numerically equal to the change in slope between A and the point in question.

When dealing with simply supported beams and overhanging beams, a certain set procedure is followed to insure a troublefree solution when applying the moment-area theorems. A procedure must also be established for cantilever beams to insure similar results. The procedure for cantilevers is simple: all tangential deviations should be taken from the horizontal tangent at the fixed support and all slopes should be calculated by adding the respective change in slope to the initial slope of zero at the point of fixity. In addition, for the vast majority of loading conditions the maximum deflection of a cantilever beam will occur at the free end.

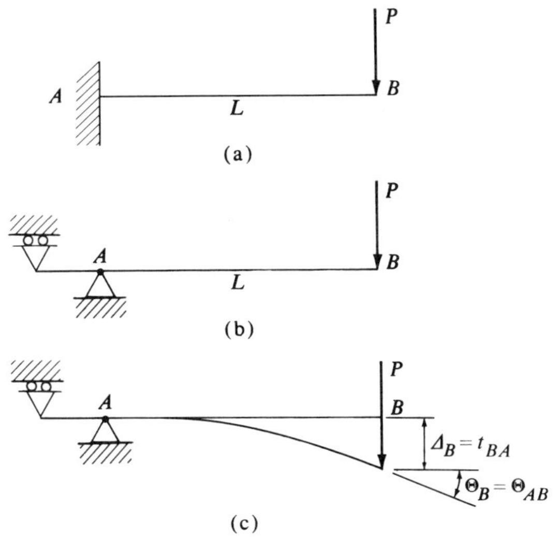

Figure 7.14

Example 7.6

Determine the displacement and slope of the free end for the cantilever beam shown in Figure 7.15a. *EI* is constant.

Solution:

1. The sketch of the elastic curve shown in Figure 7.15b reveals the following relationships:

$$\Delta_B = t_{BA}$$
$$\Theta_B = \Theta_A + \Theta_{AB} = \Theta_{AB}$$

Since more than one load is encountered, a moment diagram by parts is selected. If a moment diagram by parts is drawn with point *A* as the reference point, there is no need to determine the reactions at *A* and the areas encountered are not difficult to deal with. Generally speaking, experience has shown that when dealing with cantilever beams it is more convenient to draw a moment diagram by parts about the fixed end, i.e., use the point of fixity as the selected reference point. Figures 7.15c and 7.15d show the resulting moment diagram by parts with point *A* as the reference point.

2. Numerically, the solution proceeds as follows:

$$EIt_{BA} = [(\tfrac{1}{3})(-288)(12)](\tfrac{3}{4})(12) + [(\tfrac{1}{2})(-80)(8)](4 + \tfrac{16}{3})$$
$$= -10{,}368 - 2787$$
$$EIt_{BA} = -13{,}155 \text{ kip-ft}^3$$
$$\Delta_B = t_{BA}$$
$$\underline{\Delta_B = \frac{13{,}155}{EI} \downarrow}$$

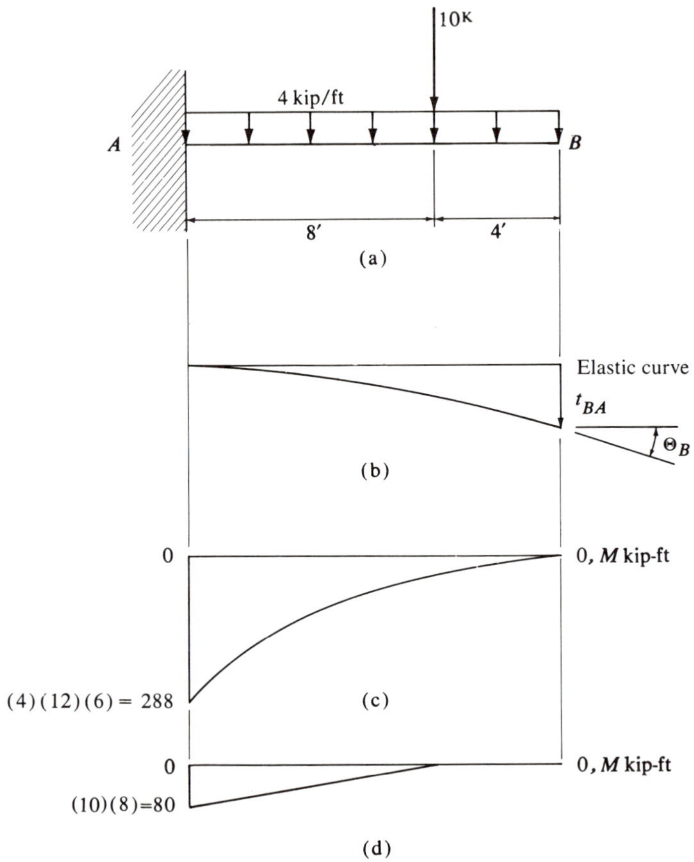

Figure 7.15

The deflection is known to be down since the negative sign associated with t_{BA} indicates that point B is below the tangent drawn at A. Continuing:

$$\Theta_B = \Theta_A + \Theta_{AB} \qquad \Theta_A = 0$$
$$\therefore \Theta_B = \Theta_{AB}$$
$$EI\Theta_{AB} = (\tfrac{1}{3})(-288)(12) + (\tfrac{1}{2})(-80)(8)$$
$$= -1472 \text{ kip-ft}^2$$
$$\Theta_B = \frac{-1472}{EI}$$

The negative sign tells us that the tangent at B is directed down and to the right as can be verified by Figure 7.15b.

7.8 Maximum Deflections

When dealing with symmetrically loaded, simply supported beams and cantilever beams, the point of maximum deflection is known. For unsymmetrically loaded, simply supported beams and for overhanging beams the point of maximum deflection is not known. Needless to say, the location of the point of maximum deflection as well as the magnitude and direction of this displacement are essential since it is these maximum limits which must be checked by the designer. To obtain the location of the point of maximum deflection, we utilize the condition that the slope of the elastic curve at the point of maximum deflection must be zero, i.e., the tangent to the elastic curve at the point of maximum displacement is a horizontal line. In other words, the slope at the point of maximum displacement is known to be zero. As with all moment–area calculations, a set procedure can again be established. Consider the simply supported beam and its corresponding elastic curve, very much exaggerated, shown in Figure 7.16. Point C, at some unknown distance of x from point A,

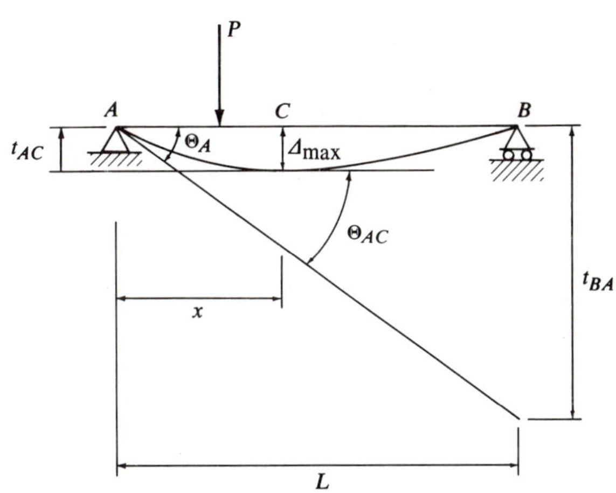

Figure 7.16

is the point of maximum deflection. At C, a horizontal tangent to the elastic curve may be drawn since it is known that the slope of the elastic curve at the point of maximum deflection is zero. Notice that the horizontal tangent at point C (see Figure 7.16) and the undeflected length AB are parallel, which means that the two angles, Θ_A and Θ_{AC}, are equal. In other words, if the slope of the elastic curve is Θ_A at point A and zero at point C, then the change in slope between points A and C must be Θ_A or Θ_A must be equal to Θ_{AC}. The quantity Θ_{AC} is known to be the area beneath the moment diagram between A and C and, hence, if we equate Θ_A to the area beneath the moment diagram over the length x, we may determine the length x which pinpoints the desired location of maximum deflection. Using Figure 7.16 and symbols, we have the following steps in the procedure:

(1) $\Theta_A = \frac{t_{BA}}{L}$, t_{BA} from application of the second moment–area theorem

(2) $\Theta_A = \Theta_{AC}$, where Θ_{AC} is a function of x. This equality yields a value for x.

(3) $\Delta_{max} = t_{AC}$, may be easily determined by again applying the second moment–area theorem since the location of C is now known.

Following this procedure, the maximum displacement occurring between two simple supports may be obtained. When dealing with an overhanging beam, however, the analyst does not know if the absolute maximum displacement of the loaded beam occurs between the supports or at the free end of the overhang. The analyst must, therefore, use the procedure outlined above to determine the maximum displacement that occurs between the supports and then determine the displacement at the free end using the procedure outlined for overhanging beams presented in Section 7.6. The absolute maximum deflection is the larger of the two.

Example 7.7

Determine the maximum deflection for the beam shown in Figure 7.17a. E is $12(10)^5$ psi and I is 30 in.^{4}

Solution:

1. The FBD, the properly labeled elastic curve, and the composite moment diagram are shown in Figures 7.17b, 7.17c, and 7.17d respectively. A composite moment diagram was utilized since there is only one applied load. Since the loading is not symmetrical, the point of maximum deflection is not known and, hence, the elastic curve shown in Figure 7.17c is labeled in accordance with the procedure established to determine the location of the point of maximum displacement.

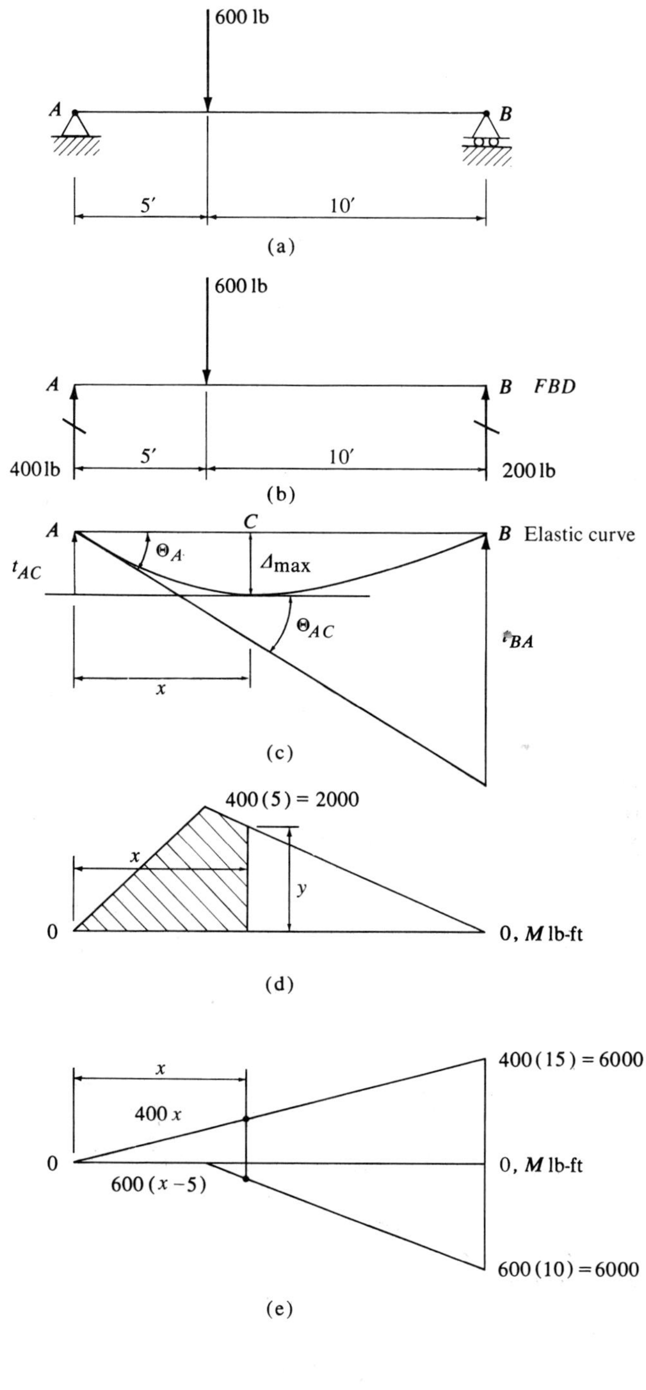
600 lb
A
B
5′
10′
(a)
600 lb
A
B FBD
400 lb
5′
10′
200 lb
(b)
A
C
B Elastic curve
Θ_A
t_{AC}
Δmax
Θ_AC
t_{BA}
x
(c)
400(5) = 2000
x
y
0
0, M lb-ft
(d)
x
400(15) = 6000
400 x
0
0, M lb-ft
600(x − 5)
600(10) = 6000
(e)

Figure 7.17

2. The procedure is as follows:

$$\theta_A = \frac{t_{BA}}{L}$$

$$EIt_{BA} = [(\tfrac{1}{2})(2000)(10)](\tfrac{2}{3})(10) + [(\tfrac{1}{2})(2000)(5)](10 + \tfrac{5}{3})$$
$$= 125{,}000 \text{ lb-ft}^3$$

$$\theta_A = \frac{125{,}000}{(15)EI} = \frac{8{,}333}{EI}$$

$$\theta_A = \theta_{AC}$$

In order to determine θ_{AC} we must set up an expression for the area beneath the moment curve between A and C. Referring to Figure 7.17d, we see that the area consists of a triangular shape and a trapezoidal shape. Such an area would be difficult to handle, particularly when we apply the second moment–area theorem. Although we could handle this area, it would be advantageous to use a moment diagram by parts such as shown in Figure 7.17e, thereby saving ourselves both time and opportunities for error by using a more tedious approach. Notice that the moment diagram by parts shown in Figure 7.17e involves only well-known triangular shapes. Continuing with the use of the moment diagram by parts we have:

$$EI\theta_{AC} = \text{Area}_{AC}$$
$$= [(\tfrac{1}{2})(400x)(x)] + [(\tfrac{1}{2})(-600)(x-5)(x-5)]$$

$$\theta_{AC} = \frac{1}{EI}(-100x^2 + 3000x - 7500)$$

$$\theta_A = \theta_{AC}$$

$$\frac{8{,}333}{EI} = \frac{1}{EI}(-100x^2 + 3000x - 7500)$$

$$x^2 - 30x + 158.3 = 0$$

Completing the square we obtain the following:

$$x^2 - 30x + (\tfrac{30}{2})^2 = (\tfrac{30}{2})^2 - 158.3$$
$$(x - 15)^2 = 66.7$$
$$x = 15 \pm \sqrt{66.7}$$

$$x = 6.83' \text{ or } 23.17'$$

An x of 6.83 ft fits the conditions of the problem. Hence, the maximum deflection occurs 6.83 ft to the right of point A.

3. The maximum displacement may now be determined as follows:

$$\Delta_{\max} = t_{AC}$$
$$EIt_{AC} = [(\tfrac{1}{2})(400x)(x)](\tfrac{2}{3})x + [(\tfrac{1}{2})(-600)(x-5)(x-5)][5 + \tfrac{2}{3}(x-5)]$$

Notice that we continue to work in terms of x since we have already established an expression for the area between A and C under the moment curve when Θ_{AC} was determined. Hence, we simply multiply these previously determined areas by their respective moment arms from A. Substituting in 6.83 ft for x, we obtain the following:

$$EIt_{AC} = 43{,}100 - 5{,}030 - 1{,}225$$
$$EIt_{AC} = 36{,}845 \text{ lb-ft}^3$$
$$t_{AC} = \frac{36{,}845 \text{ lb-ft}^3}{(12)(10)^5 \left(\frac{\text{lb}}{\text{in.}^2}\right) 30 \text{ in.}^4} \left(\frac{1728 \text{ in.}^3}{1 \text{ ft}^3}\right)$$
$$t_{AC} = 1.762 \text{ in.}$$
$$\therefore \Delta_{\max} = 1.762 \text{ in.} \downarrow$$

The maximum deflection is down since a positive result is obtained for t_{AC} indicating that point A is above a horizontal tangent @ C.

Example 7.8

Determine the maximum deflection for the beam shown in Figure 7.18a. E is $10(10)^6$ psi and I is 36 in.^{4}

Solution:

1. The FBD and the elastic curve are shown in Figures 7.18b and 7.18c. Using point B as the reference point, a moment diagram by parts is drawn as shown in Figures 7.18d and 7.18e. Using the notations used in Figure 7.18c, we may obtain the following relationships:

$$\Theta_A = \frac{t_{BA}}{L} \text{ for small angles}$$
$$EIt_{BA} = [(\tfrac{1}{2})(108)(12)](\tfrac{1}{3})(12) + [(\tfrac{1}{2})(-90)(9)](\tfrac{1}{3})(9) + [(\tfrac{1}{3})(-18)(6)](\tfrac{1}{4})(6)$$
$$= 2592 - 1215 - 54$$
$$t_{BA} = \frac{1323}{EI}$$

$$\Theta_A \times \frac{1323}{12EI} = \frac{110.2}{EI}$$

Also,
$$\Theta_A = \Theta_{AC}$$
$$EI\Theta_{AC} = \text{Area}_{AC}$$

If x is the distance from point A to point C, the point of maximum moment, a general expression for the area under the moment diagram between these two points may be obtained. Referring to Figures 7.18c and 7.18d we may proceed as follows:

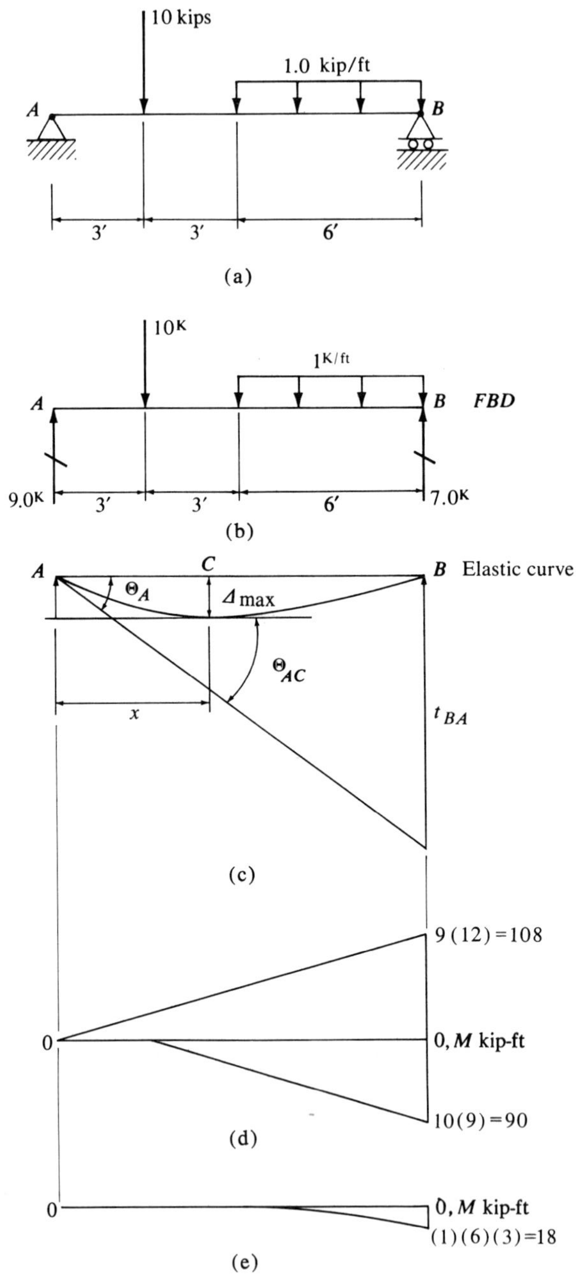
10 kips
1.0 kip/ft
A
B
3′
3′
6′
(a)
10K
1K/ft
A
B
FBD
9.0K
3′
3′
6′
7.0K
(b)
A
C
B
Elastic curve
ΘA
Δmax
ΘAC
x
tBA
(c)
9(12)=108
0
0, M kip-ft
10(9)=90
(d)
0
0, M kip-ft
(1)(6)(3)=18
(e)

Figure 7.18

$$EI\Theta_{AC} = [(\tfrac{1}{2})(9x)x] + [(\tfrac{1}{2})(-10)(x-3)(x-3)]$$
$$EI\Theta_{AC} = -0.5x^2 + 30x - 45$$

Using the equality $\Theta_A = \Theta_{AC}$, we obtain:

$$\frac{110.2}{EI} = \frac{1}{EI}(-0.5x^2 + 30x - 45)$$

Simplifying and transposing we obtain:

$$x^2 - 60x + 310.4 = 0$$

Completing the square and solving for x:

$$x^2 - 60x + (\tfrac{60}{2})^2 = (\tfrac{60}{2})^2 - 310.4$$
$$(x-30)^2 = 589.6$$
$$x = 30 \pm \sqrt{589.6}$$
$$x = 5.70' \text{ or } 54.3'$$

An x of 5.70 ft fits the conditions of the problem and, hence, the point of maximum displacement is 5.70 ft to the right of point A. Notice that when the area between A and C was initially calculated, it was calculated assuming that x was greater than 3 ft but less than 6 ft. If an x greater than 6 ft were obtained, the problem would have to be reworked since the moment area due to the uniformly distributed load (see Figure 7.18e) was not included in the first general expression for the area between A and C.

2. The maximum deflection may now be determined as follows:

$$\Delta_{max} = t_{AC}$$
$$EIt_{AC} = [(\tfrac{1}{2})(9x)(x)](\tfrac{2}{3})x + [(\tfrac{1}{2})(-10)(x-3)(x-3)][3 + (\tfrac{2}{3})(x-3)]$$

Again, we work in terms of x since we already have a general expression for the area between A and C. Substituting $x = 5.70$ ft we have:

$$EIt_{AC} = 380.2 \text{ kip-ft}^3$$
$$t_{AC} = \frac{380.2 \text{ kip-ft}^3}{(10)(10)^3\left(\frac{\text{kips}}{\text{in.}^2}\right) 36 \text{ in.}^4}\left(\frac{1728 \text{ in.}^3}{1 \text{ ft}^3}\right)$$
$$t_{AC} = 1.825 \text{ in.}$$
$$\underline{\Delta_{max} = 1.825 \text{ in.} \downarrow}$$

Since a positive sign was obtained for t_{AC}, point A is above the tangent line; therefore, the maximum deflection is downward.

7.9 The Conjugate Beam Method

The conjugate beam method, sometimes referred to as the **elastic weight method,** is very similar to the moment–area method. Briefly, the method is utilized by constructing a fictitious beam, i.e., a conjugate beam, with the following properties:

1. The length of the conjugate beam and the real, or original, beam are the same. Also, the original beam is assumed to be initially straight.
2. The load on the conjugate beam is the M/EI diagram for the real beam.
3. The supports of the conjugate beam must meet the boundary conditions of the real beam.
4. The bending moment at any section of the conjugate beam is equal to deflection of the real beam at the location of the same section.
5. The shear of any section of the conjugate beam is equal to the slope of the real beam at the same section.

In an attempt to further clarify the conjugate beam concept, the following analogy between the real beam and the conjugate may be utilized:

Real Beam	*Conjugate Beam*
M/EI Diagram	Load, w
Slope, Θ	Shear, V
Deflection, Δ	Moment, M

Using this analogy for the real beam and its counterparts in the conjugate beam, and knowing that the real and conjugate span lengths are equal, we must now set up an equivalence between the real supports and the conjugate beam supports. Recalling the analogy expressed previously, the shear and moment developed at the supports of the conjugate beam must be compatible with the slope and deflection permitted at the supports of the real beam, i.e., the support requirements of the conjugate beam are dictated by the **boundary conditions,** known deformation conditions, at the supports of the real beam. This equivalence of supports is best shown by Figure 7.19.

The conjugate beam method is relatively easy to apply when dealing with simply supported beams with a minimum number of applied loads. When the loading becomes numerous and/or complex, this results in a more complex moment diagram and, by our analogy, a more complex loading condition on the conjugate beam. Thus, the method becomes tedious and cumbersome. To reduce this problem somewhat, however, a moment diagram by parts can be used. The conjugate beam method is presented here primarily due to its uniqueness.

As with all such methods, a sign convention must be established. Positive bending moment, i.e., member tends to bend concave upward, at a section of the conjugate beam means a downward deflection of that point in the real

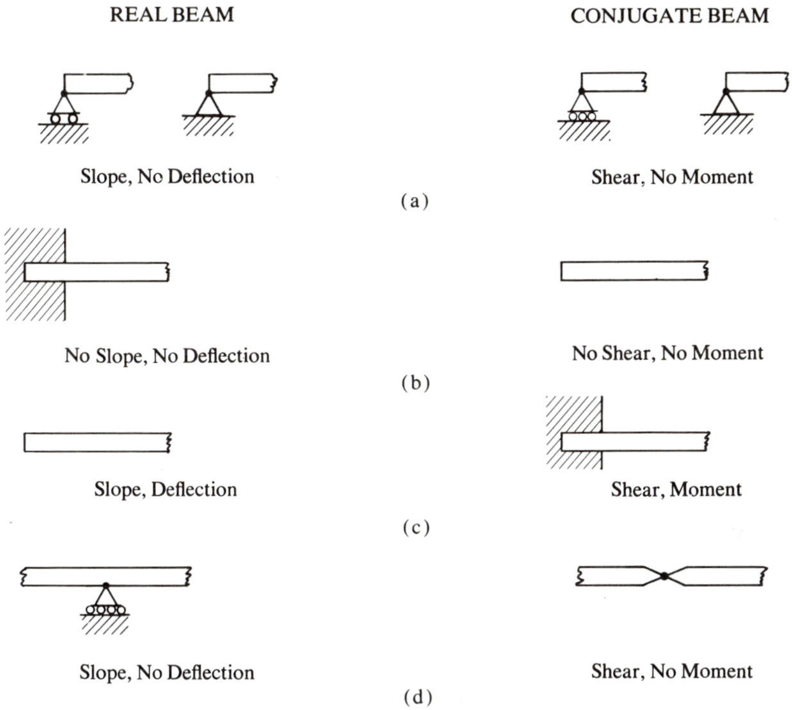

Figure 7.19

beam. Positive shear, as established in Chapter Three, indicates a negative slope (see Figure 7.3b). Positive M/EI ordinates indicate a downward loading ordinate on the conjugate beam. The following illustrative examples should help clarify the method.

Example 7.9

Using the conjugate beam method, determine the deflection and the slope of the elastic curve at the point of application of the concentrated load for the beam shown in Figure 7.20. $E = 30(10)^6$ psi, $I = 200$ in.^{4}

Solution:

1. The moment diagram for the real beam shown in Figure 7.20a is shown in Figure 7.20b. Since the moment ordinates are all positive, the corresponding load on the conjugate beam will be downward as shown in Figure 7.20c. At the supports at A and C of the real beam, there is rotation, i.e., the elastic curve possesses slope after loading, but there is no deflection. By the conjugate beam analogy, therefore, the supports at A and C of the conjugate beam must permit shear but no bending moment. The loaded conjugate beam along with its proper supports is shown in Figure 7.20c. Figure 7.19 also can be used to determine the proper supports to be used with the conjugate beam.

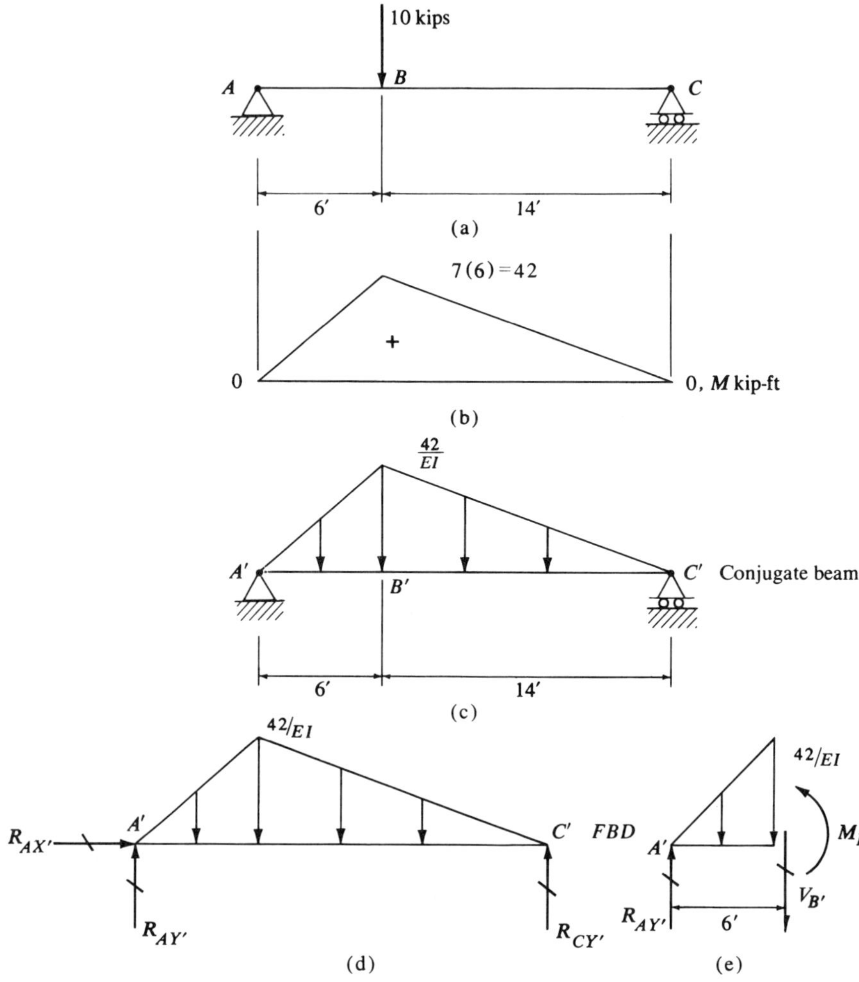

Figure 7.20

2. The conjugate beam can now be treated as any other beam, i.e., a FBD can be drawn and the beam with its loads must obey the laws of static equilibrium. Since we are after the deflection and slope at B, we must determine the shear and moment at B for the conjugate beam as indicated by our analogy. The reactions of the conjugate beam are, therefore, required. Using the FBD for the conjugate beam shown in Figure 7.20d, we may proceed as follows:

$$\Sigma F_X = 0 \rightarrow + R_{AX}' = 0$$

$$\Sigma M_B = 0 \circlearrowright +$$

$$20R_{AY}' - \left(\frac{1}{2}\right)\left(\frac{42}{EI}\right)(6)\left(14+\frac{6}{3}\right) - \left(\frac{1}{2}\right)\left(\frac{42}{EI}\right)(14)\left(\frac{2}{3}\right)(14) = 0$$

$$20R_{AY}' - \frac{4760}{EI} = 0$$

$$R_{AY}' = \frac{238}{EI} \text{ kip-ft}^2 \uparrow$$

If we cut the conjugate beam at B and draw the FBD of that portion to the left of the cut as shown in Figure 7.20e, there is no need to determine the reaction at B. Continuing, we have:

Figure 7.20e

$$\Sigma F_Y = 0 \uparrow +$$

$$R_{AY}' - \left(\frac{1}{2}\right)\left(\frac{42}{EI}\right)(6) - V_B = 0$$

$$\frac{238}{EI} - \frac{126}{EI} - V_B = 0$$

$$V_B = \frac{112}{EI} \text{ kip-ft}^2 \downarrow, \text{ or positive shear}$$

By the conjugate beam analogy,

$$\Theta_B = V_B = \frac{-112}{EI}$$

where the negative sign indicates a negative slope by our previously adopted sign convention for slope. Substituting in the appropriate values and dimensional factors we have:

$$\Theta_B = \frac{-112 \text{ kip-ft}^2}{(30)(10)^3\left(\frac{\text{kips}}{\text{in.}^2}\right)(200) \text{ in.}^4}\left(\frac{144 \text{ in.}^2}{\text{ft}^2}\right)$$

$$\underline{\Theta_B = -0.00000269 \text{ radians}}$$

Continuing the solution for M_B:

Figure 7.20e

$$\Sigma M_B = 0 \circlearrowright +$$

$$6R_{AY}' - \left[\left(\frac{1}{2}\right)\left(\frac{42}{EI}\right)(6)\right]\left(\frac{1}{3}\right)(6) - M_B = 0$$

$$(6)\left(\frac{238}{EI}\right) - \frac{252}{EI} - M_B = 0$$

$$M_B = \frac{1176}{EI} \text{ kip-ft}^3$$

By the conjugate beam analogy and recalling the sign convention we have:

$$M_B = \Delta_B = \frac{1176}{EI} \text{ kip-ft}^3 \downarrow$$

$$\Delta_B = \frac{1176 \text{ kip-ft}^3}{(30)(10)^3 \left(\frac{\text{kips}}{\text{in.}^2}\right) (200) \text{ in.}^4} \left(\frac{1728 \text{ in.}^3}{\text{ft}^3}\right)$$

$$\underline{\Delta_B = 0.338 \text{ in.} \downarrow}$$

3. These answers check with those obtained in Example 7.2 where the sample problem was worked using the moment–area method. The similarity between the two methods should be noted.

Example 7.10

Using the conjugate beam method, determine the deflection of the free end of the cantilever beam shown in Figure 7.21a. *EI* is constant.

Solution:

1. The moment diagram by parts about point *B* is shown in Figure 7.21b. This becomes the load for the conjugate beam as shown in Figure 7.21c. Notice that negative moment becomes an upward directed load on the conjugate beam. Also, there is no slope, i.e., rotation, and no deflection at *B* of the real beam; this means no shear and no moment possible at *B* for the conjugate beam. This is represented by a free end, i.e., no support. At point *A* of the real beam, both slope, i.e., rotation, and deflection occur and this is represented by a fixed support at *A* on the conjugate beam where both shear and moment are possible. Again, Figure 7.19 could have been consulted for the proper supports.
2. The FBD for the conjugate beam is shown in Figure 7.21d. By the conjugate beam analogy, the moment at *A* of the conjugate beam is the deflection at *A* for the real beam. Hence, all we must do is determine M_A' shown in Figure 7.21d. Thus:

$$\Sigma M_A' = 0 \circlearrowright +$$

$$M_A' + \left[\left(\frac{100}{EI}\right)(9)\right]\left(\frac{9}{2}\right) - \left[\left(\frac{1}{2}\right)\left(\frac{900}{EI}\right)(9)\right]\left(\frac{2}{3}\right)(9) = 0$$

$$M_A' + \frac{4050}{EI} - \frac{24{,}300}{EI} = 0$$

$$M_A' = \frac{+20{,}250}{EI}, \text{ or positive moment}$$

$$\underline{\therefore \Delta_A = \frac{20{,}250}{EI} \text{ lb-ft}^3 \downarrow}$$

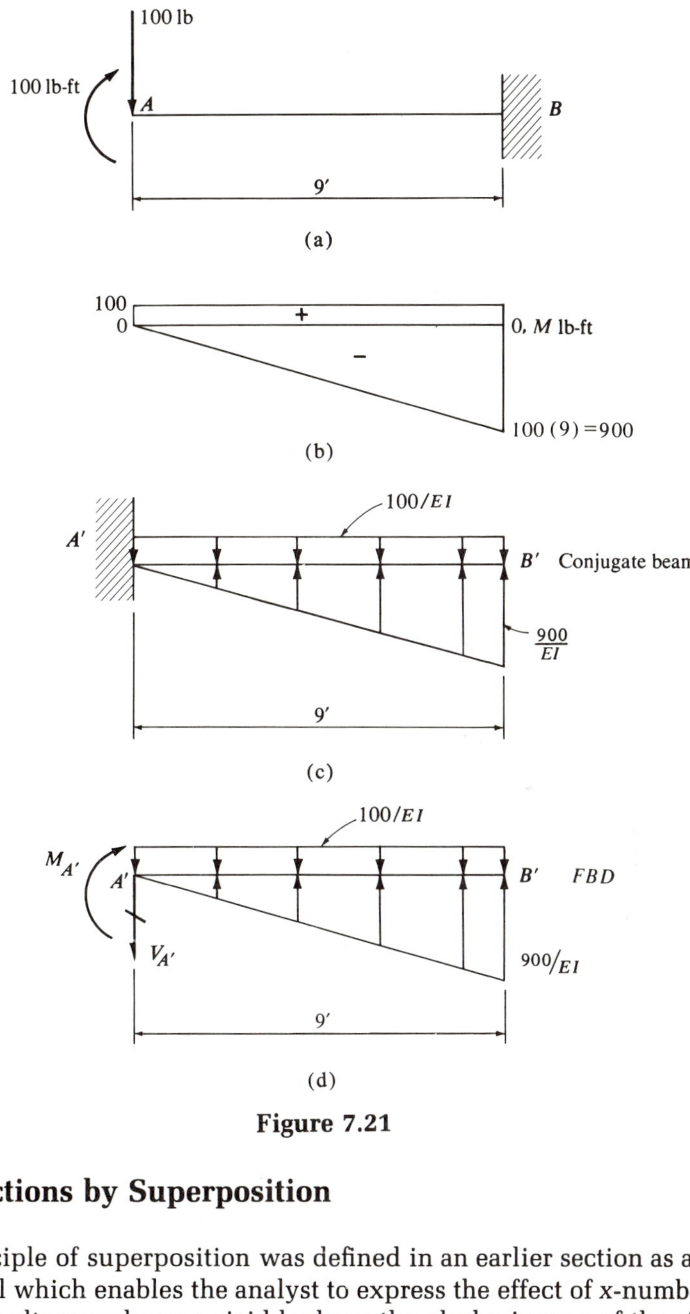

Figure 7.21

7.10 Deflections by Superposition

The principle of superposition was defined in an earlier section as a computational tool which enables the analyst to express the effect of x-number of loads acting simultaneously on a rigid body as the algebraic sum of the effect of the loads treated as if each load acts separately, provided that the load/effect relationship is linear.

There are numerous handbooks and textbooks available to the structural analyst that tabulate deflections for beams with varied loading and/or support conditions. The tabulated expressions for deflections are given in a general form and, hence, all the analyst must do is simply locate a beam with the same loading and support conditions as the one he is analyzing and substitute in the values, particular to the beam at hand, into the general, tabulated expressions for the deflection. A problem results, however, when no direct correlation can be made between the actual loading and support conditions and the tabulated conditions. Using the principle of superposition and manipulating the loading conditions, successful results can be obtained. Figure 7.22 shows some tabulated beam deflections due to various standard loadings and support systems. The following illustrative examples demonstrate the determination of deflections by superposition.

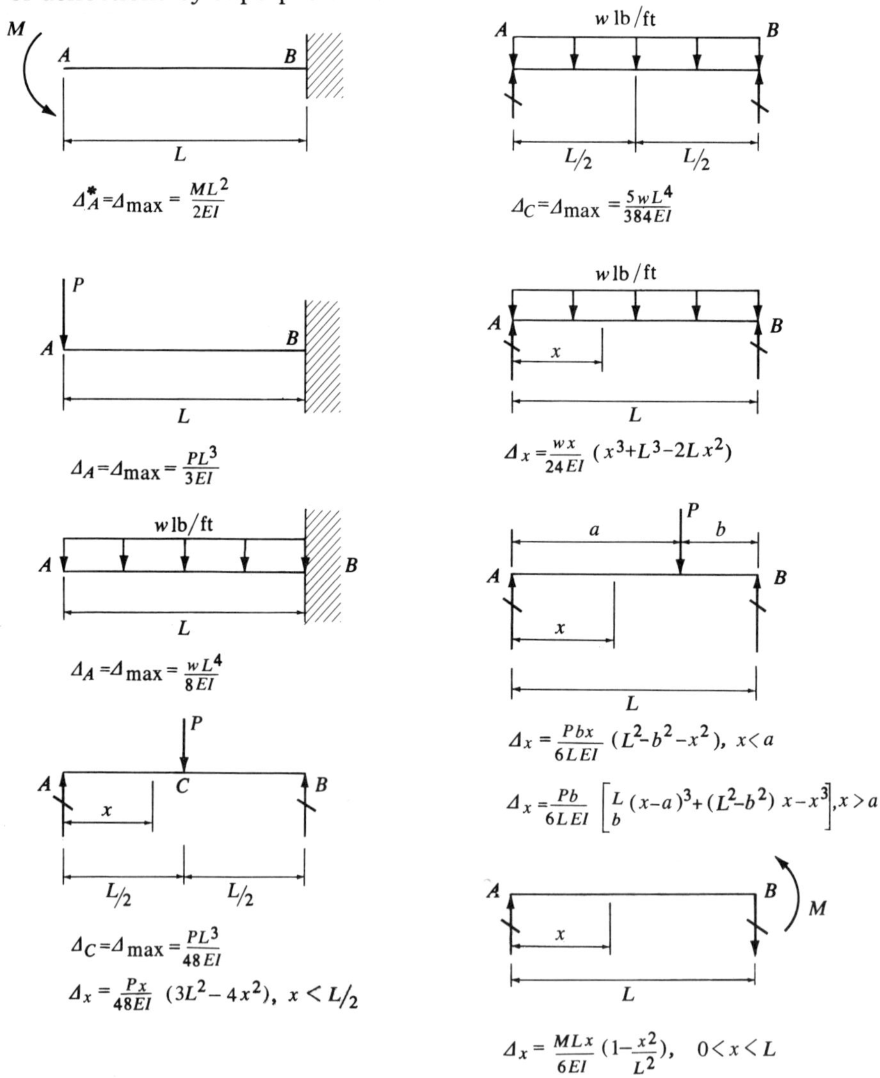

Figure 7.22

* All deflections are downward for the load and moment directions shown.

Example 7.11

Using the method of superposition, determine the deflection at the free end of the beam shown in Figure 7.23a. $E = 15(10)^6$ psi; $I = 173$ in.^{4}

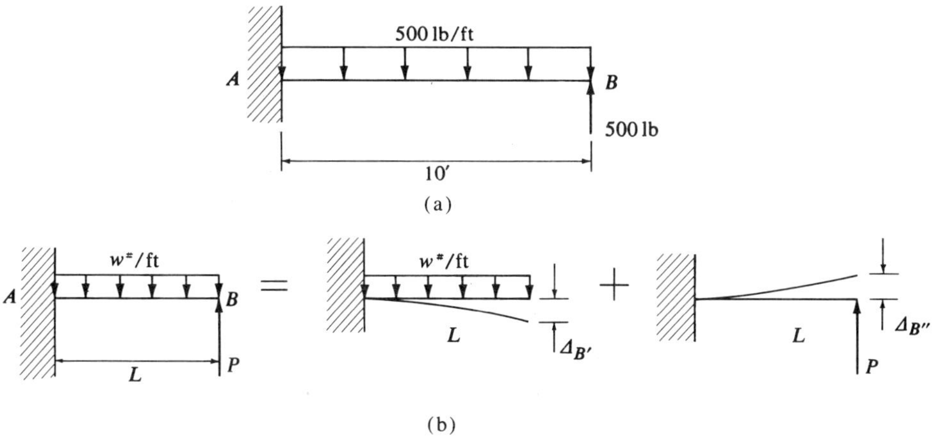

Figure 7.23

Solution:

1. Using the principle of superposition, we may represent the combined loading condition shown in Figure 7.23a by the two individual loading conditions for which deflection equations are available in Figure 7.22.

2. Thus, we may set up an expression for the deflection at B as the net result of the individual movements shown in Figure 7.23b. Using Figure 7.22 we may obtain the following:

$$\Delta_B = \Delta_B' + \Delta_B''$$

$$\Delta_B = -\frac{wL^4}{8EI} + \frac{PL^3}{3EI}$$

where downward deflections are assumed to be negative. Substituting in the appropriate values we have:

$$\Delta_B = \left[\frac{-500(120)^4}{8} + \frac{500(120)^3}{3}\right]\frac{1}{15(10)^6(173)}$$

$$\underline{\Delta_B = -4.88 \text{ in.} = 4.88 \text{ in.} \downarrow}$$

Example 7.12

Using the method of superposition, determine the midspan deflection for the beam shown in Figure 7.24a. EI is constant.

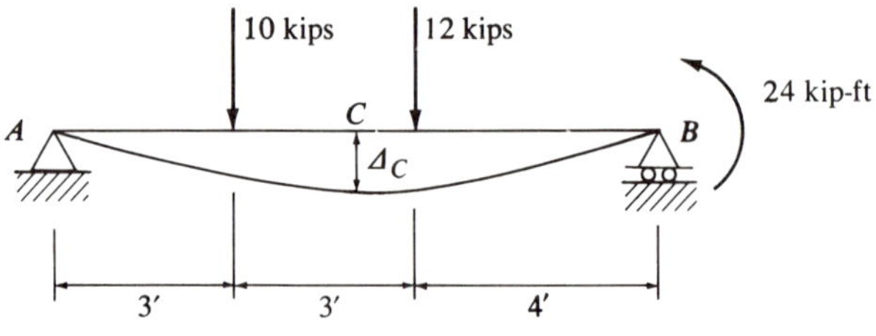

(a)

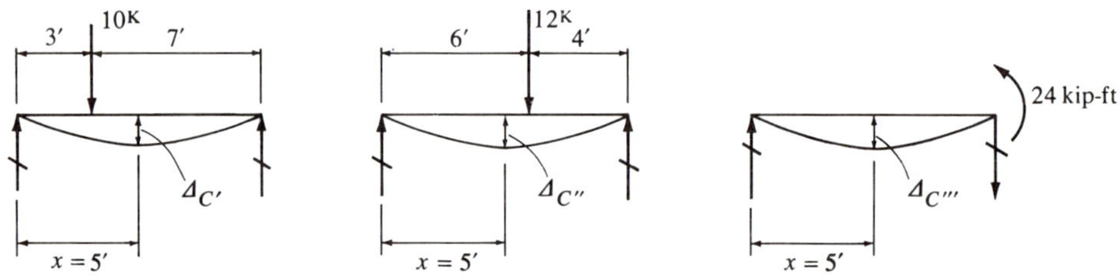

(b)

Figure 7.24

Solution:

1. The combined loading condition shown in Figure 7.24a may be represented, for deflection calculation purposes, by the individual loading conditions shown in Figure 7.24b. Using the notations of Figures 7.24a and 7.24b, we may write,

$$\Delta_C = \Delta_C' + \Delta_C'' + \Delta_C'''$$

2. Using the deflection expressed in Figure 7.22, the individual deflections may be determined as follows:

$$\Delta_C' = \frac{Pb}{6LEI}\left[\frac{L}{b}(x-a) + (L^2 - b^2)x - x^3\right]$$

where

$$\begin{aligned} a &= 3' \\ b &= 7' \\ L &= 10' \\ x &= 5' \\ p &= 10 \text{ kips} \end{aligned}$$

Substituting these values, we obtain:

$$\Delta_C' = \frac{(10)(7)}{(6)(10)EI}\left[\frac{10}{7}(5-3) + (100-49)(5) - (5)^3\right]$$

$$\Delta_C' = \frac{155}{EI} \text{ kip-ft}^3 \downarrow$$

and,

$$\Delta_C'' = \frac{Pbx}{6LEI}(L^2 - b^2 - x^2)$$

where

$$\begin{aligned} b &= 4' \\ L &= 10' \\ x &= 5' \\ P &= 12 \text{ kips} \end{aligned}$$

Substituting these values, we obtain:

$$\Delta_C'' = \frac{(12)(4)(5)}{(6)(10)EI}[(10)^2 - (4)^2 - (5)^2]$$

$$\Delta_C'' = \frac{236}{EI} \text{ kip-ft}^3 \downarrow$$

Continuing:

$$\Delta_C''' = \frac{Mx}{6EI}(L^2 - x^2)$$

where

$$\begin{aligned} M &= 24 \text{ kip-ft} \\ L &= 10' \\ x &= 5' \end{aligned}$$

Substituting, we have:

$$\Delta_C''' = \frac{(24)(5)}{6EI}[(10)^2 - (5)^2]$$

$$\Delta_C''' = \frac{1500}{EI} \text{ kip-ft}^3 \downarrow$$

Combining we obtain the desired result:

$$\Delta_C = \frac{1}{EI}(155 + 236 + 1500)$$

$$\underline{\Delta_C = \frac{1891}{EI} \text{ kip-ft}^3 \downarrow}$$

Problems

7–1 to 7–4. For the beams shown, draw the moment diagrams by parts about points A, B, and C. Perform the customary checks by summing moment ordinates at points of known moment.

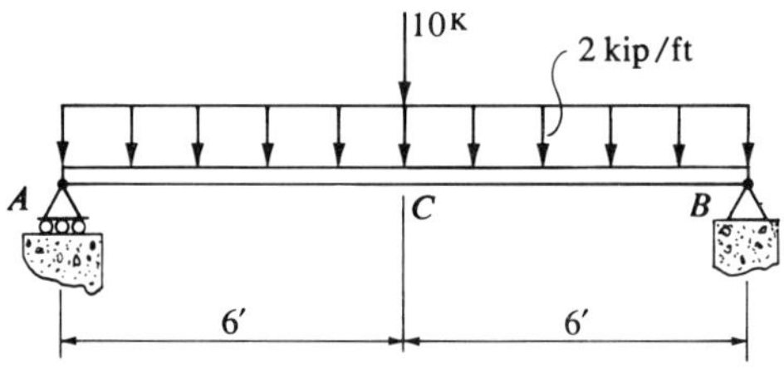

Figure P 7–1

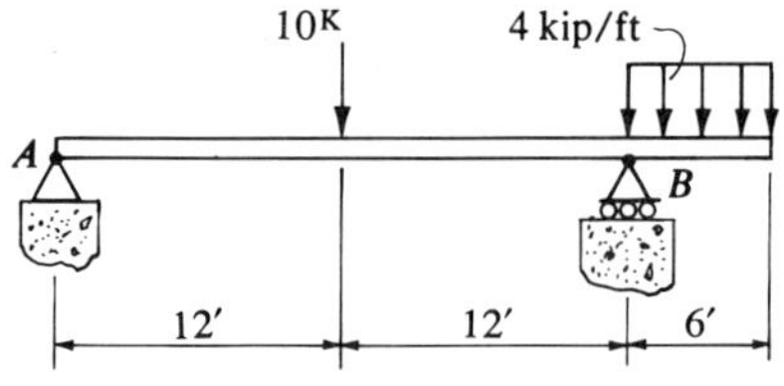

Figure P 7–2

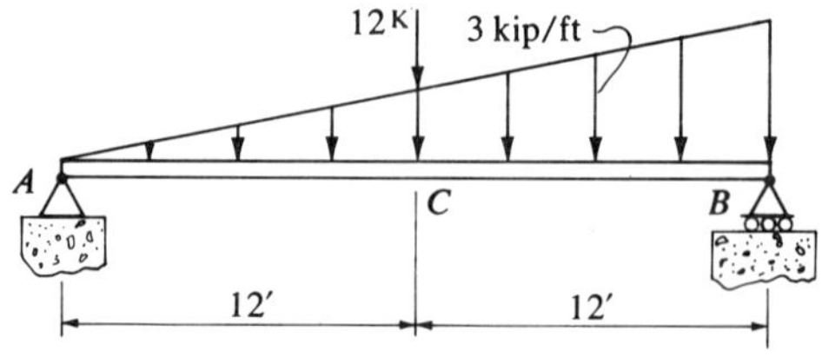

Figure P 7–3

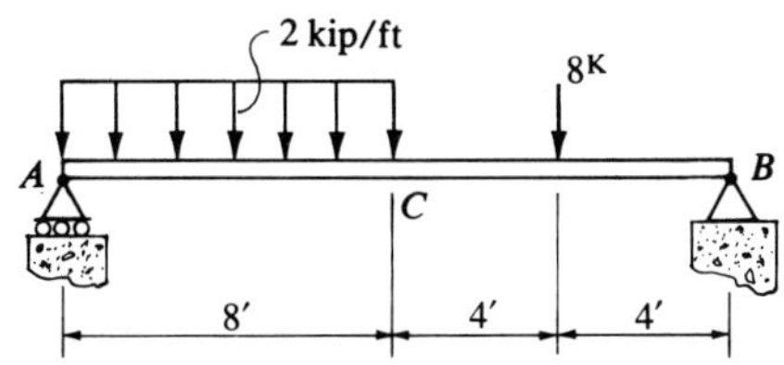

Figure P 7–4

7–5 to 7–8. Using the moment–area method, determine the deflection at the midspan of each beam shown as a function of *EI*.

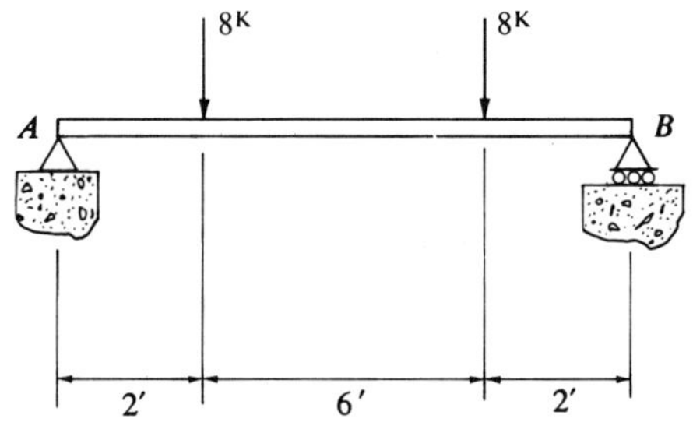

Figure P 7–5

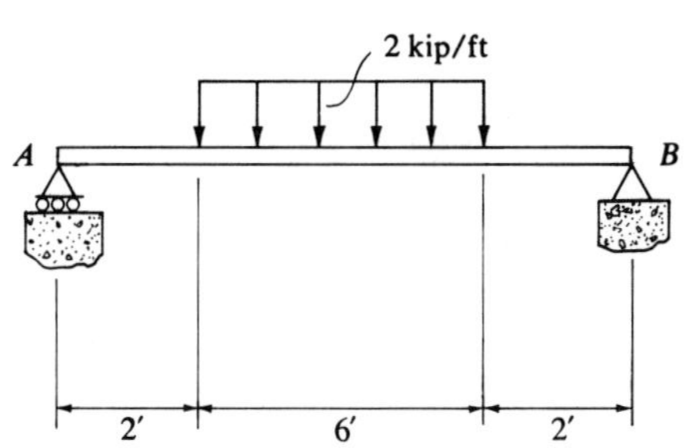

Figure P 7–6

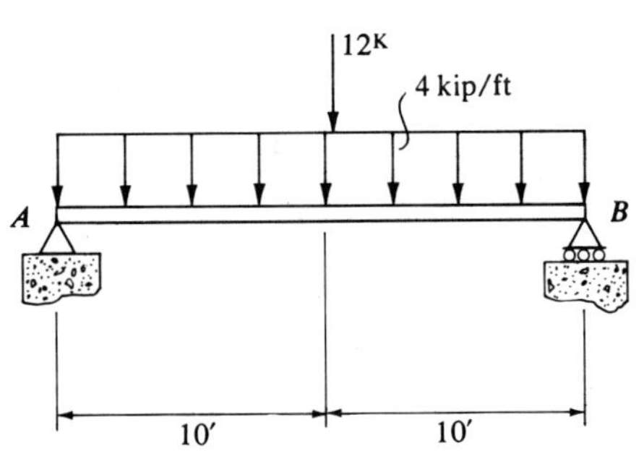

Figure P 7–7

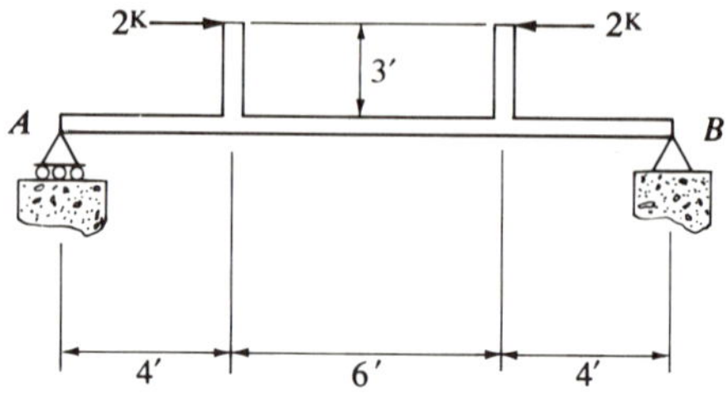

Figure P 7–8

7–9. Find the deflection at midspan for the beam shown. $E = 10(10)^6$ psi and $I = 18$ in.^{4}

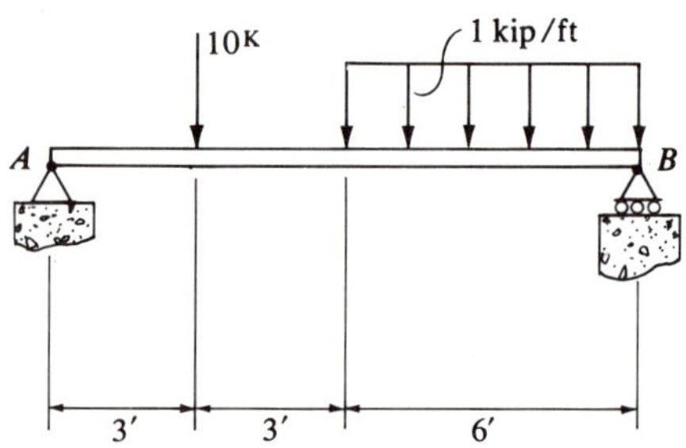

Figure P 7–9

7–10. Find the deflection under the 7-kip load for the beam shown. $E = 30(10)^6$ psi and $I = 146$ in.^{4}

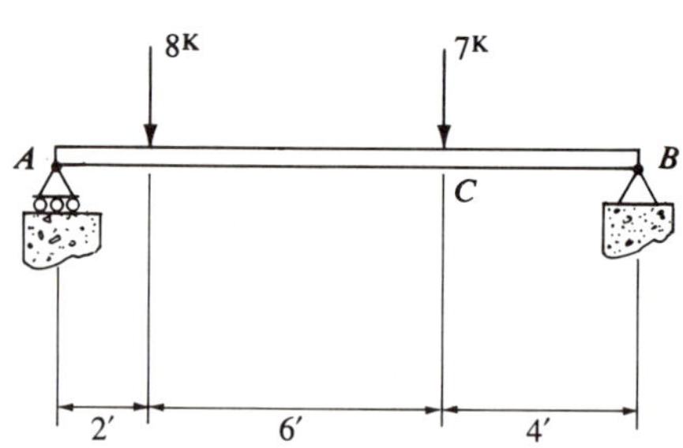

Figure P 7–10

7–11. Find the deflection of the left end of the beam shown in terms of EI.

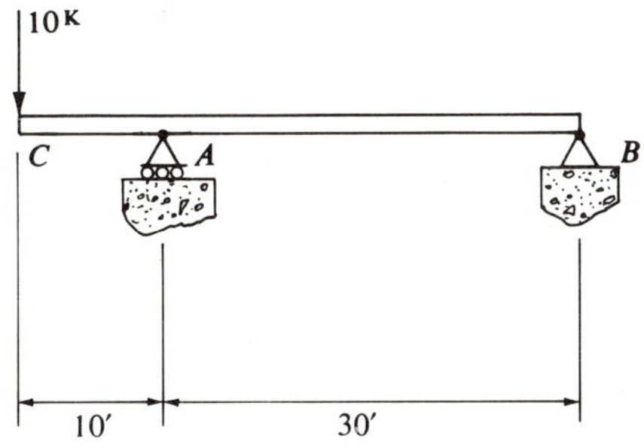

Figure P 7–11

7–12. For the beam shown, find the deflection of the right end of the beam in terms of EI.

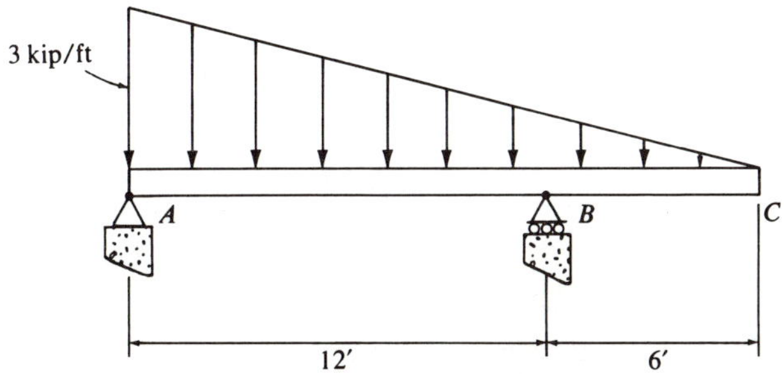

Figure P 7–12

7–13. For the beam shown, find the deflection at the right end and under the 6-kip load. $E' = 10(10)^6$ psi and $I = 13.5$ in.^{4}

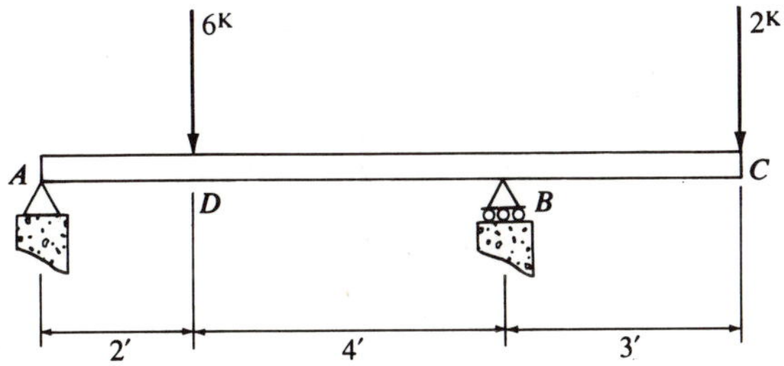

Figure P 7–13

7–14. For the beam shown, find the deflection at the free ends and at midspan in terms of EI.

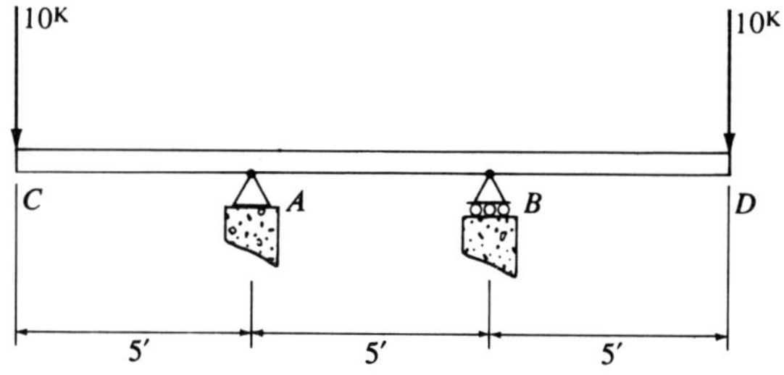

Figure P 7–14

7–15 to 7–18. For the beams shown, find the maximum deflection in terms of EI.

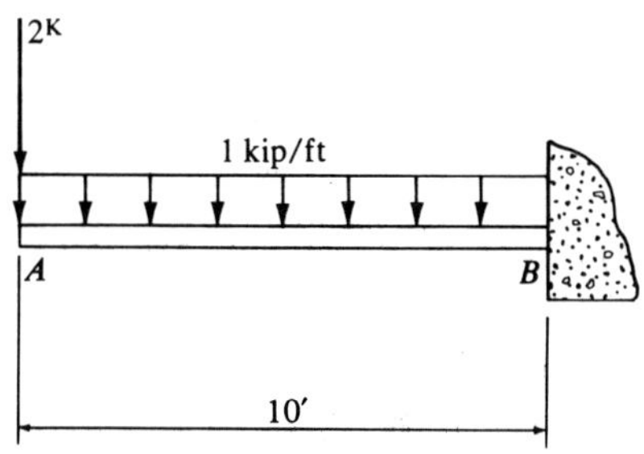

Figure P 7–15

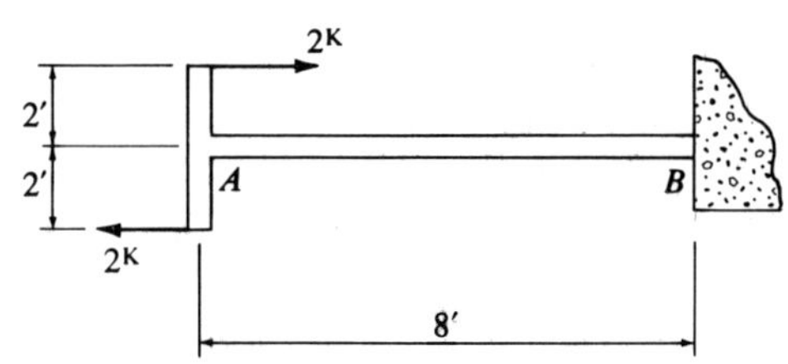

Figure P 7–16

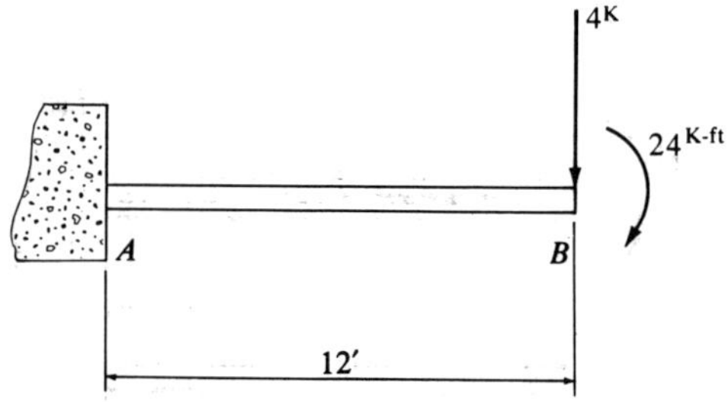

Figure P 7–17

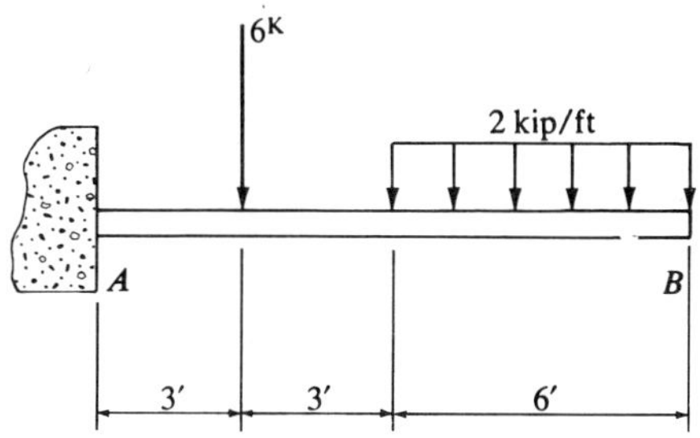

Figure P 7–18

7–19 to 7–21. For the beams shown, determine the maximum displacements.

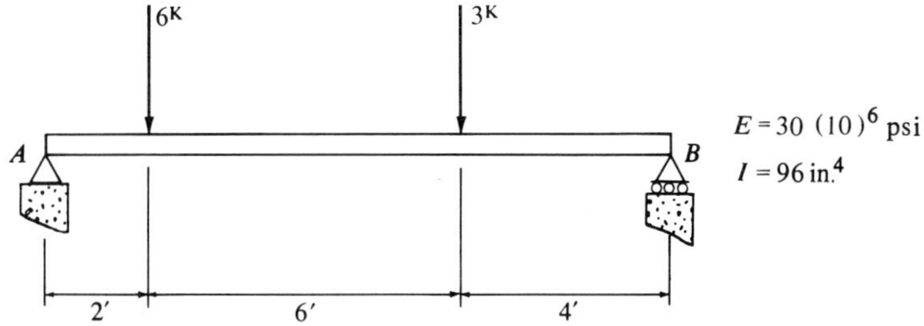

Figure P 7–19

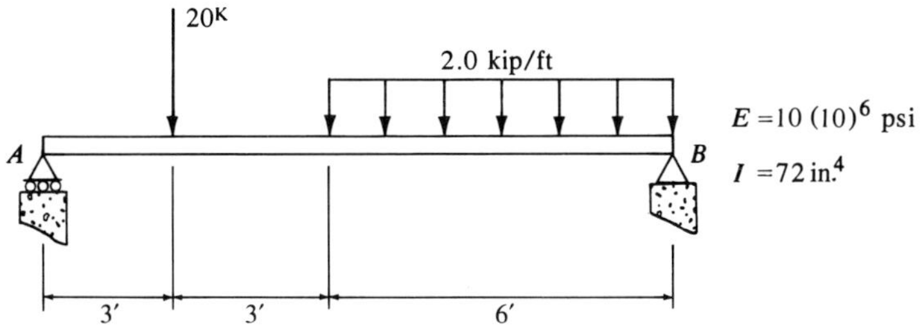

Figure P 7–20

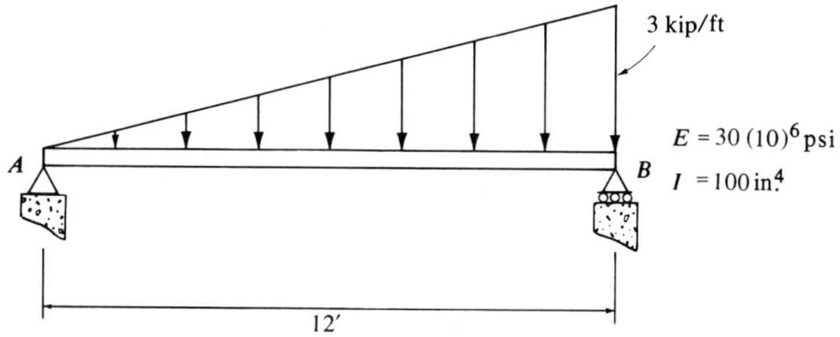

Figure P 7–21

7–22. For the beam of Problem 7–5, use the conjugate beam method to determine the deflection at midspan.

7–23. For the beam of Problem 7–8, use the conjugate beam method to determine the deflection at midspan.

7–24. Using the conjugate beam method, determine the deflection of the left end of the beam of Problem 7–11.

7–25. Using the conjugate beam method, determine the deflection of the free end of the beam of Problem 7–17.

7–26. Using the method of superposition, find the deflection at midspan for the beam shown in terms of *EI*.

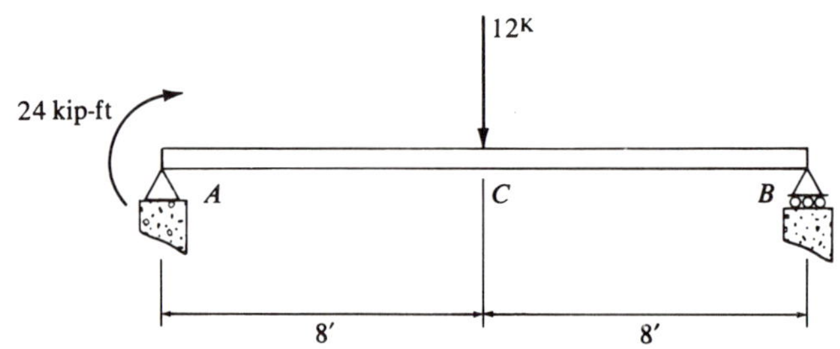

Figure P 7–26

7–27. Using the method of superposition, find the deflection of the free end of the beam shown in terms of *EI*.

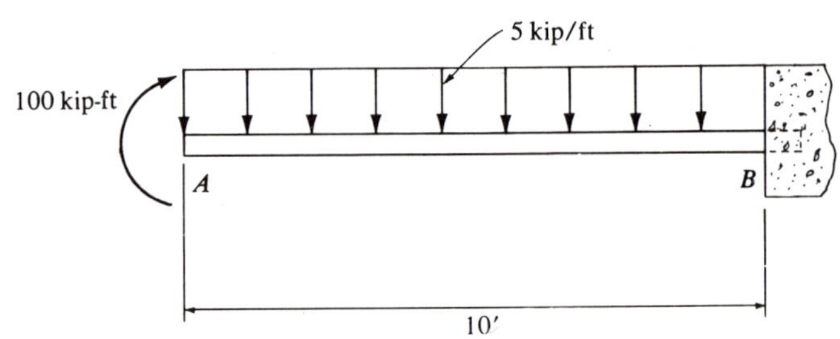

Figure P 7–27

7–28. Using the method of superposition, find the deflection at midspan for the beam shown in terms of *EI*.

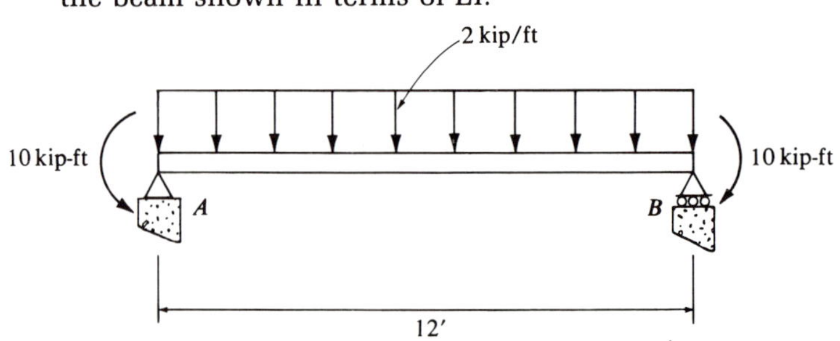

Figure P 7–28

CHAPTER EIGHT

Statically Indeterminate Structures

8.1 Introduction

A statically indeterminate structure was defined in an earlier chapter as a structure whose reactions cannot be determined by the equations of static equilibrium alone. More generally, the structure has more reaction components than are required to meet the conditions of static equilibrium. It is not unusual to wonder why such a structure is utilized, i.e., why should a structure possess more reaction components than necessary to keep it in equilibrium? The answer to this question is that statically indeterminate structures are employed to provide economy and, occasionally, for aesthetics. Consider the simply supported beam and its counterpart, the fixed end beam shown in Figure 8.1a. Suppose that each has a span of L-ft and that each carries a load of w-lb per foot. From the corresponding moment diagrams, the maximum moment for the simply supported beam is $wL^2/8$ lb-ft while the maximum moment for the fixed end beam is only $wL^2/12$ lb-ft. For the given span length and loading condition, the replacement of the simple supports by the fixed supports has produced a decrease of one-third in the simply supported maximum moment. If an interior support is added at the midspan as shown in Figure 8.1b, we have changed the simply supported beam to a **continuous beam,** i.e., the beam is continuous across an interior support. Examination of their respective moment diagrams again reveals that the addition of the interior support has produced a decrease of almost fifty percent in the simply supported maximum moment. Such moment reductions are obviously economical and, thus, it is again not unusual to wonder why statically determinate beams are used at all. To answer this, we must recall that with the addition of most economical measures something else must be sacrificed. In the case of the fixed supports, such fixity in actual

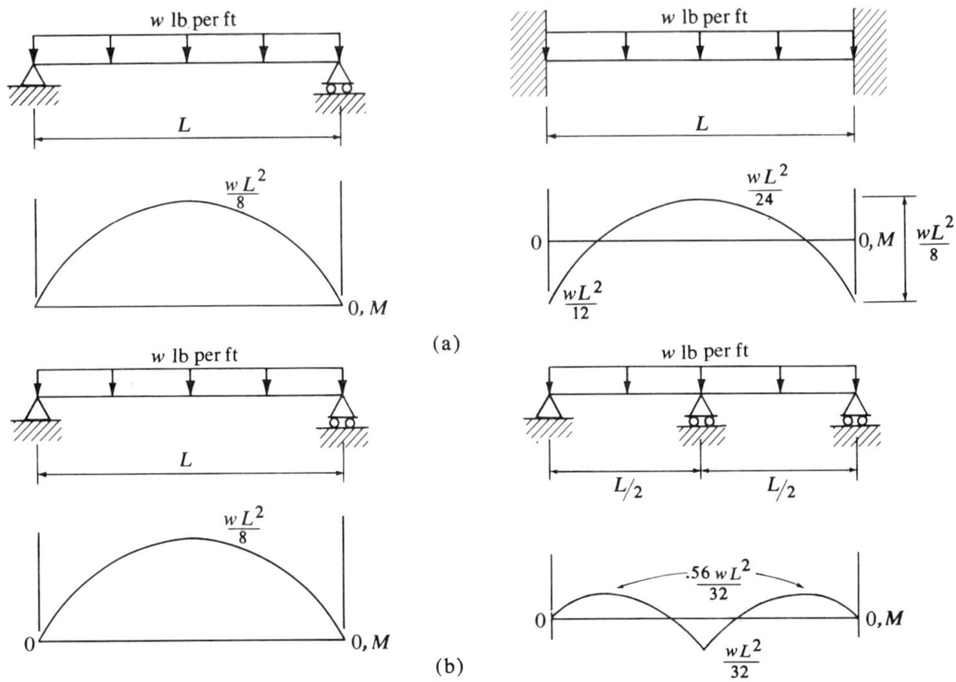

Figure 8.1

practice is difficult, at best, to attain in the field. Furthermore, both fabrication time and cost increase for the fixed supports. The addition of an interior support means that we are willing to sacrifice our original clear span of L-feet. This is rarely a prime consideration and sheer span length itself dictates the use of interior supports as both a practical and economic factor. There is a need, therefore, for both types of structures, neither type being able to fulfill all design and use considerations.

8.2 Methods of Analysis for Beams

Since the analysis of statically indeterminate structures involves more unknown reaction components than available equations of static equilibrium, we must turn to another source for additional equations. The deflected shape of the structure becomes this source of additional equations. This can be accomplished since we know specific values for the deflection and slope at various points along the span. For example, we know that if the supports are unyielding, i.e., no settlement, the vertical deflection at a hinge, roller, or fixed support must be zero. We also know that the slope of the elastic curve at a fixed support must be zero. In Chapter Seven, several methods of obtaining deflections and slopes were presented. These methods will become the tools used in this chapter to obtain the additional equations required to analyze statically indeterminate structures. This is another way of saying that we are making the deflected shape of the structure **compatible with the constraints.** Thus, the analysis of statically indeterminate structures makes use of both equilibrium equations and **compatibility equations,** i.e., equations obtained by making the deformations consistent with the constraints.

8.3 Moment–Area Method

The application of the moment–area theorems does not yield direct results for displacements or slopes along the span length. The theorems do, however, provide the analyst with a means of obtaining values for tangential deviations and changes in slope along the span. For a wide variety of beams, tangential deviations along the span, and the change in slope between points along the elastic curve are known and, hence, the moment–area theorems can provide equations to supplement the equations of equilibrium. The following illustrative examples are intended to clarify the application of the method.

Example 8.1

Determine the reactions for the beam shown in Figure 8.2a. Both E and I are constant along the span.

Solution:

1. The first step performed when determining the reactions of any type of structure is to sketch a proper FBD as shown in Figure 8.2b. A proper FBD is essential since it pinpoints the exact number and type of reaction components to be determined. Examination of the FBD reveals that there are four unknown reactions (R_A, R_B, M_A, and M_B). As before, the sense of the unknowns is assumed as shown. There are only two equations of equilibrium available ($\Sigma M = 0$ and $\Sigma F_Y = 0$) for the determination of the unknown reactions. Recall that the third remaining equation of equilibrium, $\Sigma F_X = 0$, is not applicable in this case since there are no loads in the X direction. There are, therefore, four unknowns and only two available equations. Since two additional equations are required, the structure is said to be **statically indeterminate to the second degree.**
2. We must turn to the moment–area theorems for the two additional equations. Figure 8.2c shows the exaggerated shape of the elastic curve which must be examined for geometrical relationships that can be expressed mathematically by the moment–area theorems. Notice that the slope of the elastic curve at both A and B is zero and, hence, the angle between the tangents drawn to the elastic curve at points A and B is zero, or $\Theta_{AB} = 0$. Expressed in equation form, we may have:

$$\Theta_A + \Theta_{AB} = \Theta_B$$

Since

$$\Theta_A = \Theta_B = 0$$

$$\Theta_{AB} = 0$$

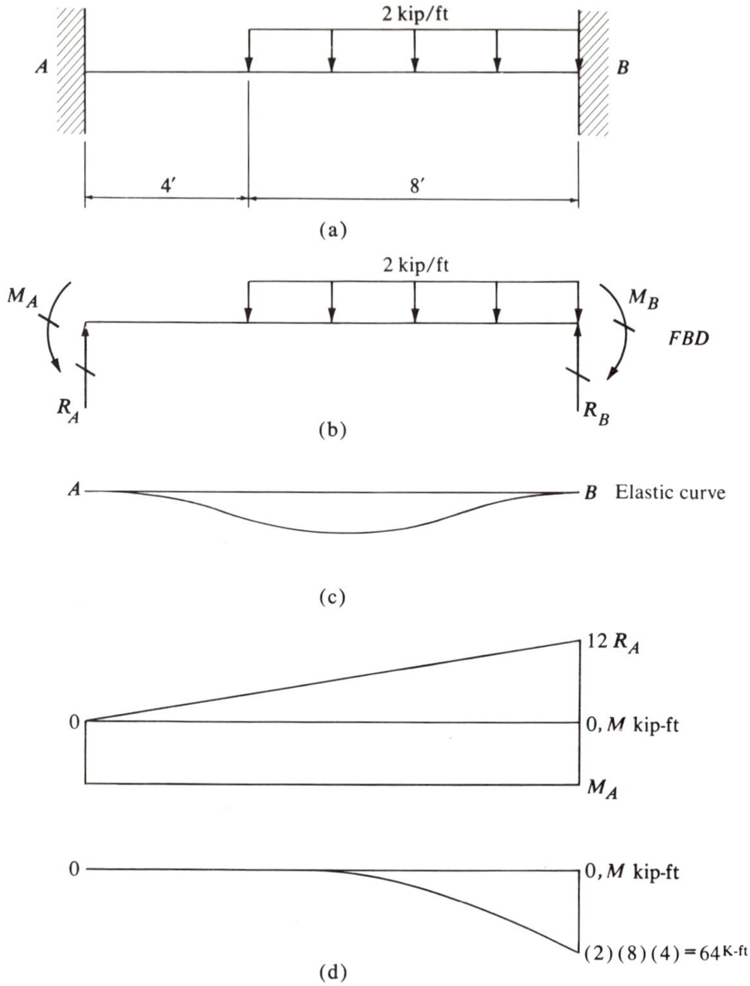

Figure 8.2

The change in slope, therefore, between A and B is zero. Applying the first moment–area theorem, the condition that $\Theta_{AB} = 0$ may be expressed by equating the area under the moment diagram between A and B to zero. Also, examination of the elastic curve reveals that the vertical distance between a tangent drawn at A to point B on the elastic curve is zero, or $t_{BA} = 0$. By the second moment–area theorem, this geometric condition reveals the fact that the first moment of all the area beneath the moment diagram between A and B about an axis through point B is equal to zero. It should be noted that the condition that $t_{AB} = 0$ could also be used instead of the condition $t_{BA} = 0$. The following four equations can now be used to determine the four unknown reactions:

$$\Sigma F_Y = 0 \qquad EI\Theta_{AB} = 0$$
$$\Sigma M = 0 \qquad EIt_{BA} = 0 \qquad (\text{or } EIt_{AB} = 0)$$

3. In order to apply the two moment–area equations, a moment diagram by parts is drawn about point B, i.e., point B is the reference point. Point B is selected as the reference point for two reasons: (1) the moment areas encountered are easy to handle, and (2) the unknowns R_B and M_B do not appear in the moment diagrams. The resulting moment diagrams are shown in Figure 8.2d. Note that the partial moment diagrams contain unknowns but the shapes and relative ordinates are known.
4. Using the moment diagram by parts and the moment area equations, we obtain:

$$EI\Theta_{AB} = 0 = (\tfrac{1}{2})(12)(12R_A) - 12(M_A) - (\tfrac{1}{3})(8)(64)$$

$$108R_A - 18M_A = 256 \qquad \textbf{(8–1)}$$

$$EIt_{BA} = 0 = (\tfrac{1}{2})(12)(12R_A)(\tfrac{1}{3})(12) - (12)(M_A)(6) - (\tfrac{1}{3})(8)(64)(\tfrac{1}{4})(8)$$

$$108R_A - 27M_A = 128 \qquad \textbf{(8–2)}$$

Solving Eqs. (8–1) and (8–2) simultaneously, the following results are obtained:

$$M_A = +14.23 \text{ kip-ft} = \underline{14.23 \text{ kip-ft}} \circlearrowleft$$

$$R_A = +4.74 \text{ kips} = \underline{4.74 \text{ kips} \uparrow}$$

The positive results obtained for both M_A and R_A indicate that their directions were correctly assumed on the FBD shown in Figure 8.2b. Continuing with the solution, the two equations of equilibrium may now be applied to the FBD shown in Figure 8.2b to obtain the remaining unknown reactions, M_B and R_B. Thus:

$$\Sigma F_Y = 0 \uparrow +$$

$$R_A - (2)(8) + R_B = 0$$

$$R_B = 16 - 4.74 = +11.26 \text{ kips}$$

$$\underline{R_B = 11.26 \text{ kips} \uparrow}$$

$$\Sigma M_B = 0 \quad (+\circlearrowright)$$

$$R_A(12) - M_A - 2(8)(4) + M_B = 0$$

$$M_B = 64 + 14.23 - 56.9 = +21.33 \text{ kip-ft}$$

$$\underline{M_B = 21.33 \text{ kip-ft}} \circlearrowright$$

5. As a check on the computations, the moment at B will be checked by summing the ordinates of the moment diagram by parts at point B. Referring to Figure 8.2d, this yields:

$$12R_A - M_A - 64 = M_B$$
$$56.9 - 14.23 - 64 = M_B$$
$$M_B = -21.33 \text{ kip-ft}$$
$$\therefore M_B = 21.33 \text{ kip-ft} \ (\text{clockwise})$$

The negative sign indicates that M_B tends to bend the member concave downward and, hence, M_B must have a clockwise sense.

Example 8.2

Determine the reactions of the beam shown in Figure 8.3a. Both E and I are constant along the span.

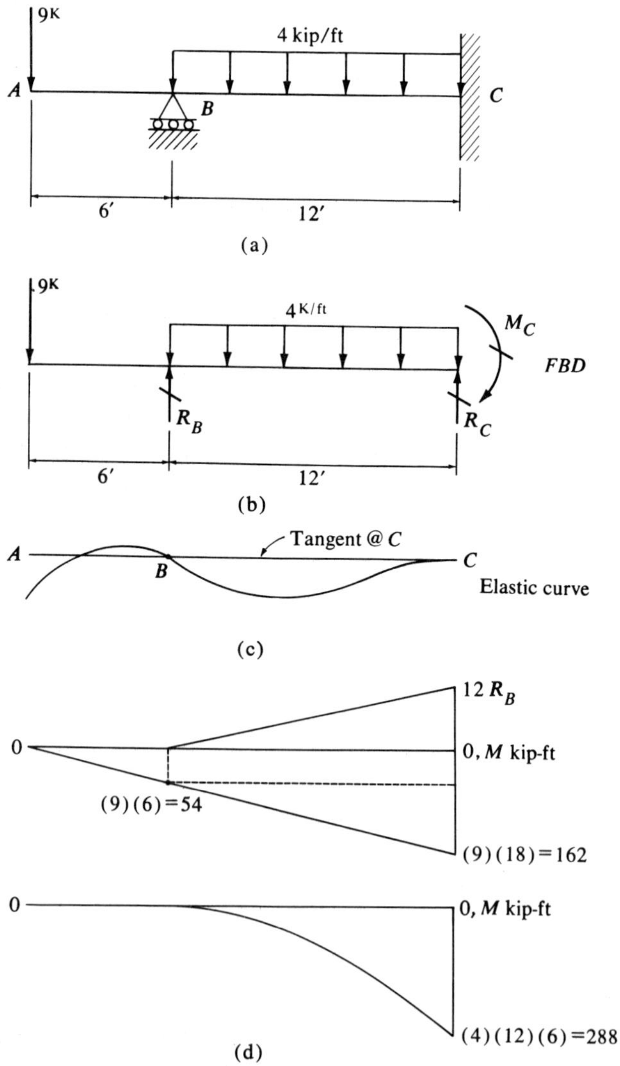

Figure 8.3

Solution:

1. The FBD sketched in Figure 8.3b reveals that there are three unknown reactions (R_B, R_C, and M_C) and only two available equations of equilibrium ($\Sigma F_Y = 0$ and $\Sigma M = 0$). Since there is one more unknown than equations available for solution, the beam is said to be **statically indeterminate to the first degree.**
2. An additional equation may be obtained by again applying the moment–area method. To successfully obtain equations by applying the moment–area theorems, we must know either something about the amount of change in slope between two points on the elastic curve or the exact vertical distance between a point on the elastic curve and a tangent drawn at another point. Examination of the much exaggerated elastic curve shown in Figure 8.3c shows that the tangent to the elastic curve at point C is horizontal, i.e., the slope of the elastic curve at C is zero. In order to obtain an additional equation by applying the first moment–area theorem, the slope must be known at least two points along the span. Since only the slope at C is known, the first moment–area theorem cannot provide an additional equation. We must now examine the deflected shape for specific knowledge about at least one tangential deviation which will enable us to apply the second moment–area theorem for the additional needed equation. The horizontal tangent drawn at point C passes through point B on the elastic curve and, thus, t_{BC} is equal to zero. Using the moment diagram by parts shown in Figure 8.3d and applying the second moment–area theorem, the following mathematical relationship is obtained:

$$EIt_{BC} = (\tfrac{1}{2})(12)(12R_B)(\tfrac{2}{3})(12) - (54)(12)(6) - (\tfrac{1}{2})(12)(162 - 54)(\tfrac{2}{3})(12) - (\tfrac{1}{3})(12)(288)(\tfrac{3}{4})(12)$$

$$0 = 2R_B - 9 - 18 - 36$$

$$\underline{R_B = +31.5 \text{ kips} = 31.5 \text{ kips} \uparrow}$$

The positive result for R_B indicates that the direction of R_B was correctly assumed on the FBD shown in Figure 8.3b. It should also be noted that drawing the moment diagram by parts about point B eliminated all unknowns from the moment plots with the exception of R_B. Thus, R_B was obtained directly and no simultaneous equations were encountered.

3. The remaining reactions may now be determined by applying the equations of equilibrium to the FBD shown in Figure 8.3b. Thus:

$$\Sigma F_Y = 0 \uparrow +$$

$$-9 + R_B - (4)(12) + R_C = 0$$

$$R_C = 57 - 31.5 = +25.5 \text{ kips}$$

$$\underline{R_C = 25.5 \text{ kips} \uparrow}$$

$$\Sigma M_C = 0 \quad (+\curvearrowright)$$

$$-(9)(18) + (31.5)(12) - (4)(12)(6) + M_C = 0$$
$$M_C = 162 + 288 - 378$$
$$\underline{M_C = +72.0 \text{ kip-ft} = 72 \text{ kip-ft}} \quad \curvearrowright$$

8.4 Theorem of Three Moments

A limitation of the moment–area method presented in the previous section is that at least one of the beam's supports must be **built-in,** i.e., fixed or prevented from rotating. This establishes a horizontal tangent to the elastic curve at the fixed support, thus insuring the knowledge that the tangential deviations from this tangent at the remaining supports are zero. Thus, the analyst is provided with the additional equations necessary for the solution of the reactions. Many times, however, the analyst will encounter the widely used continuous beam, i.e., a beam whose length is continuous across several interior supports such as that beam shown in Figure 8.4a. In the case of the continuous beam, one support is usually treated as a hinge support while the remaining supports are taken as roller supports. The one hinge support provides sufficient constraint for the beam in the horizontal direction should a horizontal load be applied to the structure either intentionally or as the result of a vertical load accidentally being applied in an inclined fashion. For this arrangement of supports, every support has only one unknown reaction, i.e., the magnitude of the vertical reaction. Hence, the following general rule for continuous beams may be stated: **the degree of statical indeterminacy for a continuous beam with only simple supports is numerically equal to the number of interior supports.** Simple supports, as used in this general rule, limit the supports to all roller types with the exception that one must be a hinge-type support. The FBD shown in Figure 8.4b shows four unknown reactions (R_A, R_B, R_C, and R_D). There are two interior supports, a roller at B and a roller at C. Hence, by the general rule for continuous beams, the beam shown in Figure 8.4a is statically indeterminate to the second degree. It should be noted that the exterior support such as at support D in Figure 8.4a, could have been a fixed support. If one, or both, of the exterior supports is fixed, then the previously stated rule for the degree of statical indeterminacy is not applicable. The analyst may then determine the degree of statical indeterminacy by the general method of subtracting the number of available equations of equilibrium from the number of reaction unknowns.

The continuous beam shown in Figure 8.4a is statically indeterminate to the second degree. In addition to the two available equations of equilibrium, $\Sigma F_Y = 0$ and $\Sigma M = 0$, two additional equations are required for the solution of the four unknown reactions. Since there are no fixed supports, there are no horizontal tangents to the elastic curve, and, therefore, nothing is known about the slope or tangential deviation at any point along the span. Using the method

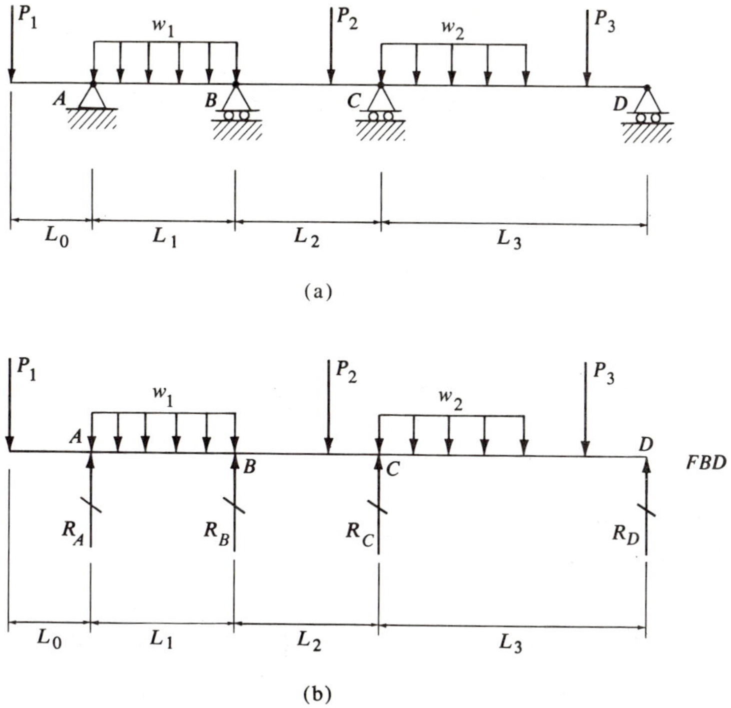

Figure 8.4

of area moments and the principle of superposition, a general method of solution for continuous beams may be developed. The elastic curve (greatly exaggerated) is sketched in Figure 8.5b. Next, a tangent to the elastic curve is drawn at support B. The angles formed by the tangent at B and the horizontal line AD are equal in magnitude. The triangles $BA'A$ and $BC'C$ are, therefore, similar triangles and the following relationship for similar triangles may be written:

$$\frac{t_{AB}}{L_1} = -\frac{t_{CB}}{L_2} \tag{8-3}$$

The negative sign on the right-hand side of the equation may be explained by referring to Figure 8.5b. By the sign convention adopted in Chapter 7, t_{AB} would be a positive value since point A lies above the tangent drawn at B. On the other hand, t_{CB} would be a negative value since point C lies below the tangent drawn at B. To insure the equality written above, a negative sign on the right side is required since a negative magnitude would be obtained for t_{CB}. Thus, positive values will be obtained on both sides of the equation.

Another tangent could be drawn at point C in Figure 8.5b and, proceeding in a similar fashion, the following would be obtained:

$$\frac{t_{BC}}{L_2} = -\frac{t_{DC}}{L_3} \tag{8-4}$$

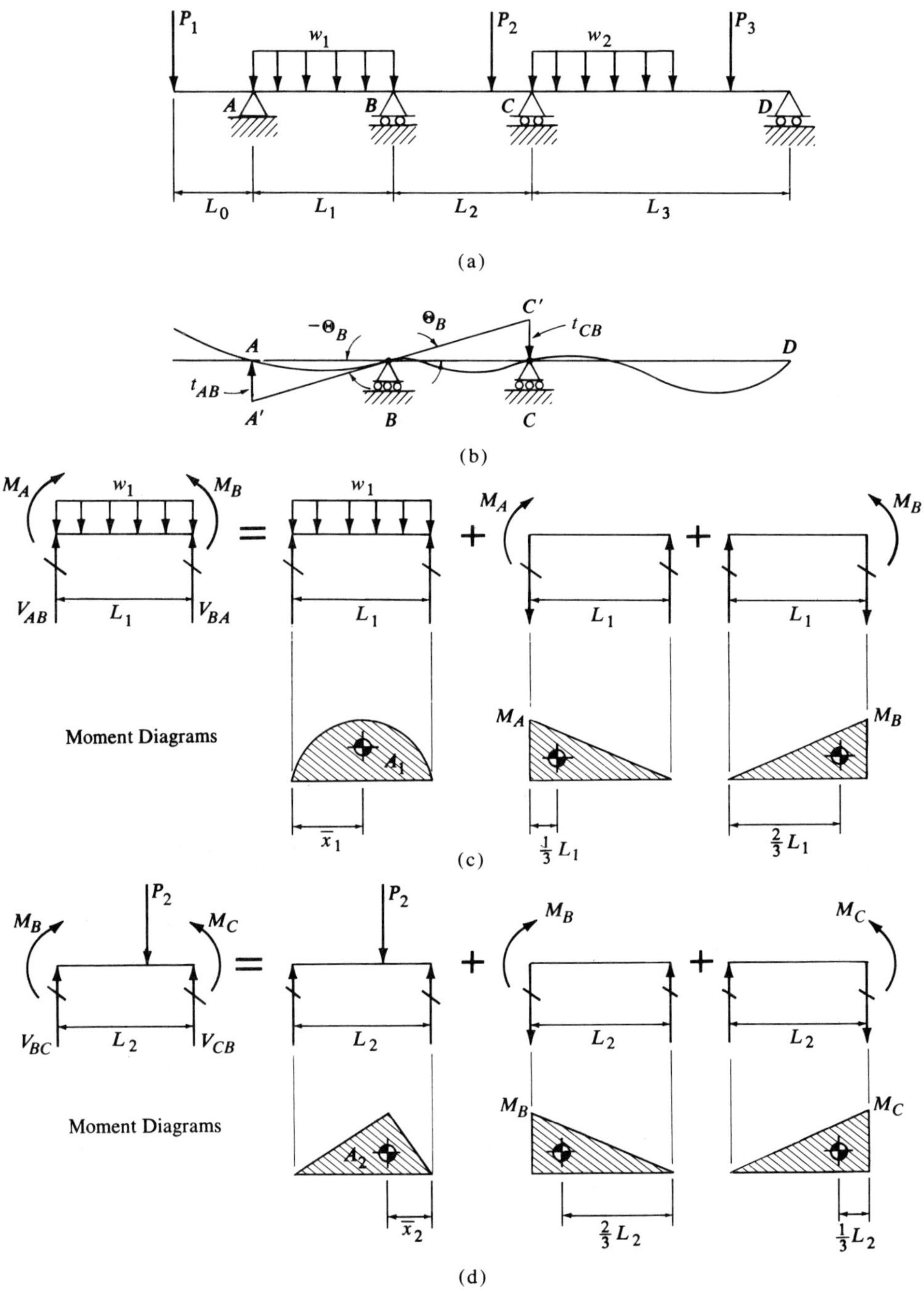
P1
w1
P2
w2
P3
A
B
C
D
L0
L1
L2
L3
(a)
C'
−ΘB
ΘB
tCB
A
D
tAB
A'
B
C
(b)
MA
w1
MB
VAB
L1
VBA
Moment Diagrams
A1
x̄1
1/3 L1
2/3 L1
(c)
P2
MB
MC
VBC
L2
VCB
Moment Diagrams
A2
x̄2
2/3 L2
1/3 L2
(d)

Figure 8.5

The expressions in Eqs. (8–3) and (8–4) provide the two additional equations necessary for the solution of the unknown reactions. It should be emphasized that these expressions can be written only at the interior supports since these expressions are only really another way of stating that the magnitude of the slope of the elastic curve is equal on both sides of the interior support, i.e., angle $C'BC$ is numerically equal to angle $A'BA$ in Figure 8.5b. The problem now becomes one of solving for the tangential deviations. Solving first for t_{AB} and t_{CB}, spans L_1 and L_2 are isolated by cutting the member immediately adjacent to the supports as shown in Figures 8.5c and 8.5d. In the FBDs for spans L_1 and L_2, the moments M_A, M_B, and M_C were assumed to be positive, i.e., they tend to bend the segments in a positive fashion (concave upward). Each reaction for span L_1 in Figure 8.5c is represented by a two-letter subscript since these reactions are actually the shears just to the right and left of supports A and B respectively. For example, V_{AB} indicates the shear at support A on the side toward support B and V_{BA} identifies the shear at support B on the side toward support A. It should be noted that these shears should not be confused with the unknown reactions at both supports A and B. The relationship between the shears adjacent to the supports and the reactions at the supports can best be seen by removing a small segment of beam with cuts taken immediately to the left and right of, say, the support at B as shown in Figure 8.6. Using the notations of Figure 8.6:

$$\Sigma F_Y = 0 \uparrow +$$
$$R_B - V_{BA} - V_{BC} = 0$$
$$R_B = V_{BA} + V_{BC}$$

Referring again to the FBDs shown in Figures 8.5c and 8.5d, the applied loads and reactions for spans L_1 and L_2 are replaced by three simply supported beams by the principle of superposition. The first of the equivalent three simple beams includes **all** the applied loads while the remaining two show the effect of the moments acting at the ends of the span. Also shown are the corresponding complete moment diagrams for each of the simple beams. Using the notations of Figure 8.5c and the second moment–area theorem, we obtain,

$$EI_1 t_{AB} = A_1\bar{x}_1 + (1/2)M_A L_1(L_1/3) + (1/2)M_B L_1(2L_1/3)$$

where A_1 denotes the area under the moment diagram for span L_1 considered as a simply supported beam under the action of the actual loads on this span and x_1 is the centroidal distance of area A_1 from the left reaction of span L_1, and E is the modulus of elasticity. I_1 is the moment of inertia for span L_1. Treating span L_2 in a similar manner and using the notations of Figure 8.5d, the second moment-area theorem yields:

$$EI_2 t_{CB} = A_2\bar{x}_2 + (1/2)M_B L_2(2L_2/3) + (1/2)M_A L_2(L_2/3)$$

where A_2 denotes the area under the moment diagram for span L_2 considered as a simply supported beam under the action of the actual applied loads on this span and x_2 is the centroidal distance of area A_2 from the right support of span L_2, I_2 is the moment of inertia for span L_2. Rewriting Eq. (8–3):

$$\frac{t_{AB}}{L_1} = -\frac{t_{CB}}{L_2}$$

Substitution of the expressions for t_{AB} and t_{CB} in the above yields:

$$M_A L_1 + 2M_B\left(L_1 + L_2\frac{I_1}{I_2}\right) + M_C L_2\frac{I_1}{I_2} = -\frac{6A_1\bar{x}_1}{L_1} - \frac{6A_2\bar{x}_2}{L_2}\frac{I_1}{I_2} \qquad \textbf{(8–5)}$$

Equation (8–5) is based upon the assumption that the modulus of elasticity, E, is constant throughout the entire beam, i.e., the beam is fabricated entirely from the same material. Also, the moment of inertia is constant along each span but may differ from span to span. When the moment of inertia is constant for both spans, Eq. (8–5) reduces to

$$M_A L_1 + 2M_B(L_1 + L_2) + M_C L_2 = -\frac{6A_1\bar{x}_1}{L_1} - \frac{6A_2\bar{x}_2}{L_2} \qquad \textbf{(8–6)}$$

Equation (8–6) is known as the **theorem of three moments.** Equation (8–6) was obtained by the continuity condition that:

angle $A'BA$ = angle $C'BC$ (see Figure 8.5b)

A similar continuity condition may also be utilized at support C and a similar **equation of three moments** may be written following the format of Eq. (8–6). This is really a simplified way of writing Eq. (8–4). Since an equation of three moments can be written at each interior support and since the degree of statical indeterminacy is numerically equal to the number of interior supports for a simply supported continuous beam, it follows that a sufficient number of three-moment equations can be written to supplement the equilibrium equations and, hence, render a solution for the reactions. Notice that the three moment equations contain only moments as unknowns. Referring again to Figure 8.5a, the moments at the exterior supports A and D are known, while the moments at the interior supports B and C are not known. Since an equation of three moments can be written for each interior support, the analyst will have at his disposal two three-moment equations with two unknowns, M_B and M_C. It should be remembered that the number of three-moment equations available always equals the number of unknown bending moments, i.e., the analyst can always write a sufficient number of three-moment equations to directly obtain the bending moment at the interior supports. Knowing the moments at the supports, the shears at the supports may be determined from the FBDs as shown in Figures 8.5c and 8.5d, and, lastly, the external reactions may be obtained using the relationship demonstrated in Figure 8.6.

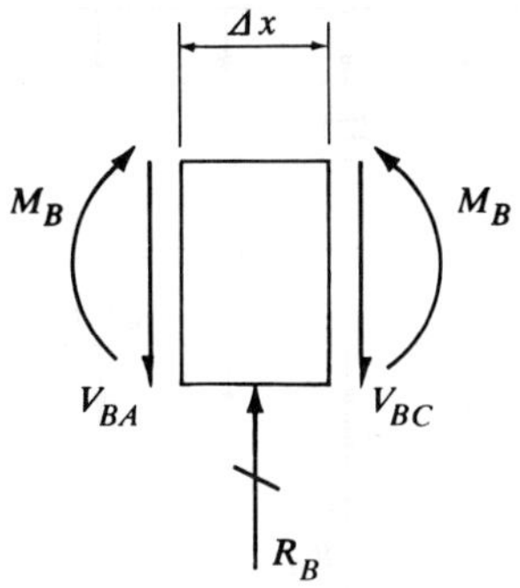

Figure 8.6

Before proceeding to some illustrative examples, it should be noted that the support moments used in the derivations were assumed positive, i.e., tend to bend the member concave upward. A negative result obtained for a support moment simply indicates a reverse condition.

Example 8.3

Determine the reactions for the continuous beam shown in Figure 8.7a. Both E and I are constant along the beam's entire length.

Solution:

1. The FBD shown in Figure 8.7b shows that there are three unknown reactions. Since there are only two available equations of equilibrium ($\Sigma F_Y = 0$ and $\Sigma M = 0$), the beam is statically indeterminate to the first degree. Also, there is one interior support and, hence, the degree of statical indeterminacy is one.
2. The additional equation required can be provided by applying the theorem of three moments at the interior support. Thus, from Eq. (8–6):

$$M_A L_1 + 2M_B(L_1 + L_2) + M_C L_2 = -\frac{6A_1\bar{x}_1}{L_1} - \frac{6A_2\bar{x}_2}{L_2}$$

The moment at both supports A and C is zero, and, hence, we have:

$$2M_B(L_1 + L_2) = -\frac{6A_1\bar{x}_1}{L_1} - \frac{6A_2\bar{x}_2}{L_2}$$

In order to find $A_1\bar{x}_1$ and $A_2\bar{x}_2$, the moment diagrams for the applied loads are constructed considering spans L_1 and L_2 as simply supported as shown in Figure 8.7c. From Figure 8.7c:

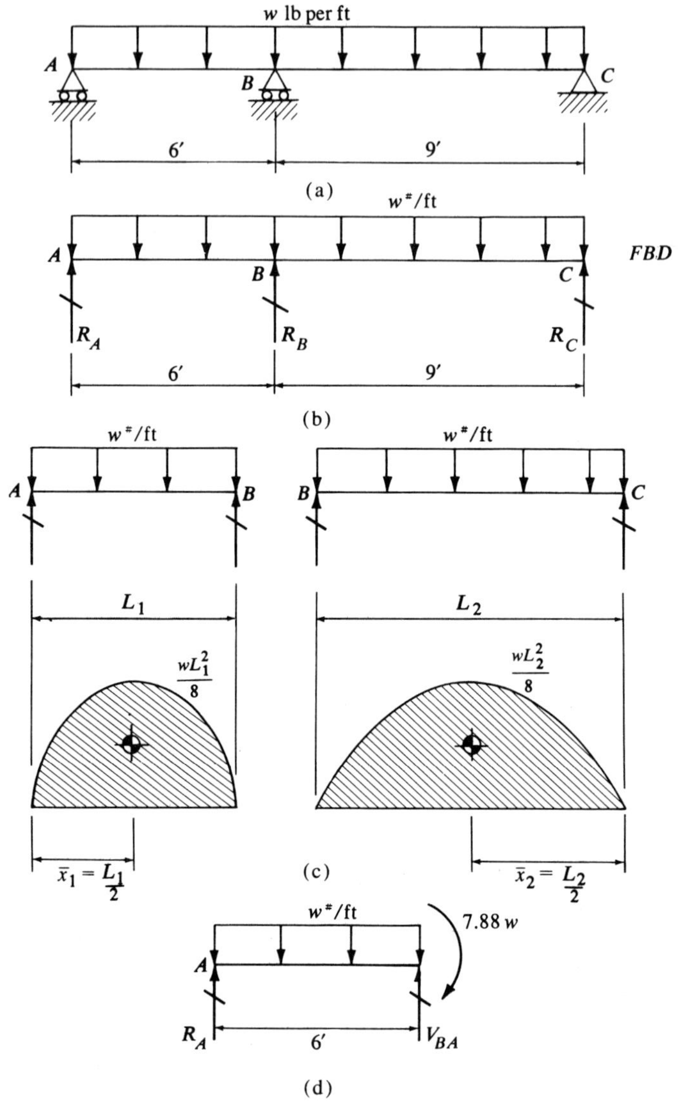

Figure 8.7

$$A_1\bar{x}_1 = \left[\left(\frac{2}{3}\right)\left(\frac{wL_1^2}{8}\right)(L_1)\right]\left[\frac{L_1}{2}\right] = \frac{wL_1^4}{24}$$

$$A_2\bar{x}_2 = \left[\left(\frac{2}{3}\right)\left(\frac{wL_2^2}{8}\right)(L_2)\right]\left[\frac{L_2}{2}\right] = \frac{wL_2^4}{24}$$

Substituting these values into the three-moment equation yields

$$2M_B(L_1 + L_2) = -\frac{wL_1^3}{4} - \frac{wL_2^3}{4}$$

Notice the special form of the right-hand side of the equation when uniformly distributed loads are encountered over each span. Reference will be made later for this special case. Continuing:

$$2M_B(6+9) = -\frac{w}{4}(6)^3 - \frac{w}{4}(9)^3$$

$$M_B = -\frac{w}{120}(216+729)$$

$$M_B = -7.875\ w \text{ lb-ft}$$

The negative sign indicates negative moment at B.

3. With M_B known, the reaction at A can be determined from the FBD of Figure 8.7d. Notice the direction of M_B since it was found to be negative. From Figure 8.7d:

$$\Sigma M_B = 0 \quad (\curvearrowright +)$$

$$6R_A + 7.88w - w(6)(\tfrac{6}{2}) = 0$$

$$\underline{R_A = 1.687w \uparrow}$$

With R_A now known, the FBD shown in Figure 8.7b contains only two unknowns. Applying the equations of equilibrium to the FBD in Figure 8.7b we obtain:

$$\Sigma M_C = 0 \quad (\curvearrowright +)$$

$$1.687w(15) + R_B(9) - w(15)(\tfrac{15}{2}) = 0$$

$$\underline{R_B = 9.69w \uparrow}$$

$$\Sigma M_A = 0 \quad (\curvearrowright +)$$

$$w(15)(\tfrac{15}{2}) - 6(9.69w) - 15R_C = 0$$

$$\underline{R_C = 3.66w \uparrow}$$

Example 8.4

Determine the reactions for the continuous beam shown in Figure 8.8a. E and I are constant along the beam's length.

Solution:

1. The FBD shown in Figure 8.8b indicates that there are four unknown reactions. Since only two equilibrium equations are applicable ($\Sigma F_Y = 0$ and $\Sigma M = 0$), the beam is statically indeterminate to the second degree. There is only one interior support, i.e., support B, and, hence, only one equation

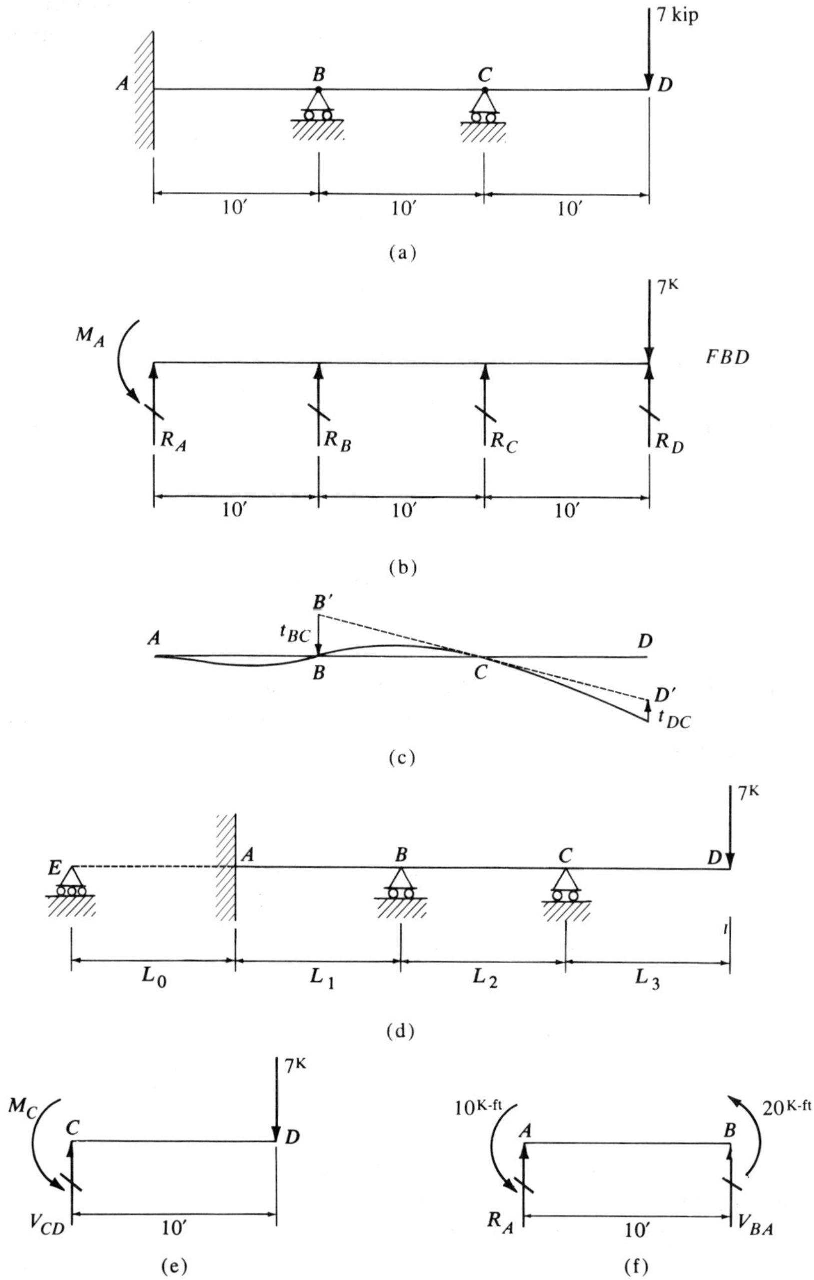

Figure 8.8

of three moments can be written. Recall that the general rule equating the degree of statical indeterminacy to the number of interior supports is not applicable in this instance since the support at A is fixed.

2. With only one equation of three moments available, we are still one equation short of being able to solve the problem. It has been the author's experience that many students erroneously attempt to write a three-moment equation for the support at C. Recall that the three-moment equation is simply a standard way of expressing a condition of continuity at a support, i.e., the slopes of the elastic curves at an interior support of a continuous beam must be equal. The slopes to the left and right of a common interior support were equated and, thus, the theorem of three moments was derived. For example, in Figure 8.8c:

$$\text{angle } B'CB = t_{BC}/\overline{BC}$$

However:

$$\text{angle } D'CD \neq t_{DC}/\overline{CD}$$

and hence, even so the relationship

$$\text{angle } B'CB = \text{angle } D'CD$$

is valid at support C, the expression

$$t_{BC}/\overline{BC} \neq t_{DC}/\overline{CD}$$

is a non-equality. This expression must be an identity as it is the basic form of the three-moment equation [see Eqs. (8–3) and (8–4)]. Since the slope at C cannot be represented by $t_{DC}/\overline{CD}$, the three-moment equation cannot be applied when one span, such as CD, in Figure 8.8a, is an overhang.

3. Still being an equation shy, let us turn our attention to Figure 8.8d. Due to the fixed support, the slope of the elastic curve at A is known to be zero. Let us further assume that the fixed support creates another span beyond it having a length of zero but having an infinitely large moment of inertia at all sections. These assumptions are valid since the length of the imaginary span is indeed zero and the fixed support is, in fact, infinitely stiff, i.e., no rotation permitted. In addition, the condition of continuity of slope is met since the slope of the elastic curve is zero considering either the A end of span AB or the A end of span AE. Also, t_{BA} and t_{EA} are both zero, and, therefore, the basic form of the three-moment equation is maintained [see Eqs. (8–3) and (8–4)]. Writing the equation of three moments for support A [in the form of Eq. (8–5) because of the different moments of inertia] we have:

$$M_E L_0 + 2M_A\left(L_0 + L_1\frac{I_0}{I_1}\right) + M_B L_2\frac{I_0}{I_1} = 0$$

where I_0 and I_1 are the moments of inertia of spans L_0 and L_1 respectively. The right side of the three-moment equation is zero since no applied loads are encountered on either span. Rewriting the above:

$$M_E L_0\left(\frac{I_1}{I_0}\right) + 2M_A\left(L_0\frac{I_1}{I_0} + L_1\right) + M_B L_1 = 0$$

Since I_0 is infinitely large:

$$M_E L_0\left(\frac{I_1}{\infty}\right) + 2M_A\left(L_0\frac{I_1}{\infty} + L_1\right) + M_B L_1 = 0$$

Noting that $I_1/\infty = 0$ and $M_E = 0$ in the above, we obtain:

$$2M_A L_1 + M_B L_1 = 0$$

The remaining additional equation has thus been established.

4. It has been shown that if an end support of a continuous beam is fixed, a special form of the three-moment equation can be utilized. In fact, the equation of three moments can be applied directly using the imaginary span introduced by the fixed support. For example, using Figure 8.8d:

$$M_E L_0 + 2M_A(L_0 + L_1) + M_B L_1 = 0$$

Since $L_0 = 0$ and $M_E = 0$

$$2M_A L_1 + M_B L_1 = 0 \tag{8–7}$$

The above equation verifies that obtained previously.

5. Writing the three-moment equation at support B:

$$M_A L_1 + 2M_B(L_1 + L_2) + M_C L_2 = 0$$

From Figure 8.8e, a cut taken to the right of C yields M_C.

$$\Sigma M_C = 0 \quad (\curvearrowright +)$$

$$7(10) - M_C = 0$$

$$M_C = 70 \text{ kip-ft} \circlearrowleft$$

By our sign convention, M_C is negative. Thus:

$$M_A(10) + 2M_B(10 + 10) + (-70)(10) = 0$$

$$M_A + 4M_B = 70 \tag{8–8}$$

From our earlier work:

$$2M_A(10) + M_B(10) = 0 \tag{8–7}$$

Solving Eqs. (8–7) and (8–8) simultaneously:

$$\underline{M_A = -10.0 \text{ kip-ft}} \qquad \underline{M_B = 20.0 \text{ kip-ft}}$$

Knowing the moments at all the supports, the reactions may now be determined. Referring to the FBD shown in Figure 8.8f:

$$\Sigma M_B = 0 \quad (\curvearrowright +)$$

$$R_A(10) - 10 - 20 = 0$$

$$\underline{R_A = 3 \text{ kips} \uparrow}$$

Note that span AB was isolated initially since R_A could be determined directly. If any other span had been selected, end shears would have been obtained, thus requiring additional calculations for the reactions. With R_A known, we can apply the equations of equilibrium to the FBD shown in Figure 8.8b:

$$\Sigma M_B = 0 \quad (\curvearrowright +)$$

$$-10 + (3)(10) - R_C(10) + 7(20) = 0$$

$$\underline{R_C = 16 \text{ kips} \uparrow}$$

$$\Sigma M_A = 0 \quad (\curvearrowright +)$$

$$-10 - 10R_B - 20(16) + 30(7) = 0$$

$$R_B = -12 \text{ kips}$$

$$\underline{R_B = 12 \text{ kips} \downarrow}$$

Example 8.5

Draw the complete shear and moment diagrams for the beam shown in Figure 8.9a. Both E and I are constant along the beam.

Solution:

1. To fulfill the requirements of this problem, i.e., draw the shear and moment diagrams, we must determine the reactions. The given beam is statically indeterminate to the second degree. With only two equilibrium equations available, two additional equations will be obtained by applying the equation of three moments at supports A and B. Because the support at A is

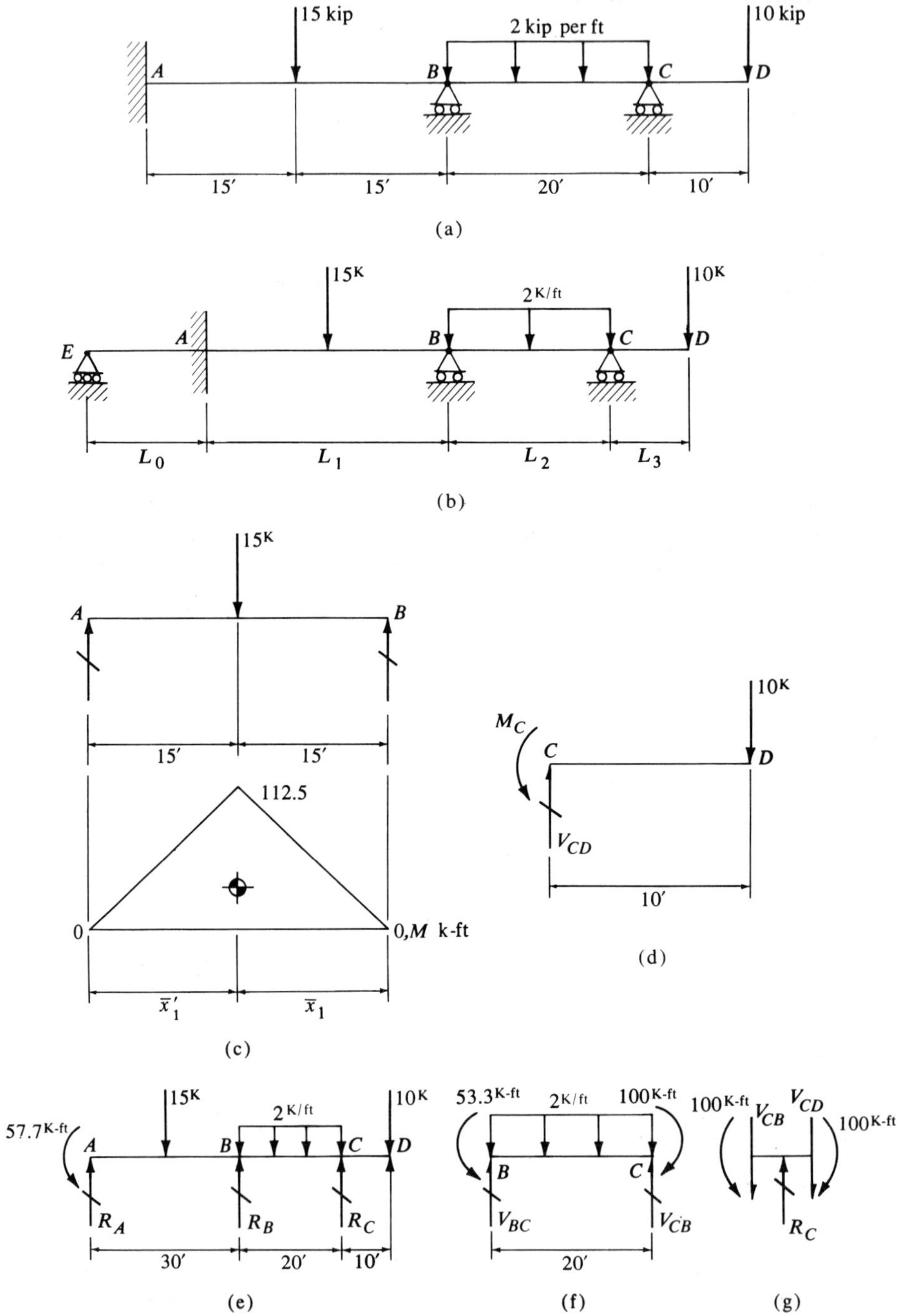
15 kip
2 kip per ft
10 kip
A
B
C
D
15′
15′
20′
10′
(a)
15K
2K/ft
10K
E
A
B
C
D
L0
L1
L2
L3
(b)
15K
A
B
15′
15′
112.5
0
0,M k-ft
x̄′1
x̄1
(c)
10K
MC
C
D
VCD
10′
(d)
57.7K-ft
15K
2K/ft
10K
A
B
C
D
RA
RB
RC
30′
20′
10′
(e)
53.3K-ft
2K/ft
100K-ft
B
C
VBC
VCB
20′
(f)
100K-ft
VCB
VCD
100K-ft
RC
(g)

Figure 8.9

fixed, an imaginary span L_0 is added as shown in Figure 8.9b. Starting at support A we have:

$$M_E L_0 + 2M_A(L_0 + L_1) + M_B L_1 = -\frac{6A_1\bar{x}_1}{L_1}$$

$$0 + 2M_A(0 + 30) + M_B(30) = -\frac{6A_1\bar{x}_1}{30}$$

$$60M_A + 30M_B = -\frac{A_1\bar{x}_1}{5}$$

In order to determine the $A_1\bar{x}_1$ term, span AB along with its loads is placed on simple supports and the corresponding moment diagram is drawn as shown in Figure 8.9c. Thus, we have from Figure 8.9c:

$$A_1 = (\tfrac{1}{2})(30)(112.5) = 1687.50 \text{ kip-ft}^2$$
$$\bar{x}_1 = 15 \text{ ft} \quad (\bar{x}_1 \text{ is measured from support } B)$$

Substituting and simplifying in the three-moment equation yields:

$$6M_A + 3M_B = -506.3 \tag{8-9}$$

Continuing at support B:

$$M_A L_1 + 2M_B(L_1 + L_2) + M_C L_2 = -\frac{6A_1\bar{x}_1}{L_1} - \frac{6A_2\bar{x}_2}{L_2}$$

$$M_A(30) + 2M_B(30 + 20) + (-100)20 = -\frac{6(1687.5)(15)}{30} - \frac{(2)(20)^3}{4}$$

where M_C was obtained from a cut taken just to right of C in Figure 8.9a and shown by the FBD in Figure 8.9d, A_1 and $\bar{x}_1'$ were obtained from the simple beam moment diagram shown in Figure 8.9c (in this case, $\bar{x}_1'$ is measured from support A), and the term containing A_2 takes the special form of $-wL^3/4$ for the case of a uniformly distributed load acting over the entire span. Simplifying this last three-moment equation we have:

$$3M_A + 10M_B = -706.3 \tag{8-10}$$

We now have two equations, Equations (8–9) and (8–10), in two unknowns, M_A and M_B. Solving these equations simultaneously we obtain:

$$\underline{M_A = -57.73 \text{ kip-ft}} \qquad \underline{M_B = -53.31 \text{ kip-ft}}$$

2. Knowing the moment at supports A and B, we may proceed to solve for the reactions. Recall that this continuous beam is statically indeterminate to the second degree, but knowing the moment at two points, point A and B

in this case, now makes a conventional solution possible. If we deal with a FBD of the entire beam as shown in Figure 8.9e, we encounter three unknown reactions, even though we have shown M_A as a known quantity. In order to utilize the fact that M_B is a known quantity, we may isolate span BC by taking cuts just to the right and left of supports B and C respectively as shown in Figure 8.9f. This segment is statically determinate since M_B is known from the three-moment equations and M_C is known from the FBD shown in Figure 8.9d. The problem now becomes one of determining the shear to the left and right of support C, V_{CD} in Figure 8.9d, and V_{CB} in Figure 8.9f. Knowing these shears, R_C may be determined by isolating an infinitesimally small segment of beam immediately above the reaction at C as shown in Figure 8.9g. The solution proceeds as follows:

Figure 8.9d:

$$\Sigma F_Y = 0 \uparrow +$$
$$V_{CD} - 10 = 0$$
$$V_{CD} = 10^K \uparrow \text{ on } CD$$
$$\therefore V_{CD} = 10^K \downarrow \text{ on segment shown in Figure 8.9g}$$

Figure 8.9f:

$$\Sigma M_B = 0 \circlearrowright +$$
$$-53.3 + 2(20)(10) + 100 - 20V_{CD} = 0$$
$$V_{CB} = 22.34^K \uparrow \text{ on } CB$$
$$\therefore V_{CB} = 22.34^K \downarrow \text{ on segment shown in Figure 8.9g}$$

Figure 8.9g:

$$\Sigma F_Y = 0 \uparrow +$$
$$R_C - 22.34 - 10 = 0$$
$$\underline{R_C = 32.34^K \uparrow}$$

With R_C known, the FBD in Figure 8.9e may be utilized as follows:

$$\Sigma M_A = 0 \circlearrowright +$$
$$-57.7 + 15(15) - 30R_B + 2(20)(40) - 50(32.34) + 10(60) = 0$$
$$\underline{R_B = 25.1^K \uparrow}$$

$$\Sigma F_Y = 0 \uparrow +$$
$$R_A - 15 + 25.1 - 2(20) + 32.34 - 10 = 0$$
$$\underline{R_A = 7.65^K \uparrow}$$

3. The shear and moment diagrams are shown in Figure 8.10.

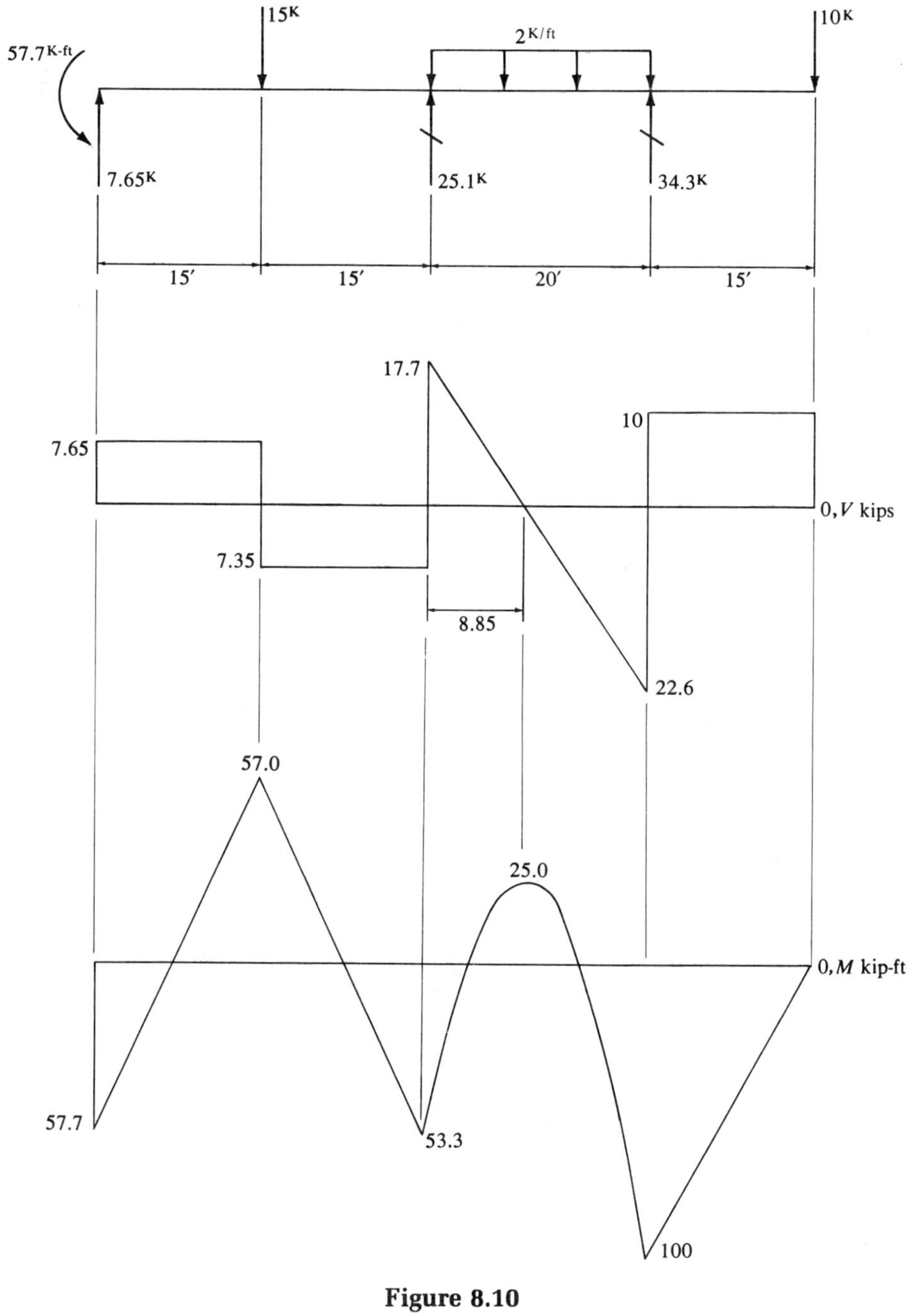

Figure 8.10

8.5 The Conjugate Beam Method

The conjugate beam method is well suited for the solution of statically indeterminate beams. Recall that in the conjugate beam method, an analogy is drawn between the real beam and a fictitious conjugate beam. In the analogy, the *M/EI* diagram for the real beam becomes the loading diagram for the conjugate beam while the supports of the conjugate beam are taken to insure conformity to the slope and deflection introduced by their counterparts in the real beam. The conjugate beam is identical to the actual beam with regard to the length of the beam. Thus, with this analogy and with properly prescribed support conditions for the conjugate beam, the following results may be obtained: (1) the slope at a given section of the real beam equals the shear at the corresponding section of the conjugate beam; and (2) the deflection at a given section of the real beam equals the bending moment at the corresponding section of the conjugate beam. The reader may refer to the previous chapter where a complete explanation of the method is given.

For the case of statically indeterminate beams, the conjugate beam method takes on a very unique form. For example, let us turn our attention to the real statically indeterminate beams, along with their conjugate beam counterparts, shown in Figure 8.11. The relationship between the real and conjugate support systems can be briefly summarized as follows: (1) A fixed support on the real beam permits no rotation and no deflection, and, hence, its counterpart on the conjugate beam must allow no shear and no moment to occur. Thus, a real fixed support is represented by a free end on the conjugate beam. (2) At a simple support on a real beam rotation can occur but no deflection is permitted. The corresponding conjugate beam support must, therefore, permit only shear and no moment. A real simple support has a simple support as its conjugate counterpart. (3) At an interior support of a real beam there is rotation with no deflection, and therefore, its corresponding conjugate support must allow shear to exist without bending moment. A real interior support becomes, therefore, an interior hinge on the conjugate beam. A re-examination of the conjugate beams shown in Figure 8.11 should reveal this important and helpful fact: **a conjugate beam is never statically indeterminate, although it may be unstable in itself.** For instance, all the conjugate beams in Figure 8.11 are unstable beams. Once loaded, however, the conjugate beams will be maintained in a condition of unstable equilibrium. The conjugate beam shown in Figure 8.11a,

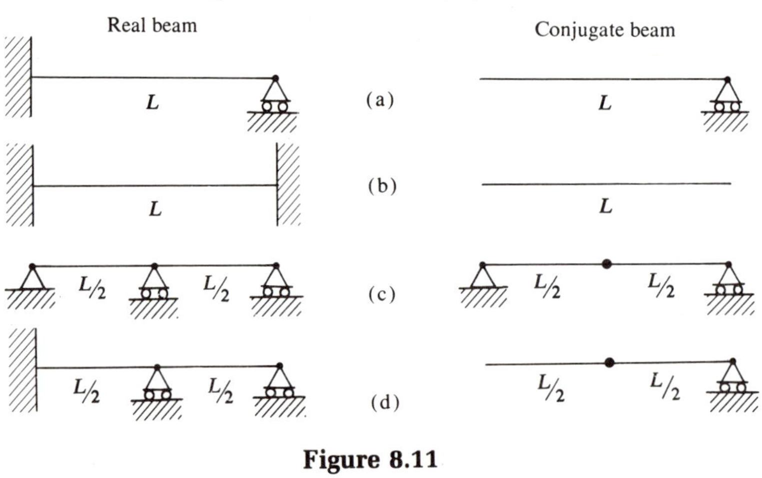

Figure 8.11

for example, appears to be completely unstable. Once loaded, however, with the M/EI diagram of the real beam, these loads, along with the conjugate beam reaction at the right end, will maintain the required equilibrium. Since the real beam is statically indeterminate, the magnitude of the moment diagram ordinates will be unknown but the shape of the moment diagram will be known. With the shape of the loading diagram acting on the conjugate beam known, i.e., the same shape as the moment diagram of the real beam, and because the conjugate beam is always statically determinate, application of the conventional equations of equilibrium to the conjugate beam will provide a solution for the loading ordinates on the conjugate beam and thus a solution for the real moment ordinates which can provide the reactions for the real beam. Therein lies the application of the conjugate beam method to statically indeterminate beams. The following illustrative examples are presented to further clarify the application of the method.

Example 8.6

Determine the reactions for the beam shown in Figure 8.12a. E and I are both constant along the span.

Solution:

1. Because the load acting on the conjugate beam is identical to the M/EI for the real beam, a moment diagram for the real beam is required initially. The FBD shown in Figure 8.12b shows that there are four unknown reactions, thereby making the given beam statically indeterminate to the second degree. A moment diagram by parts about point B, as shown in Figure 8.12c, is selected because all the moment shapes encountered are such that their properties, i.e., areas and centroids, are well known. Selection of the type of moment diagram is important since simple moment diagram shapes will decrease the complexity of the load diagrams acting on the conjugate beam. Generally, a moment diagram by parts is well adapted for the conjugate beam method.
2. The fixed supports of the real beam allow neither rotation nor deflection to occur in the real beam. The conjugate beam, therefore, must permit no shear or moment to occur at the corresponding support points. Thus, the real beam's fixed supports become free ends on the conjugate beam. The loaded conjugate beam is shown in Figure 8.12d. Positive bending moment on the real beam is downward loading on the conjugate beam, and negative bending moment on the real beam corresponds to upward loading on the conjugate beam. On the FBD shown in Figure 8.12b, the senses, or directions, of the reactions were assumed. The moment diagram and, thus, the conjugate beam loading must be consistent with these initial assumptions as shown in Figures 8.12c and 8.12d. An incorrect assumption will show up as a negative magnitude for that particular reaction. As before, no adverse results will be obtained if an incorrect sense is assumed for a reaction.
3. The equations of static equilibrium can be applied to the statically determinate conjugate beam shown in Figure 8.12d. It should be noted that the

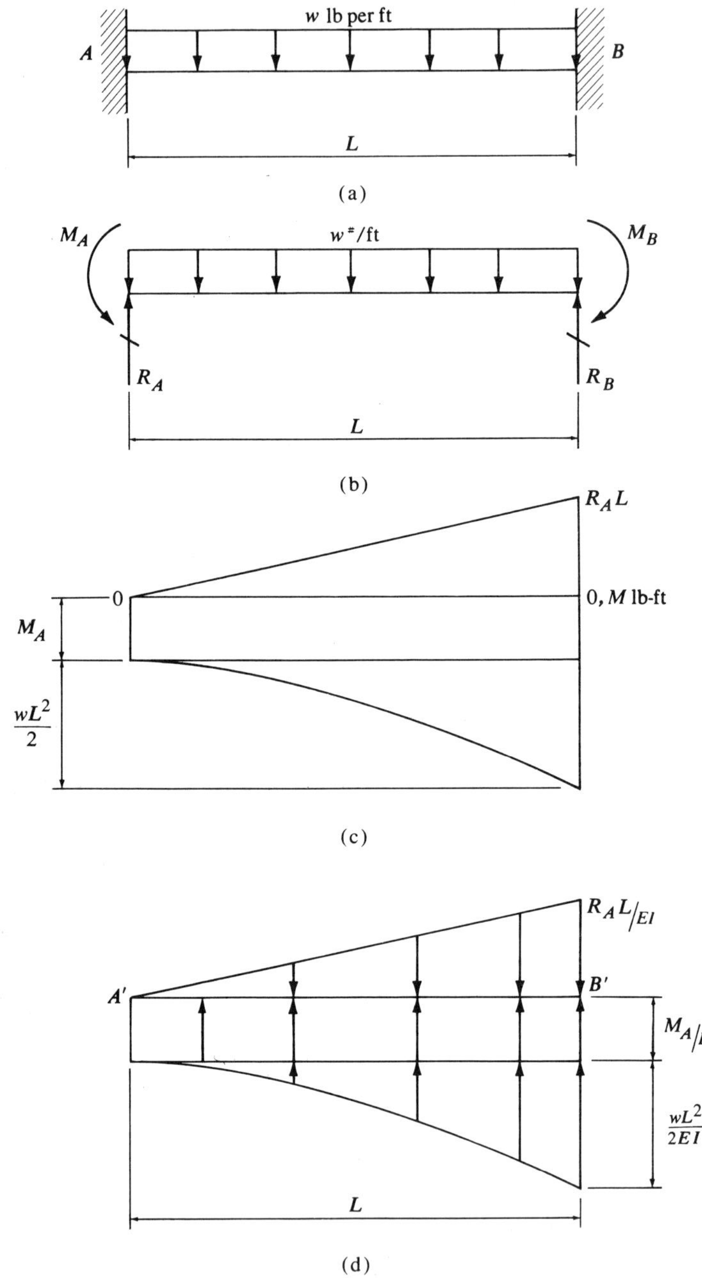

Figure 8.12

applied loads alone keep the conjugate beam in equilibrium. Also, there are two equilibrium equations available ($\Sigma F_Y = 0$ and $\Sigma M = 0$) and only two unknown reactions (R_A and M_A) involved in the loading diagrams. Referring to Figure 8.12d, the solution may be accomplished as follows:

$$\Sigma F_Y = 0 \uparrow +$$

$$\left(\frac{1}{2}\right)\left(\frac{R_A L}{EI}\right)L - \left(\frac{M_A}{EI}\right)L - \left(\frac{1}{3}\right)\left(\frac{wL^2}{2EI}\right)L = 0$$

$$R_A L - 2M_A - \frac{wL^2}{3} = 0 \qquad (8\text{–}11)$$

$$\Sigma M_{B'} = 0 \ \circlearrowright +$$

$$-\left(\frac{1}{2}\right)\left(\frac{R_A L}{EI}\right)(L)\frac{L}{3} + \left(\frac{M_A}{EI}\right)(L)\frac{L}{2} + \left(\frac{1}{3}\right)\left(\frac{wL^2}{2EI}\right)(L)\frac{L}{4} = 0$$

$$R_A L - 3M_A - \frac{wL^2}{4} = 0 \qquad (8\text{–}12)$$

Solving Eqs. (8.11) and (8.12) simultaneously:

$$\underline{M_A = +\frac{wL^2}{12}\ \text{lb-ft}} \qquad \underline{R_A = +\frac{wL}{2}\ \text{lb}}$$

The positive results indicate that the directions assumed on the FBD in Figure 8.12b are correct. With R_A and M_A known, we can refer to the FBD in Figure 8.12b for the remainder of the solution.

$$\Sigma F_Y = 0 \uparrow +$$

$$\frac{wL}{2} + R_B - wL = 0$$

$$R_B = +\frac{wL}{2}$$

$$\underline{R_B = \frac{wL}{2}\ \text{lb} \uparrow}$$

$$\Sigma M_A = 0 \ \circlearrowright +$$

$$-\frac{wL}{12} + w(L)\left(\frac{L}{2}\right) - \left(\frac{wL}{2}\right)(L) + M_B = 0$$

$$\underline{M_B = +\frac{wL}{12}\ \text{lb-ft} = \frac{wL}{12}\ \text{lb-ft} \circlearrowright}$$

Example 8.7

Determine the reactions for the two-span, continuous beam shown in Figure 8.13a. The beam is of one material, i.e., E is constant.

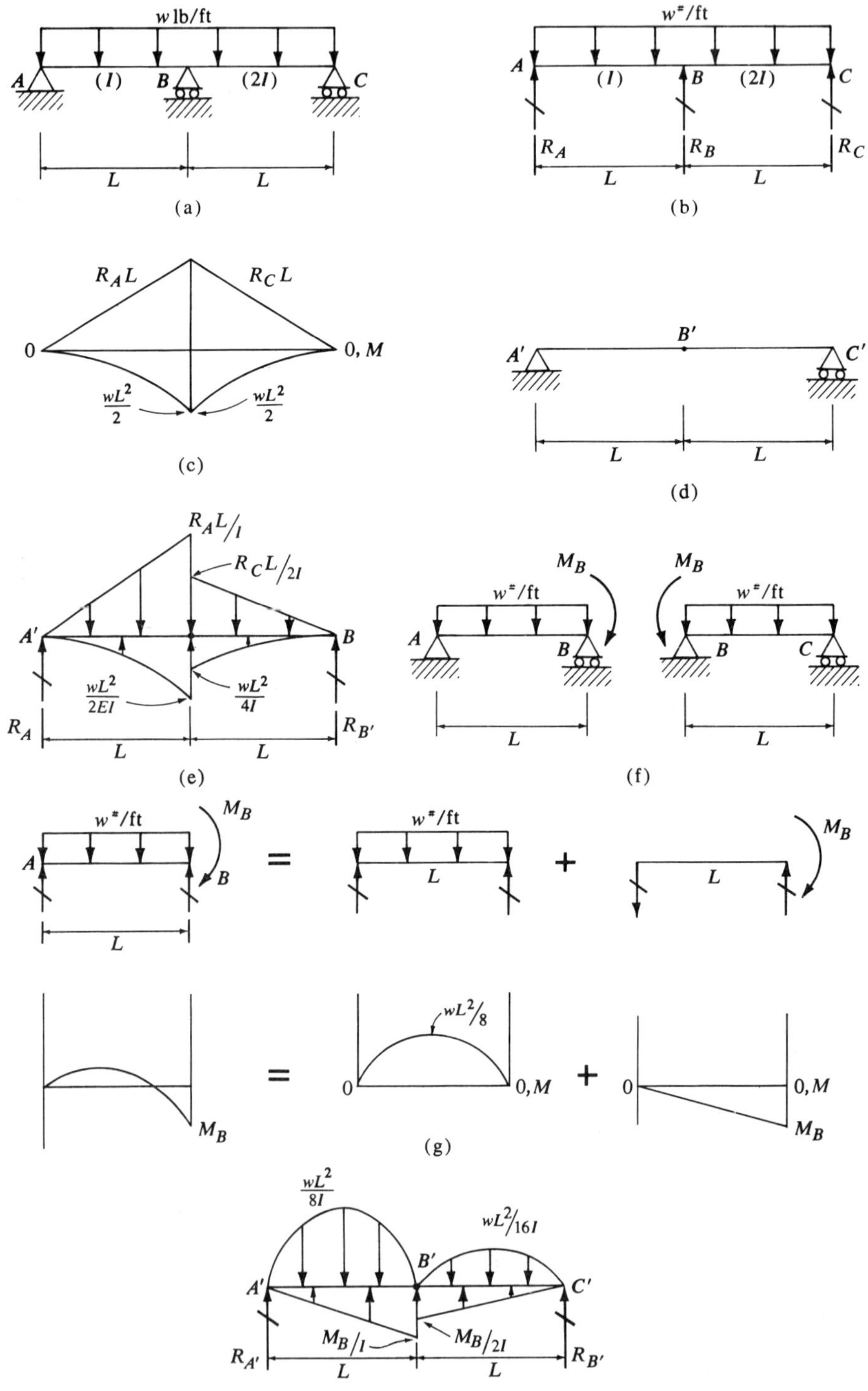
w lb/ft
A
(I)
B
(2I)
C
L
L
(a)
w#/ft
A
(I)
B
(2I)
C
R_A
R_B
R_C
L
L
(b)
R_A L
R_C L
0
0, M
wL²/2
wL²/2
(c)
B′
A′
C′
L
L
(d)
R_A L/I
R_C L/2I
A′
B
wL²/2EI
wL²/4I
R_A
R_B′
L
L
(e)
M_B
w#/ft
A
B
M_B
w#/ft
B
C
L
L
(f)
w#/ft
M_B
A
B
L
=
w#/ft
L
+
M_B
L
M_B
=
wL²/8
0
0, M
+
0
0, M
M_B
(g)
wL²/8I
wL²/16I
B′
A′
C′
M_B/I
M_B/2I
R_A′
R_B′
L
L
(h)

Figure 8.13

Solution:

1. A FBD of the given continuous beam along with a moment diagram by parts about point B are shown in Figures 8.13b and 8.13c respectively. Point B was selected only because of the convenient arrangement of the simple moment areas. The conjugate beam itself is shown in Figure 8.13d. The exterior simple supports of the real beam are analogous to the exterior simple supports of the conjugate beam. In the real beam, rotation but no deflection is permitted at the interior support B and, hence, the analogous conjugate beam support must allow only shear and no bending moment. This is accomplished by using an interior hinge which is incapable of transmitting moment.
2. The loaded conjugate beam is shown in Figure 8.13e. Since each span has a different moment of inertia, care must be exerted so as not to neglect this factor. The loading for the conjugate beam is the M/EI diagram of the real beam and, therefore, dividing the real moment ordinates by the respective EI terms automatically compensates for the different moments of inertia. Examination of the loaded conjugate beam reveals that there are four unknowns included among the reactions and applied loads ($R_{A'}$, $R_{B'}$, R_A, and R_B). The loaded conjugate beam can be analyzed just as any equilibrium problem; hence, we have at our disposal the two equations of equilibrium ($\Sigma F_Y = 0$ and $\Sigma M = 0$) and the condition that the moment at B' is zero. This leaves us one equation, or condition, shy of determining the four unknowns. It would appear that the conjugate beam is statically indeterminate to the first degree had we not earlier stated that a conjugate beam, along with its loads and reactions, is never statically indeterminate. In order to overcome such an obstacle, the conjugate beam method must be modified slightly for the case of continuous beams. The problem originates because the real moment diagram contains two unknowns, R_A and R_B, while the real beam is only statically indeterminate to the first degree. These two unknowns, along with the two unknown conjugate beam reactions, appear to make the conjugate beam statically indeterminate. We cannot alter the number of conjugate beam reactions, but we can alter the real moment diagram, and, thus, the conjugate loading diagram, to contain only one unknown.
3. The original beam shown in Figure 8.13a can be cut at B and represented by the two beams shown in Figure 8.13f. Restraint against translation in the x and y directions, as well as rotational restraint, is provided at B; thus, the beams shown in Figure 8.13f are equivalent to the original beam. Using the principle of superposition, the FBDs and corresponding moment diagrams are shown for the left span in Figure 8.13g. The moment diagram for the continuous beam may, therefore, be represented by the combined effects of that produced by the applied loads and that produced by the moment at the interior support. Identical results can be obtained for the right span. The loading diagram for the conjugate beam, as shown in Figure 8.13h, now contains only one unknown, M_B. Note that the moment ordinates were divided by the appropriate moments of inertia. The following

modification must, therefore, be made when analyzing a continuous beam by the conjugate beam method: **the moment diagram must be drawn in terms of the applied loads and the moment at the interior supports.**

4. The equations of equilibrium can now be applied to the beam shown in Figure 8.13h as follows:

$$\Sigma M_{A'} = 0 \quad (+\curvearrowright) \quad \text{(Considering only span } A'B'\text{)}$$

$$\left(\frac{2}{3}\right)\left(\frac{wL^2}{8I}\right)(L)\left(\frac{L}{2}\right) - \left(\frac{1}{2}\right)\left(\frac{M_B}{I}\right)(L)\left(\frac{2L}{3}\right) - V_{B'}(L) = 0$$

$$V_{B'} = \frac{wL^3}{24I} - \frac{M_B L}{3I} \uparrow \text{ on } A'B'$$

$$\Sigma M_{C'} = 0 \quad (+\curvearrowright) \quad \text{(Considering only span } B'C'\text{)}$$

$$-V_{B'}(L) - \left(\frac{2}{3}\right)\left(\frac{wL^2}{16I}\right)(L)\left(\frac{L}{2}\right) + \left(\frac{1}{2}\right)\left(\frac{M_B}{2I}\right)(L)\left(\frac{2}{3}\right)L = 0$$

$$-\left(\frac{wL^3}{24I} - \frac{M_B L}{3I}\right) - \frac{wL^3}{48I} + \frac{M_B L}{6I} = 0$$

$$\underline{M_B = +\frac{wL^2}{8} \text{ lb-ft}}$$

The positive sign indicates that the direction of M_B was correctly assumed in Figure 8.13f. With M_B known, Figure 8.13f may be utilized to determine R_A:

$$\Sigma M_B = 0 \quad (+\curvearrowright) \quad \text{(Considering only the left span)}$$

$$R_A(L) - wL(L/2) + \frac{wL^2}{8} = 0$$

$$\underline{R_A = (\tfrac{3}{8})wL \uparrow}$$

$$\Sigma M_B = 0 \quad (+\curvearrowright) \quad \text{(Considering only the right span)}$$

$$(wL)(L/2) - R_C L - \frac{wL^2}{8} = 0$$

$$\underline{R_C = (\tfrac{3}{8})wL \uparrow}$$

Using Figure 8.13b,

$$\Sigma F_Y = 0 \uparrow +$$
$$(\tfrac{3}{8})wL + (\tfrac{3}{8})wL - w(2L) + R_B = 0$$
$$\underline{R_B = \tfrac{5}{4}wL \uparrow}$$

The conjugate beam method is not a particularly convenient method when dealing with continuous beams and the analyst will generally prefer using the three-moment equations. The method is, however, well suited for the solution of fixed-end beams.

8.6 The Slope Deflection Method

The slope deflection method can be used to analyze all types of statically indeterminate beams or **rigid** frames. Rigid, in this case, is taken to indicate that the angles between intersecting members of a frame do not change in value as the loads are applied, i.e., if two members of a frame are at right angles at a joint before loading, they will remain at right angles after loading. For example, the frame shown in Figure 8.14 deflects as shown under the loading. Although there is rotation at joint *B*, the angle between members *BA* and *BC* at joint *B* remains at 90 degrees. Note that continuity of slope exists at joint *B*. All members meeting at a common joint must rotate an equal amount. Thus, in Figure 8.14a, $\Theta_{BA} = \Theta_{BC} = \Theta_B$, i.e., there is only one rotation (Θ_B) at any rigid joint such as joint *B*.

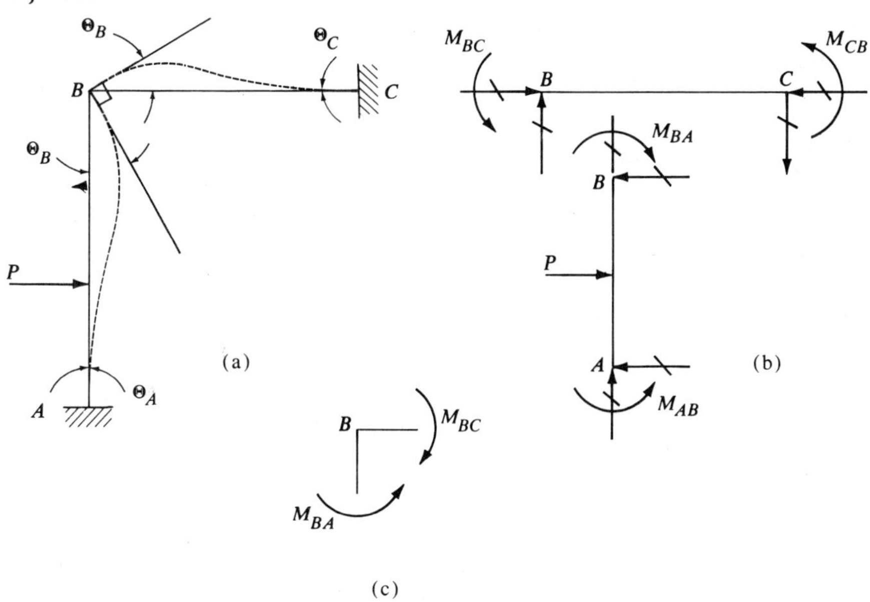

(c)

Figure 8.14

The bases of the slope deflection method are the **slope deflection equations,** which express the end moments of each member in terms of the end displacements of that member. Referring to Figure 8.14, let us suppose that we could express the end moments of each member as a function of the end displacements of each member and the load on the member as follows:

$$M_{CB} = f_1\ (\Theta_B, \Theta_C, \text{load on span})^*$$
$$M_{BC} = f_2\ (\Theta_B, \Theta_C, \text{load on span})$$
$$M_{BA} = f_3\ (\Theta_B, \Theta_A, \text{load on span})$$
$$M_{AB} = f_4\ (\Theta_B, \Theta_A, \text{load on span})$$

These four equations contain seven unknowns, M_{CB}, M_{BC}, M_{BA}, M_{AB}, Θ_A, Θ_B, and Θ_C. Since the supports at A and C are fixed, Θ_A and Θ_C are known to be zero, thus leaving five unknowns in the form:

$$M_{CB} = f_1(\Theta_B), \text{ (no load on span } CB)$$
$$M_{BC} = f_2(\Theta_B), \text{ (no load on span } BC)$$
$$M_{BA} = f_3\ (\Theta_B, P)$$
$$M_{AB} = f_4\ (\Theta_B, P)$$

All four end moments can, therefore, be expressed as different functions of Θ_B. For equilibrium, the summation of all moments acting on each joint must be zero. Using this joint condition and referring to the FBD of joint B in Figure 8.14c (only moments are shown) we have:

$$M_{BC} + M_{BA} = 0$$

Since both expressions, that for M_{BC} and that for M_{BA}, contain only one unknown, Θ_B, we have a solution for Θ_B. All end moments can then be found by substituting Θ_B into the expressions $f_1(\Theta_B)$, $f_2(\Theta_B)$, $f_3(\Theta_B, P)$, and $f_4(\Theta_B, P)$. These expressions are known as the **slope deflection equations** for the end moments. We now know the moments acting at supports A and C as well as the moment at joint B. The six unknown reactions are, therefore, reduced to four with the moments at A and C known. The four remaining unknown reactions can be obtained with the three equilibrium equations along with the condition that the moment at joint B is known.

The slope deflection equations can provide the additional equations necessary for the solution of a statically indeterminate beam or frame. The problem now becomes one of deriving the slope deflection equations. It should be noted that the only end displacements considered thus far have been rotations. It is possible for the joints not only to rotate, but, in some instances, translate. For example, joints A, B, and C in the frame shown in Figure 8.14 are not free to move in any direction with the possible exception of that negligible amount due to axial stress. The joints at B and C for the frames shown in Figure 8.15 are free to move under action of the applied loads and such relative movement between, say, the joints at A and B must be taken into account by the slope deflection equations.

---

*Recall that this equation is read, M_{CB} **is a function of** Θ_B, Θ_C, **and the load on span CB.**

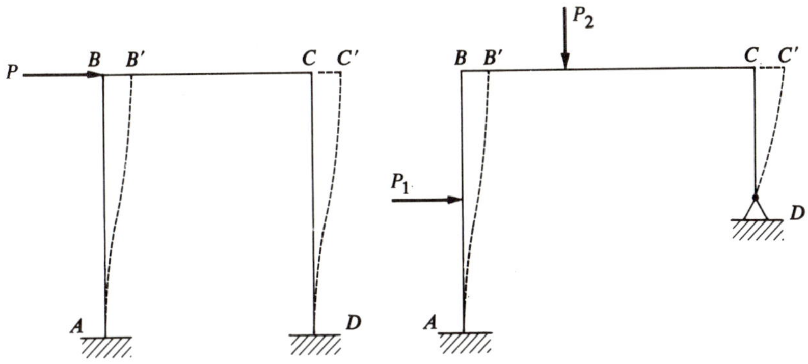

Figure 8.15

8.7 Slope Deflection Equations—No Joint Translation

The slope deflection equations express the end moments acting at the ends of a member in terms of end displacements and applied loads acting on the member. Initially, only end rotations will be considered, i.e., no joint or end translation permitted. Later, a modification will be made in the slope deflection equations to account for end translations. Consider member *AB* shown in Figure 8.16a, which is isolated from a loaded indeterminate beam or rigid frame (not shown). The member is deformed, as shown by the dashed line, with end rotations Θ_A and Θ_B. Our objective is to express the end moments, M_{AB} and M_{BA}, in terms of the end rotations and applied loads P_0 and P_1. Let us first establish the following sign convention for the slope deflection equations:

1. End moments acting on the end of a member are positive when clockwise.
2. End rotations are positive when the tangent to the elastic curve rotates clockwise from its original position.

By the concept of superposition, we may represent the combined loading acting on member *AB* by the three separated beams shown in Figure 8.16b. Thus, by superposition

$$\begin{aligned} \Theta_A &= \Theta_{A1} - \Theta_{A2} \\ \Theta_B &= -\Theta_{B1} + \Theta_{B2} \end{aligned} \tag{8–13}$$

where clockwise rotations are taken as positive, and

$$\begin{aligned} M_{AB} &= M_{FAB} + M_{A'} \\ M_{BA} &= M_{FBA} + M_{B'} \end{aligned} \tag{8–14}$$

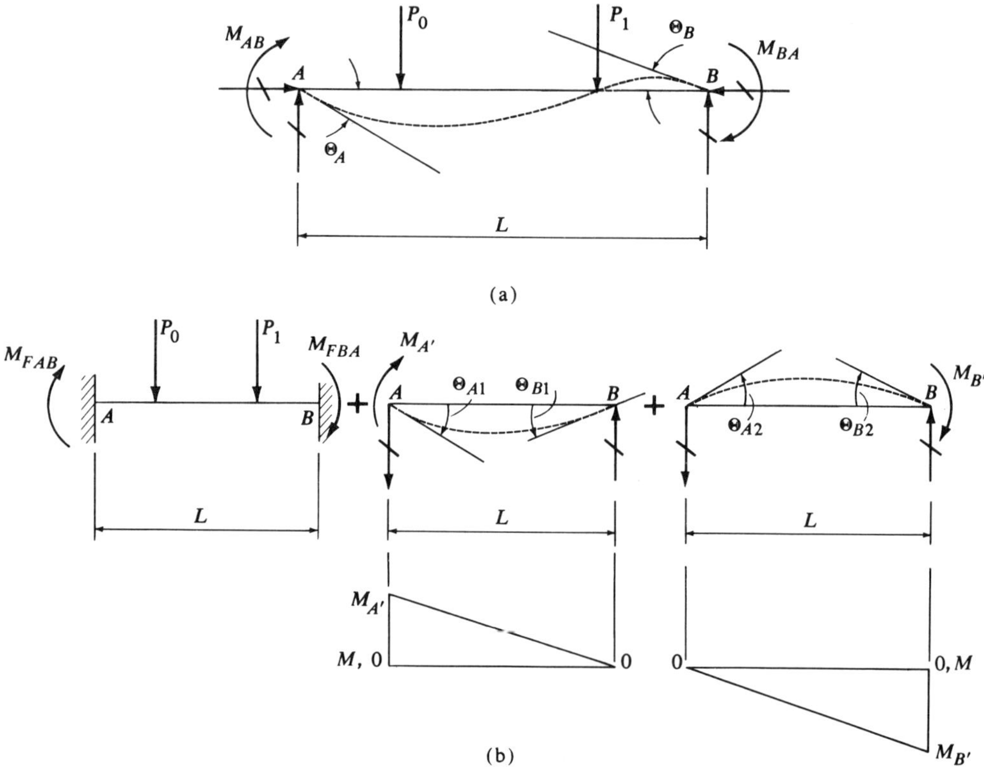

Figure 8.16

where

M_{FAB}, M_{FBA} = the end moments acting at ends A and B respectively due to the applied loads and considering ends A and B fixed. These moments are called **fixed-end moments.**

$M_{A'}$, $M_{B'}$ = the additional moments acquired to cause rotations of Θ_A and Θ_B respectively.

Equations (8–14) are, in crude form, the slope deflection equations. The fixed-end moments can be easily determined by using, for example, the conjugate beam method. The remaining moments, $M_{A'}$ and $M_{B'}$, can be determined in terms of the end rotations and substituted back into Eq. (8–14) to obtain the final form of the slope deflection equations. Referring to Figure 8.16b, we may apply the second moment–area theorem to obtain the rotations Θ_{A1}, Θ_{A2}, Θ_{B1}, and Θ_{B2}. Thus:

$$\Theta_{A1} = (t_{BA}/L)$$

$$= \left(\frac{1}{L}\right)\left(\frac{1}{2}\right)\frac{(M_{A'})(L)}{EI}\left(\frac{2L}{3}\right)$$

$$\Theta_{A1} = \frac{M_{A'}L}{3EI}$$

$$\Theta_{B1} = (t_{AB}/L)$$

$$= \left(\frac{1}{L}\right)\left(\frac{1}{2}\right)\frac{(M_{A'})(L)}{EI}\left(\frac{L}{3}\right)$$

$$\Theta_{B1} = \frac{M_{A'}L}{6EI}$$

In a similar fashion

$$\Theta_{A2} = \frac{M_{B'}L}{6EI} \qquad \Theta_{B2} = \frac{M_{B'}L}{3EI}$$

Using the concept of superposition and referring to Figure 8.16 we can again write Eqs. (8–13) taking clockwise rotations as positive:

$$\Theta_A = \Theta_{A1} - \Theta_{A2}$$
$$\Theta_B = -\Theta_{B1} + \Theta_{B2}$$

Substituting, we now have:

$$\Theta_A = \frac{M_{A'}L}{3EI} - \frac{M_{B'}L}{3EI}$$

$$\Theta_B = -\frac{M_{A'}L}{6EI} + \frac{M_{B'}L}{3EI}$$

Solving these equations simultaneously for $M_{A'}$ and $M_{B'}$ we obtain:

$$M_{A'} = \frac{2EI}{L}(2\Theta_A + \Theta_B)$$

$$M_{B'} = \frac{2EI}{L}(2\Theta_B + \Theta_A)$$

Again by superposition, or rewriting Eqs (8–14), we have the final form of the slope deflection equations (assuming no joint translation):

$$M_{AB} = M_{FAB} + \frac{2EI}{L}(2\Theta_A + \Theta_B) \tag{8–15}$$

$$M_{BA} = M_{FBA} + \frac{2EI}{L}(2\Theta_B + \Theta_A) \tag{8–16}$$

We have now expressed the end moments of any indeterminate member in terms of the applied loads and end displacements. The procedure of analyzing any statically indeterminate beam or frame by the slope deflection method is as follows:

1. Determine the fixed-end moments at the ends of each span or member using the conjugate beam method or the formulas shown in Figure 8.17.
2. Apply the slope deflection Eq. (8–15) or Eq. (8–16) to each end of each span or member.
3. Applying the equations of equilibrium and using the condition of continuity of slope, obtain a sufficient number of simultaneous equations with the end rotations as unknowns.
4. Solve the simultaneous equations for the unknown rotations and substitute the rotations back into the slope deflection equations for the end moments.
5. With the end moments known, determine the reactions using the conventional laws of static equilibrium.

All the preceding steps are straightforward with the possible exception of step 3. The simultaneous equations mentioned in step 3 may be obtained, for example, by applying the condition that the sum of the end moments acting on the ends of members meeting at a support or joint must be zero as shown at joint B in Figure 8.14c. Also, continuity of slope must exist at an interior support of a continuous beam, i.e., $\Theta_{BA} = \Theta_{BC}$ at interior support B of a continuous beam (not shown). In some instances, specific knowledge about an end rotation may be known, such as zero rotation at a fixed support. Summarizing, all possibilities must be examined to provide the simultaneous equations necessary for the solution of the unknown rotations. The following illustrative examples should further clarify the slope deflection method.

Example 8.8

Determine the reactions for the continuous beam shown in Figure 8.18a. Both E and I are constant along the length.

Solution:

1. The fixed-end moments for spans AB and BC can be determined by using formulas given in Figures 8.17a and b:

$$M_{FAB} = -\frac{PL}{8} = -\frac{15(30)}{8} = -56.3 \text{ kip-ft}$$

$$M_{FBA} = +\frac{PL}{8} = +56.3 \text{ kip-ft}$$

$$M_{FBC} = -\frac{wL^2}{12} = -\frac{(2)(20)^2}{12} = -66.7 \text{ kip-ft}$$

$$M_{FCB} = +\frac{wL^2}{12} = +66.7 \text{ kip-ft}$$

2. The slope deflection equations can now be written using the fact that $\Theta_{BA} = \Theta_{BC} = \Theta_B$ and $\Theta_{CB} = \Theta_{CD} = \Theta_C$, i.e., continuity of slope exists at the supports.

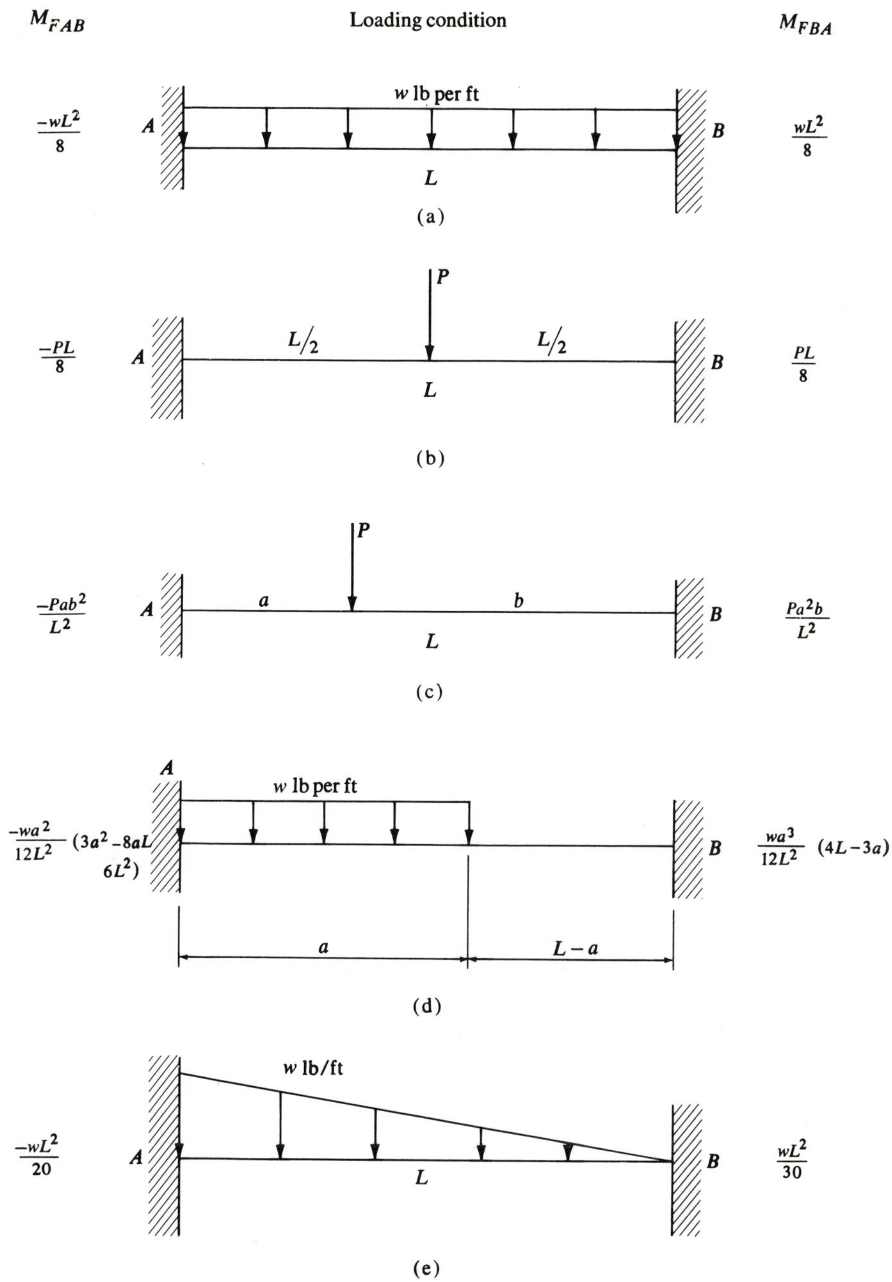
M_{FAB}
Loading condition
M_{FBA}
$\frac{-wL^2}{8}$
A
w lb per ft
L
B
$\frac{wL^2}{8}$
(a)
$\frac{-PL}{8}$
A
P
L/2
L/2
L
B
$\frac{PL}{8}$
(b)
$\frac{-Pab^2}{L^2}$
A
P
a
b
L
B
$\frac{Pa^2b}{L^2}$
(c)
A
w lb per ft
$\frac{-wa^2}{12L^2}(3a^2 - 8aL$
$6L^2)$
B
$\frac{wa^3}{12L^2}(4L - 3a)$
a
L − a
(d)
w lb/ft
$\frac{-wL^2}{20}$
A
L
B
$\frac{wL^2}{30}$
(e)

Figure 8.17

$$M_{AB} = M_{FAB} + \frac{2EI}{L_{AB}}(2\Theta_A + \Theta_B)$$

$$= -56.3 + \frac{2EI}{30}(2\Theta_A + \Theta_B)$$

$$M_{BA} = M_{FBA} + \frac{2EI}{L_{AB}}(2\Theta_B + \Theta_A)$$

$$= 56.3 + \frac{2EI}{30}(2\Theta_B + \Theta_A)$$

$$M_{BC} = M_{FBC} + \frac{2EI}{L_{BC}}(2\Theta_B + \Theta_C)$$

$$= -66.7 + \frac{2EI}{20}(2\Theta_B + \Theta_C)$$

$$M_{CB} = M_{FCB} + \frac{2EI}{L_{BC}}(2\Theta_C + \Theta_B)$$

$$M_{CB} = 66.7 + \frac{2EI}{20}(2\Theta_C + \Theta_B)$$

The coefficients of the rotation terms are rather cumbersome to carry along during the remaining calculations. We can replace these coefficients with relative values with no loss of generality or accuracy. Thus, we have:

$$K_{AB} = \frac{2EI_{AB}}{L_{AB}} = \frac{2EI}{30}$$

$$K_{BC} = \frac{2EI_{BC}}{L_{BC}} = \frac{2EI}{20}\left(K_{BC} = \frac{3}{2}K_{AB};\ K_{AB} = \frac{2}{3}K_{BC}\right)$$

If we let K_{AB} equal unity, i.e., $K_{AB} = 1$, then $K_{BC} = 1.5$ since only relative values of K are needed. We could have just as easily let $K_{BC} = 1$, then K_{AB} would become 2/3. We now rewrite the slope deflection equations in simpler form utilizing the relative K coefficients:

$$M_{AB} = -56.3 + 2\Theta_A + \Theta_B$$
$$M_{BA} = 56.3 + 2\Theta_B + \Theta_A$$
$$M_{BC} = -66.7 + 3\Theta_B + (\tfrac{3}{2})\Theta_C$$
$$M_{CB} = 66.7 + 3\Theta_C + (\tfrac{3}{2})\Theta_B$$

3. Note that the slope deflection equations contain three unknown rotations. We must now examine all equilibrium and deformation conditions for possible equations necessary for the solution of the unknown rotations. As a first source of equations, we can easily check to see if the slope at points A, B, or C is known. By inspection, therefore, $\Theta_A = 0$. The interior support at B can now be isolated as shown in Figure 8.18b. Summing moments, this joint condition yields:

$$M_{BA} + M_{BC} = 0$$

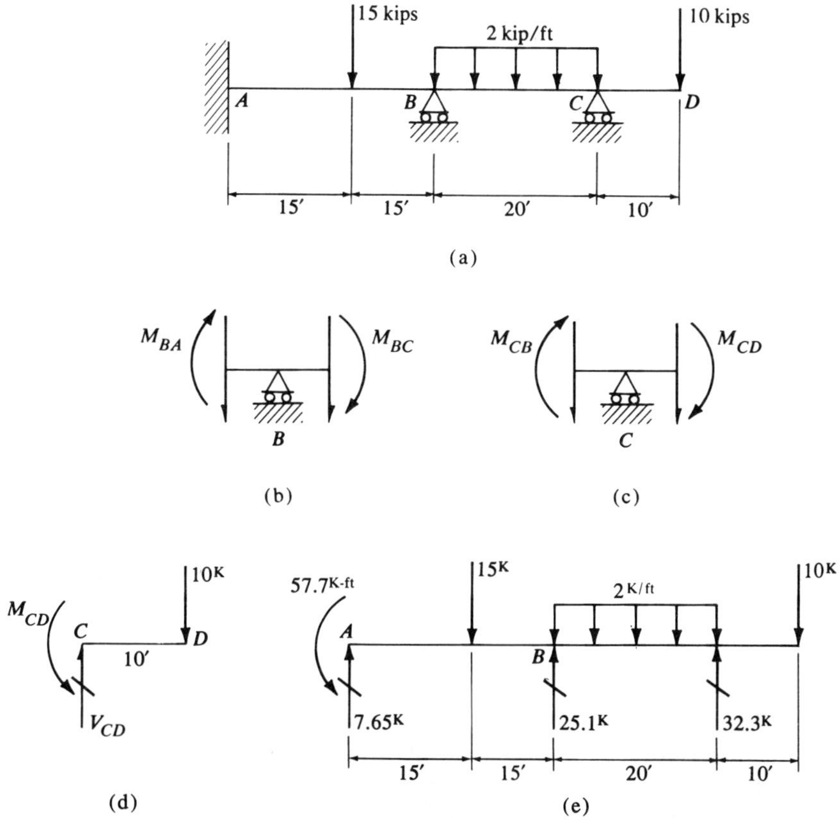

Figure 8.18

The joint at C can also be isolated as shown in Figure 8.18c. We note, however, that M_{CD} is a known quantity as can be seen from the FBD of segment CD shown in Figure 8.18d. We have, therefore,

$$M_{CD} = 100 \text{ kip-ft} \quad \text{at joint } C \text{ (Figure 8.18c)}$$

$$M_{CB} = 100 \text{ kip-ft} \quad \text{at joint } C \text{ (Figure 8.18c)}$$

$$\therefore M_{CB} = 100 \text{ kip-ft} \quad \text{on span } BC$$

Since M_{CB} acts clockwise on the C end of span BC, M_{CB} is +100 kip-ft by the adopted sign convention. Thus, by our second joint condition we have:

$$M_{CB} = +100 \text{ kip-ft}$$

Substitution of the slope deflection expressions into the joint moment equations yields:

$$\left(56.3 + 2\Theta_B + \Theta_A\right) + \left(-66.7 + 3\Theta_B + \frac{3\Theta_C}{2}\right) = 0$$

$$\left(66.7 + 3\Theta_C + \frac{3\Theta_B}{2}\right) = +100$$

Recalling that $\Theta_A = 0$ and simplifying we obtain:

$$10\Theta_B + 3\Theta_C = 20.8$$
$$3\Theta_B + 6\Theta_C = 66.7$$

4. Solving these joint equations simultaneously yields:

$$\Theta_B = -1.478$$
$$\Theta_C = 10.53$$

Substitution of these rotations into the slope equations yields the following end moments:

$$M_{AB} = -57.77 \text{ kip-ft}$$
$$M_{BA} = +53.3 \text{ kip-ft}$$
$$M_{BC} = -53.3 \text{ kip-ft}$$
$$M_{CB} = +100 \text{ kip-ft}$$

where positive moments act clockwise on the end of a member.

5. With the moments now known at supports A and B, the reactions can be determined using the procedure followed when using the three-moment equations (see Example 8.5). A FBD of the given beam is shown in Figure 8.18d.

Example 8.9

Determine the reactions for the rigid frame shown in Figure 8.19a. E is constant throughout the frame.

Solution:

1. The initial step is to determine all the fixed-end moments. From 8.17b:

$$M_{BC} = -\frac{PL}{8} = -\frac{(32)(10)}{8} = -40.0 \text{ kip-ft}$$

$$M_{CB} = +40.0 \text{ kip-ft}$$

Note that even though no moment exists at support C, a fixed-end moment at the C end of span BC was determined. A slope deflection equation for M_{CB} may be written despite the fact that $M_{CB} = 0$. Thus, the fixed-end moment at the C end of span BC is required for the slope deflection equation.

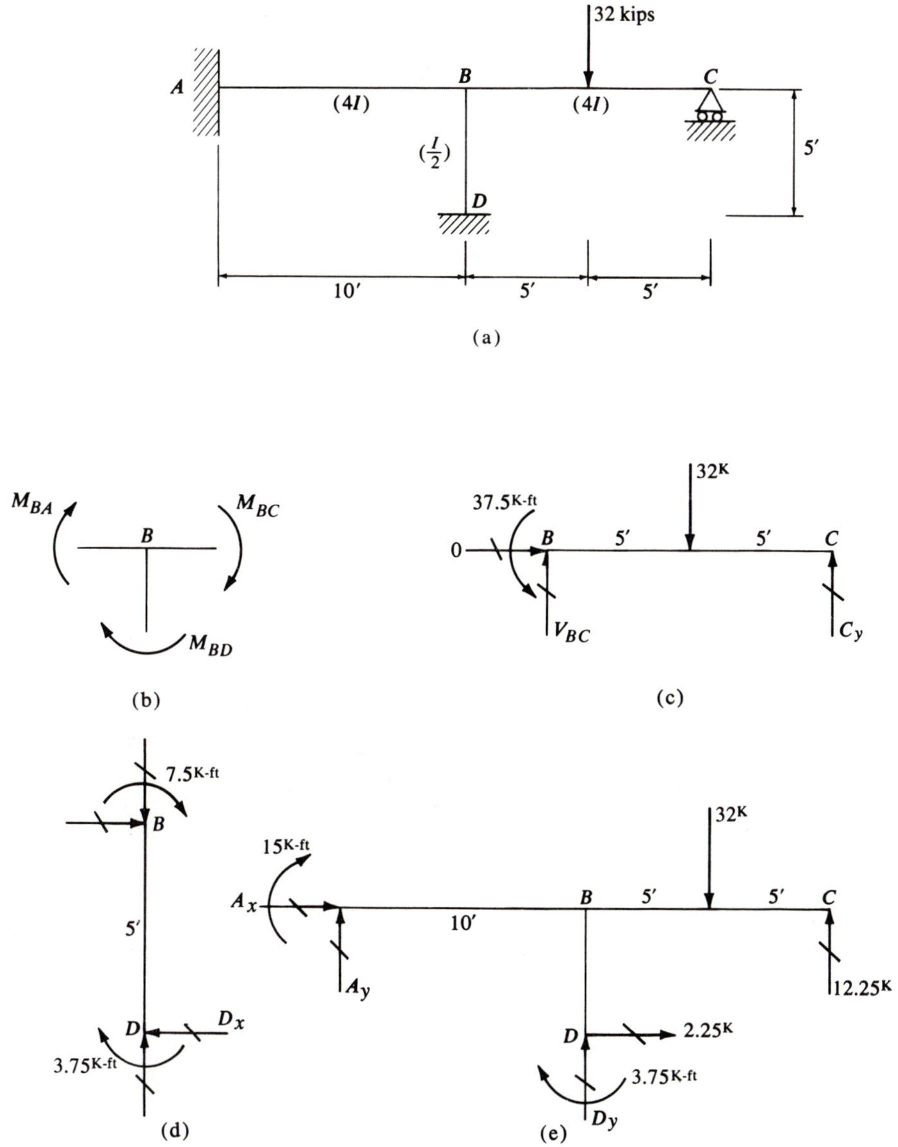

Figure 8.19

2. The slope deflection equations may now be written as follows:

$$M_{AB} = \frac{2E(4I)}{10}(2\Theta_A + \Theta_B)$$

$$M_{BA} = \frac{2E(4I)}{10}(2\Theta_B + \Theta_A)$$

$$M_{BC} = -40.0 + \frac{2E(4I)}{10}(2\Theta_B + \Theta_C)$$

$$M_{CB} = 40.0 + \frac{2E(4I)}{10}(2\Theta_C + \Theta_B)$$

$$M_{BD} = \frac{2E(I/2)}{5}(2\Theta_B + \Theta_D)$$

$$M_{DB} = \frac{2E(I/2)}{5}(2\Theta_D + \Theta_B)$$

Note that there exists no relative translation between any of the joints in the given frame. Thus, only end rotational displacements occur and the above written slope deflection equations are adequate. The use of the following relative K-values simplifies the calculations:

If $$K_{BD} = \frac{2E(I/2)}{5} = \frac{2EI}{10}, \text{ use } 1$$

then $$K_{AB} = K_{BC} = \frac{2E(4I)}{10}, \text{ use } 4\ (K_{AB} = K_{BC} = 4K_{BD})$$

Rewriting the slope deflection equations:

$$\begin{aligned} M_{AB} &= 8\Theta_A + 4\Theta_B \\ M_{BA} &= 8\Theta_B + 4\Theta_A \\ M_{BC} &= -40 + 8\Theta_B + 4\Theta_C \\ M_{CB} &= 40 + 8\Theta_C + 4\Theta_B \\ M_{BD} &= 2\Theta_B + \Theta_D \\ M_{DB} &= 2\Theta_D + \Theta_B \end{aligned}$$

3. Simultaneous equations must now be written in order to determine the unknown rotations. Looking to the deformed structure, we may write by inspection:

$$\Theta_A = \Theta_B = 0$$

This reduces the number of unknown rotations to two. We now need only two equations, and, hence, we turn our attention to the joint conditions. Referring to Figure 8.19b, which shows only the moments acting at B, the sum of the moments acting at the joint must be zero for equilibrium. Thus:

$$M_{BA} + M_{BC} + M_{BD} = 0 \text{ (all assumed positive on member)}$$

The moment at support C is zero, thereby giving us the second required equation:

$$M_{CB} = 0$$

The slope deflection relationships can be substituted into these joint equations:

$$(8\Theta_B) + (-40 + 8\Theta_B + 4\Theta_C) + (2\Theta_B) = 0$$
$$40 + 8\Theta_C + 4\Theta_B = 0$$

Simplifying:

$$18\Theta_B + 4\Theta_C = 40$$
$$\Theta_B + 2\Theta_C = -10$$

4. Solving these equations simultaneously yields:

$$\Theta_B = 3.75 \qquad \Theta_C = -4.375$$

Substitution of these rotations into the slope deflection equations will yield the member end moments:

$$M_{AB} = +15 \text{ kip-ft}$$
$$M_{BA} = +30.0 \text{ kip-ft}$$
$$M_{BC} = -37.5 \text{ kip-ft}$$
$$M_{CB} = 0$$
$$M_{BD} = +7.5 \text{ kip-ft}$$
$$M_{DB} = +3.75 \text{ kip-ft}$$

5. With the end moments known, the reactions can be determined. (Note that positive end moments act clockwise at the member ends.)

Figure 8.19c:

$$\Sigma M = 0 \quad (+\curvearrowright)$$

$$-35.5 + 32(5) = C_Y(10) = 0$$
$$\underline{C_Y = 12.25 \text{ kips} \uparrow}$$

Figure 8.19d:

$$\Sigma M_B = 0 \quad (+\curvearrowright)$$

$$7.5 + 3.75 + D_X(5) = 0$$
$$\underline{D_X = -2.25^K = 2.25 \text{ kips} \rightarrow}$$

Figure 8.19e:

$$\Sigma M_A = 0 \quad (+\curvearrowright)$$

$$15 + 3.75 - 10D_Y - 2.25(5) + 32(15) - 12.25(20) = 0$$
$$\underline{D_Y = 24.25 \text{ kips} \uparrow}$$

$$\Sigma F_Y = 0 \uparrow +$$
$$A_Y + 24.25 - 32 + 12.25 = 0$$
$$\underline{A_Y = -4.50 = 4.50 \text{ kips} \downarrow}$$

$$\Sigma F_X = 0 \rightarrow +$$
$$A_X + 2.25 = 0$$
$$\underline{A_X = -2.25 = 2.25 \text{ kips} \leftarrow}$$

When writing the joint equations, confusion sometimes arises when trying to determine the directions of the moments acting at the joints. When the direction of a moment acting at a joint is not known, it is assumed to act positive, i.e., clockwise, on the member end. Thus, the direction of the moments acting at joing B in Figure 8.19b were obtained by assuming that the moments act positively on the ends of the respective members. In some instances, however, the directions of the moments acting at a joint are known. See, for example, joint C in example 8.8.

8.8 Slope Deflection—With Joint Translation

Recall that the basis of the slope deflection method lies in the fact that the end moments of each member can be expressed in terms of the end displacements of that member and the loads on that member. Thus far, the only end displacements encountered were end rotations. The slope deflection equations presented so far are limited to those structures having no relative joint translations. The slope deflection equations will now be modified for the case when relative joint translations occur. The moment acting at the end of any member of a statically indeterminate frame or beam must now be written in terms of the loads on the member, the end rotations, and the relative end displacement, i.e., the movement of one end with respect to the other. The slope deflection equations will take the following form for any single member AB:

$$M_{AB} = M_{FAB} + \frac{2EI}{L}(2\Theta_A + \Theta_B) + M''_{AB}$$

$$M_{BA} = M_{FBA} + \frac{2EI}{L}(2\Theta_B + \Theta_A) + M''_{BA}$$

where

M_{FAB}, M_{FBA} = the fixed-end moment which represents the contribution of the loads on the member to the final end moment.

$\frac{2EI}{L}(2\Theta_A + \Theta_B)$, $\frac{2EI}{L}(2\Theta_B + \Theta_A)$ = the contribution to the final end moment due to the rotations at the member ends.

M''_{AB}, M''_{BA} = the contribution to the final end moment due to the relative translation of the member ends.

In order to determine this final contribution, let us consider the fixed-end beam shown in Figure 8.20a. The problem becomes one of simply determining the moment introduced at the ends if there is a relative translation of the member ends as shown in Figure 8.20b. To solve this problem, the moment-area method will be utilized. The beam in Figure 8.20b is statically indeterminate to the second degree but the following two geometric facts are known.

$$\Theta_{AB} = 0$$
$$t_{BA} = -\Delta$$

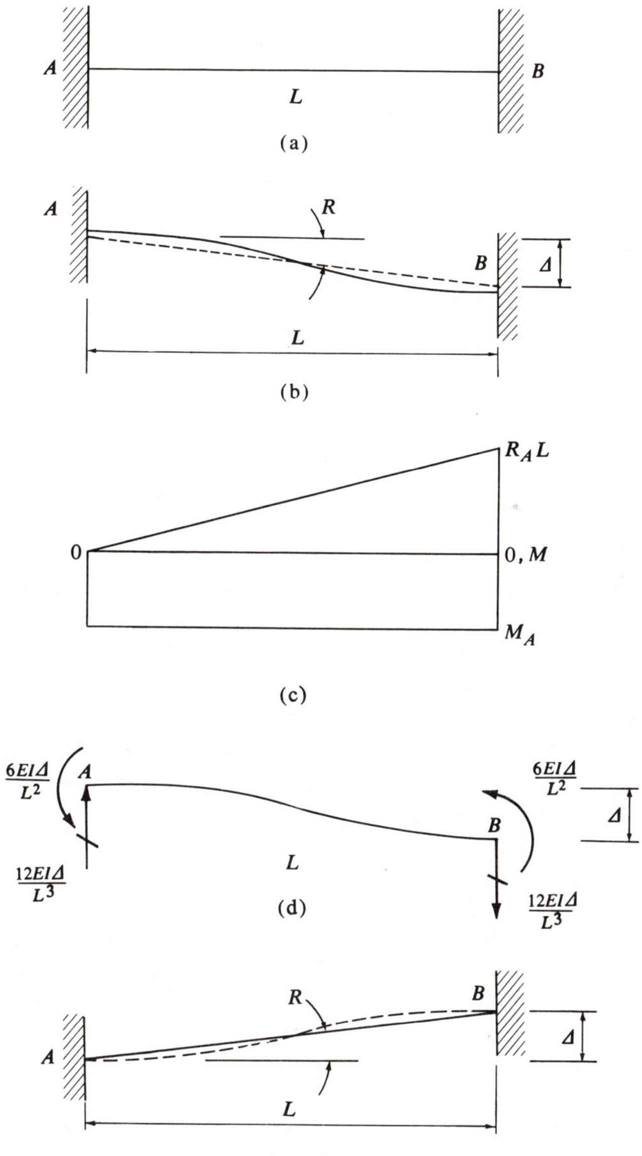

Figure 8.20

Using the first and second moment-area theorems and the moment diagram by parts about point B shown in Figure 8.20c, we have:

$$EI\Theta_{AB} = 0 = (\tfrac{1}{2})(R_A L)L - M_A(L)$$

$$\frac{R_A L^2}{2} - M_A L = 0$$

and

$$-EI\Delta = \frac{R_A L^2}{2}(L/3) - M_A L(L/2)$$

$$\frac{R_A L^3}{6} - \frac{M_A L^2}{2} = -EI\Delta$$

Solving these two equations simultaneously for R_A and M_A we obtain

$$M_A = +\frac{6EI\Delta}{L^2} \qquad R_A = +\frac{12EI\Delta}{L^3}$$

The remaining reactions at B can now be determined. The final FBD of the beam is shown in Figure 8.20d. The end moments at A and B may be written in the form:

$$M_A = \frac{2EI}{L}\left(\frac{3\Delta}{L}\right) \qquad M_B = \frac{2EI}{L}\left(\frac{3\Delta}{L}\right)$$

for easy insertion within the parentheses of the slope deflection equations. First, however, a sign convention must be established. In this instance, support B moved downward with respect to support A. Using the previously adopted sign convention, i.e., clockwise end moments positive, both M_A and M_B are negative. If support B moves upward with respect to support A as shown in Figure 8.20e, both M_A and M_B would be positive. To establish a sign convention for relative joint translation, the angle R will be introduced as shown in Figures 8.20b and 8.20e. Where:

$$R = \Delta/L$$

Thus, for a clockwise R, i.e., a clockwise rotation of a line joining the member ends as shown in Figure 8.20b, the resulting end moments are negative, i.e., counterclockwise. For a counterclockwise R, the end moments are positive, i.e., clockwise on the member ends. For the beam shown in Figure 8.20b the end moments become

$$M_A = -\frac{2EI}{L}(3R) \qquad M_B = -\frac{2EI}{L}(3R)$$

and for the beam shown in Figure 8.20e we have:

$$M_A = -\frac{2EI}{L}(-3R) \qquad M_B = -\frac{2EI}{L}(-3R)$$

The modified form of the slope deflection equations now becomes

$$M_{AB} = M_{FAB} + \frac{2EI}{L}(2\Theta_A + \Theta_B - 3R)$$

$$M_{BA} = M_{FBA} + \frac{2EI}{L}(2\Theta_B + \Theta_A - 3R)$$

A positive result for M_{AB}, or M_{BA}, indicates a clockwise moment acting at the member end. The following illustrative example is intended to demonstrate the application of the modified slope deflection equations.

Example 8.10

Analyze the rigid frame shown in Figure 8.21a by the slope deflection method.

Solution:

1. The fixed-end moments can be obtained using Figure 8.17b and 8.17c:

$$M_{FAB} = -\frac{Pab^2}{L^2} = -\frac{6(9)(6)^2}{(15)^2} = -8.64 \text{ kip-ft}$$

$$M_{FBA} = \frac{Pa^2b}{L^2} = \frac{6(9)^2(6)}{(15)^2} = +12.96 \text{ kip-ft}$$

$$M_{FBC} = -\frac{PL}{8} = -\frac{12(12)}{8} = -18.0 \text{ kip-ft}$$

$$M_{FCB} = +18.0 \text{ kip-ft}$$

2. Before writing the slope deflection equations, the relative K values can be determined:

$$\text{if} \quad K_{CD} = \frac{2EI}{2(10)} = \frac{EI}{10}, \text{ use } 3 \quad \left(\frac{EI}{10} \times \frac{30}{EI} = 3\right)$$

$$\text{then } K_{AB} = \frac{2EI}{2(15)} = \frac{EI}{15}, \text{ use } 2 \quad \left(\frac{EI}{15} \times \frac{30}{EI} = 2\right)$$

$$\text{and } K_{BC} = \frac{2EI}{12} = \frac{EI}{6}, \text{ use } 5 \quad \left(\frac{EI}{6} \times \frac{30}{EI} = 5\right)$$

In addition, under the action of the applied loads the frame will deform as shown in Figure 8.21b. Notice that the top end of each vertical member moves horizontally relative to the base. This horizontal movement is known as **sidesway.** For frames, sidesway occurs due to the action of the lateral (horizontal) loads and due to unsymmetrical loading or geometry. This lateral movement of the top of the vertical member with respect to the base introduces moments at the vertical member ends. The modified slope deflection equations can account for sidesway through the use of the R angles. The R angle is the angle of rotation of a line joining the ends of the member due to the relative end motion as shown in Figure 8.21b. For both vertical members, AB and DC, a line joining the member ends rotates clockwise during the relative end displacement Δ. Thus:

$$R_{AB} = R_{BA} = +\frac{\Delta}{15} = 2R$$

$$R_{DC} = R_{CD} = +\frac{\Delta}{10} = 3R \quad (R_{DC} = \tfrac{3}{2} R_{AB})$$

$$R_{BC} = R_{CB} = 0$$

Again, relative values may be utilized for the R angles. Before writing the slope deflection equations, the continuity of slope at a joint must be reviewed as follows:

$$\Theta_{BA} = \Theta_{BC} = \Theta_B$$
$$\Theta_{CB} = \Theta_{CD} = \Theta_C$$

That is to say, all the members meeting at a joint rotate the same amount. The slope deflection equations are:

$$M_{AB} = -8.64 + 2(2\Theta_A + \Theta_B - 2R)$$
$$M_{BA} = 12.96 + 2(2\Theta_B + \Theta_A - 2R)$$
$$M_{BC} = -18 + 5(2\Theta_B + \Theta_C)$$
$$M_{CB} = 18 + 5(2\Theta_C + \Theta_B)$$
$$M_{CD} = 3(2\Theta_C + \Theta_D - 3R)$$
$$M_{DC} = 3(2\Theta_D + \Theta_C - 3R)$$

Note that since relative R values are used, the constant coefficient three for the R terms in the slope deflection equations may be omitted because it appears for all R values.

3. By inspection:

$$\Theta_A = \Theta_D = 0$$

At joints B and C we have the following joint moment conditions (see Figure 8.21c):

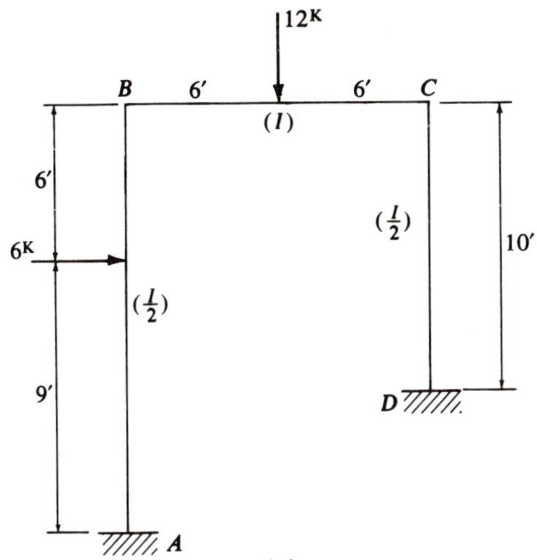

(a)

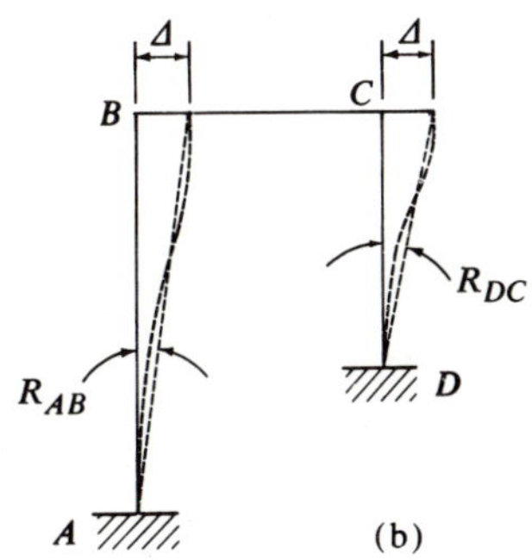

(b)

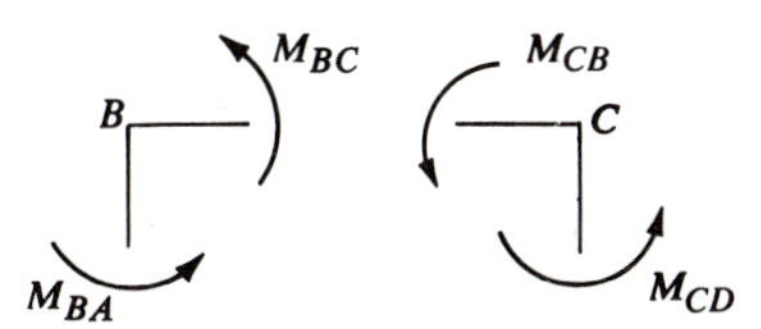

(c)

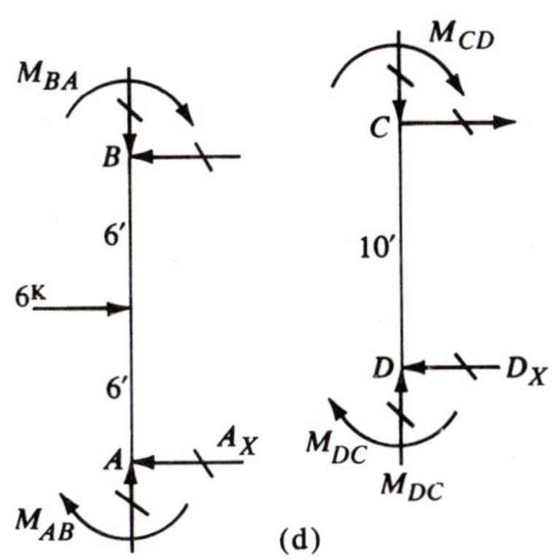

(d)

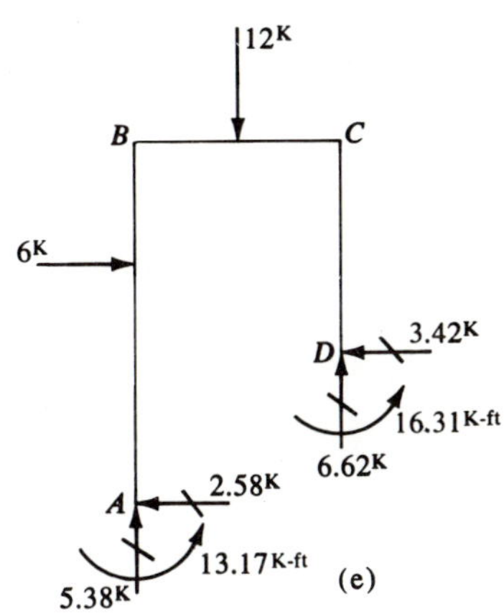

(e)

Figure 8.21

$$M_{BA} + M_{BC} = 0$$
$$M_{CB} + M_{CD} = 0$$

The joint conditions provide two equations but the slope deflection equations contain three unknowns: Θ_B, Θ_C, and R. A third condition, or equation, must be secured. For the case of frames we go to the equilibrium of the structure itself for this condition. By applying the equation $\Sigma F_X = 0$ to the entire frame shown in Figure 8.21a, we have:

$$A_X + B_X + 6 = 0$$

By isolating the vertical members we can obtain both A_X and B_X in terms of the member end moments. From Figure 8.21d:

$$\Sigma M_B = 0 \quad (\curvearrowright +)$$

$$M_{AB} + M_{BA} + 15A_X - 6(6) = 0$$
$$A_X = \frac{36 - M_{AB} - M_{BA}}{15} \rightarrow$$

$$\Sigma M_C = 0 \quad (\curvearrowright +)$$

$$M_{CD} + M_{DC} + 10B_X = 0$$
$$B_X = -\frac{(M_{CD} + M_{DC})}{10} = \frac{(M_{CD} + M_{DC})}{10} \rightarrow$$

Summing horizontal forces on the whole frame once again:

$$\Sigma F_X = 0 \rightarrow +$$
$$-\frac{(36 - M_{AB} - M_{BA})}{15} + \frac{M_{CD} + M_{DC}}{10} + 6 = 0$$

This last equation is referred to as the **shear equation** or the **bent equation** and it provides the necessary condition, corresponding to the additional unknown R. Substitution of the slope deflection relationships into the joint and bent equations yields:

$$12.96 + 4\Theta_B - 4R - 18 + 10\Theta_B + 5\Theta_C = 0$$
$$14\Theta_B + 5\Theta_C - 4R = 5.04 \qquad \textbf{(8–17)}$$
$$18 + 10\Theta_C + 5\Theta_B + 6\Theta_C - 9R = 0$$
$$5\Theta_B + 16\Theta_C - 9R = -18 \qquad \textbf{(8–18)}$$
$$(-2.11 + .4\Theta_B - .533R) + (.9\Theta_C - 1.8R) + 6 = 0$$
$$.4\Theta_B + .9\Theta_C - 2.33R = -3.89 \qquad \textbf{(8–19)}$$

4. Solving these equations simultaneously we obtain:

$$\Theta_B = 1.01$$
$$\Theta_C = -0.520$$
$$R = 1.639$$

Substitution into the slope deflection equations:

$$M_{AB} = -13.17 \text{ kip-ft}$$
$$M_{BA} = +10.46$$
$$M_{BC} = -10.46$$
$$M_{CB} = +17.88$$
$$M_{CD} = -17.88$$
$$M_{DC} = -16.31$$

5. With the end moments known, the structure's reactions can be determined using the equations of equilibrium. The FBD of the entire frame is shown in Figure 8.21a. For frames subject to sidesway, the slope deflection method must be modified somewhat. When establishing the equations necessary for the solution of the unknown end displacements, the bent equation must be used in addition to the joint equations. Again, relative values may be utilized for the R values but care must be exercised to determine the sign associated with each R.

Problems

8–1 to 8–8. Determine the reactions for the beams shown using the moment-area method.

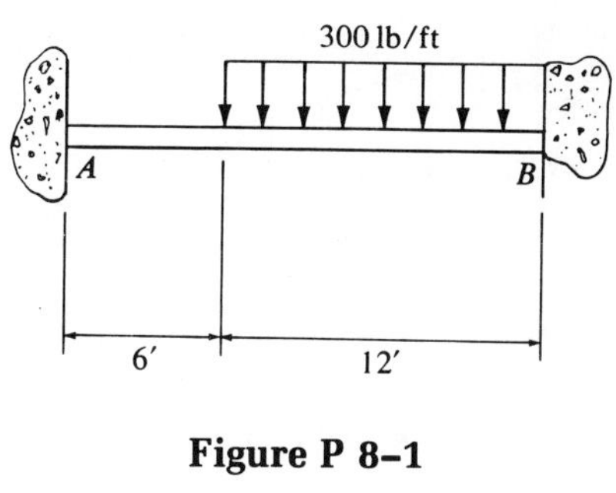

Figure P 8–1

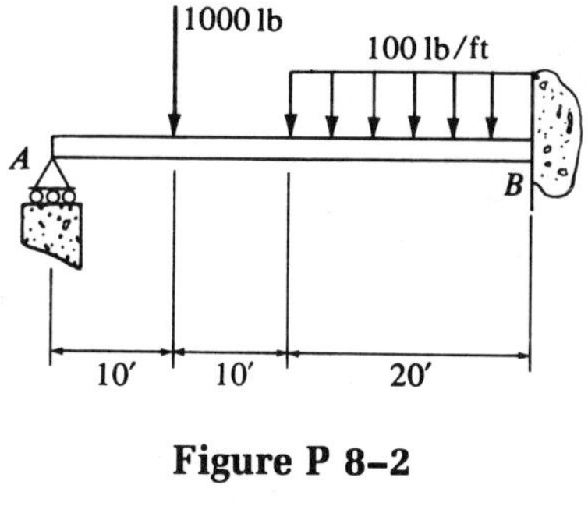

Figure P 8–2

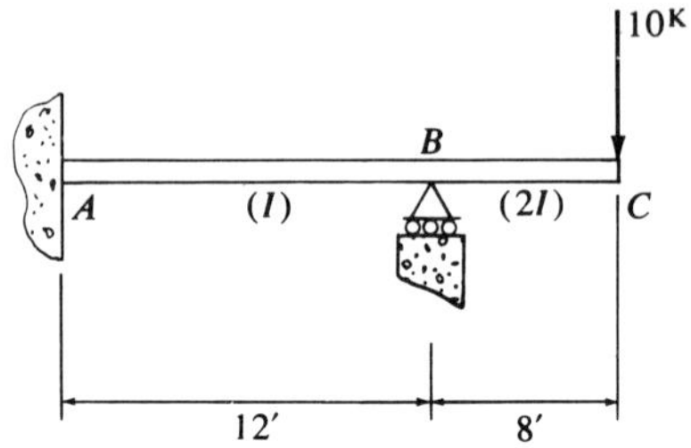

Figure P 8–3

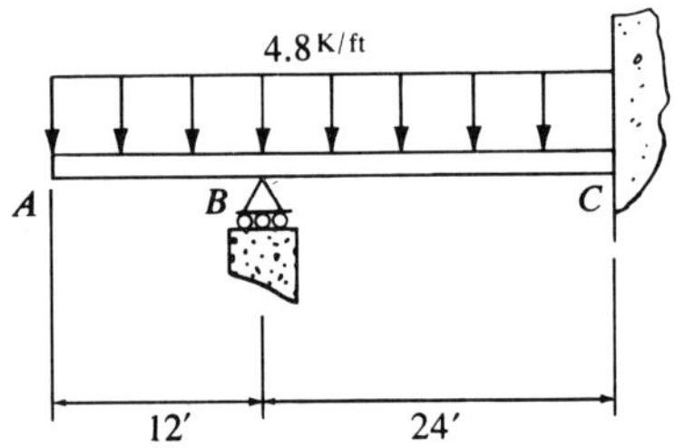

Figure P 8–4

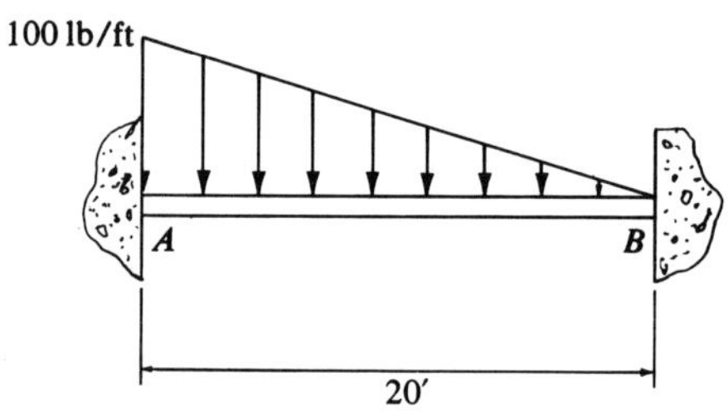

Figure P 8–5

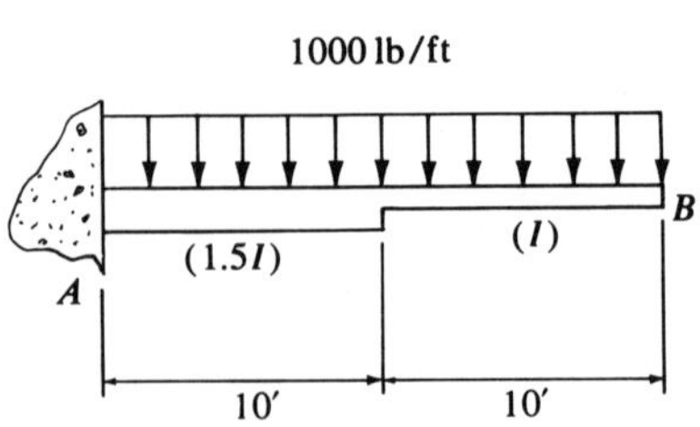

Figure P 8–6

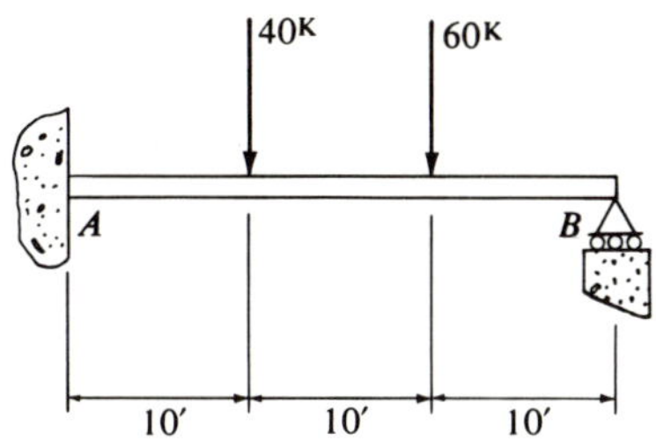

Figure P 8–7

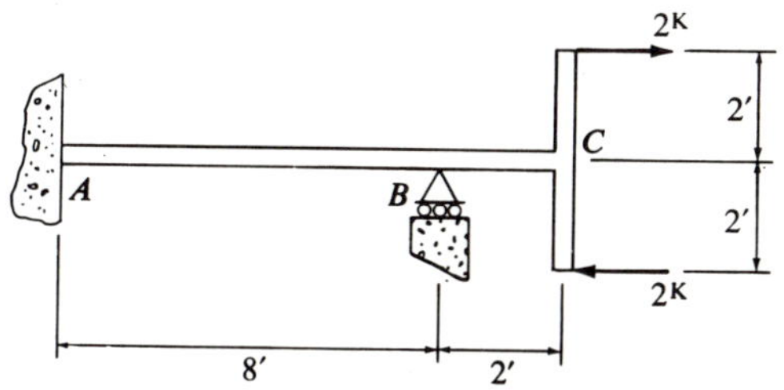

Figure P 8–8

8–9 to 8–16. Determine the reactions for the beams shown using the theorem of three moments.

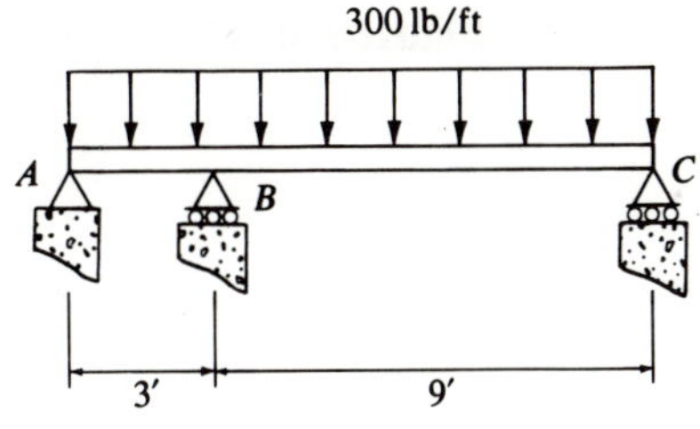

Figure P 8–9

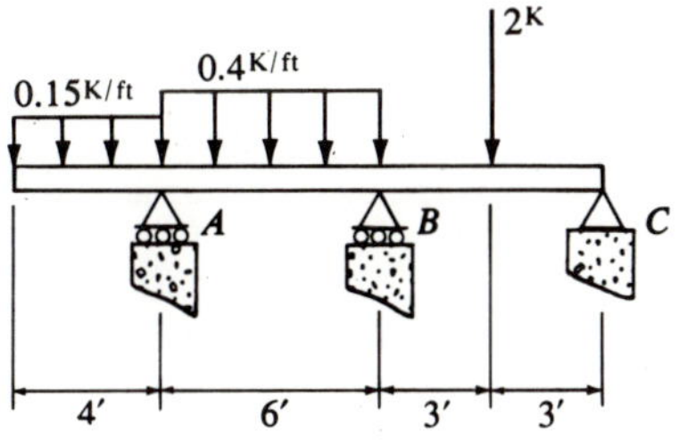

Figure P 8–10

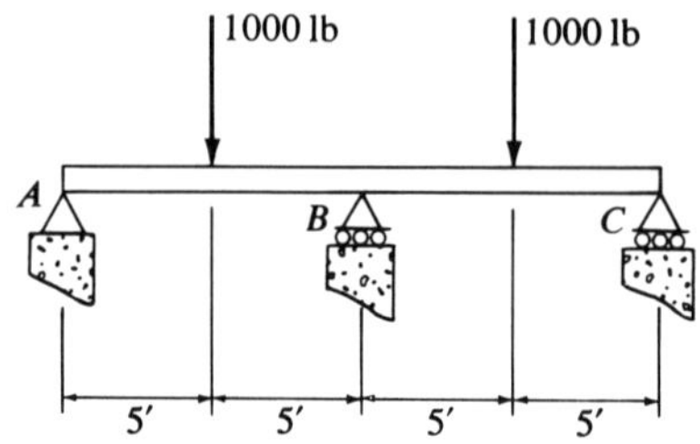

Figure P 8–11

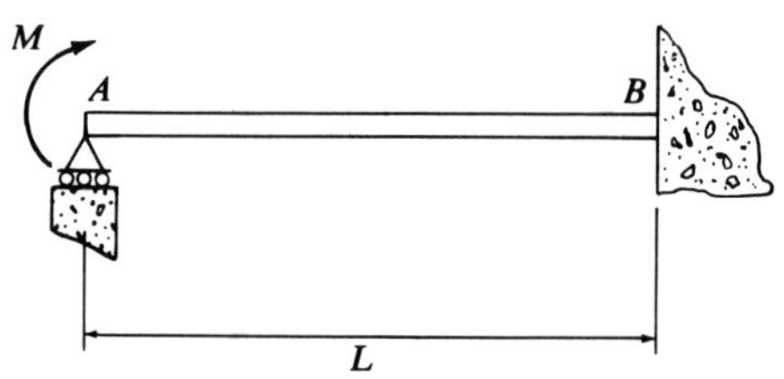

Figure P 8–12

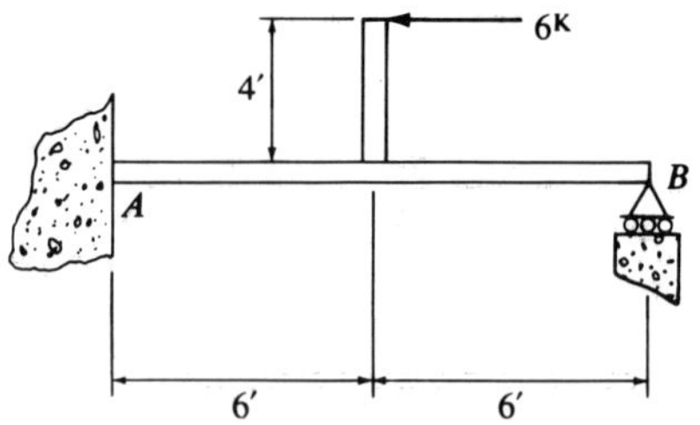

Figure P 8–13

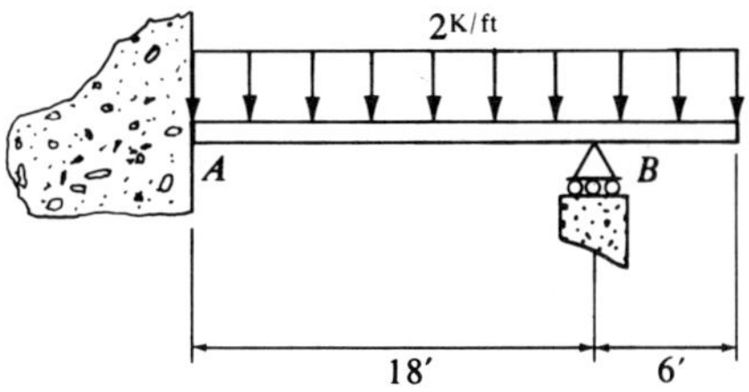

Figure P 8–14

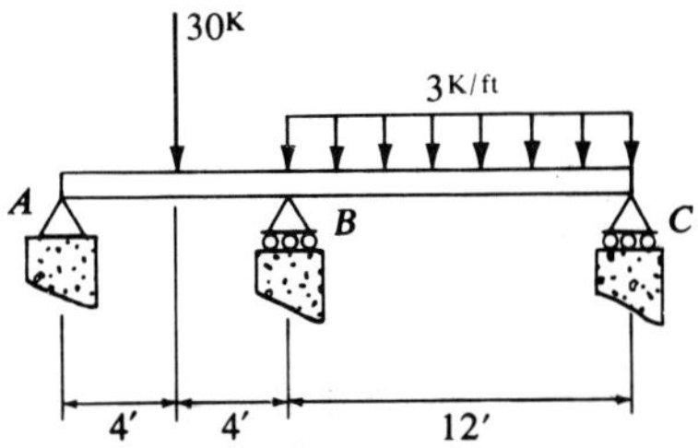

Figure P 8–15

Figure P 8–16

8–17. Determine the shear (V_{BA}) just to the left of the support at B of the beam shown.

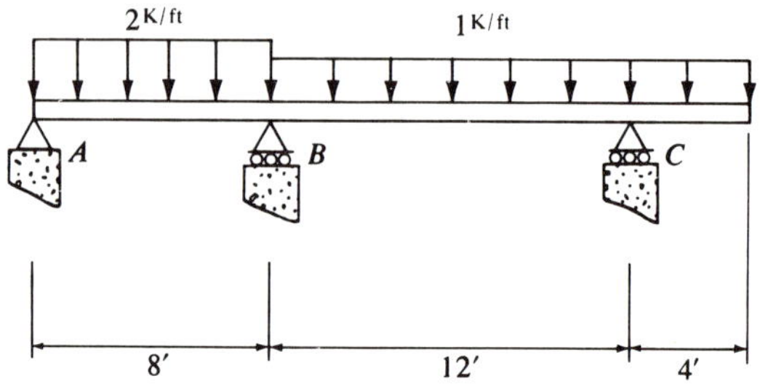

Figure P 8–17

8–18 to 8–21. Using the theorem of three moments, determine the moment at each support for each beam shown.

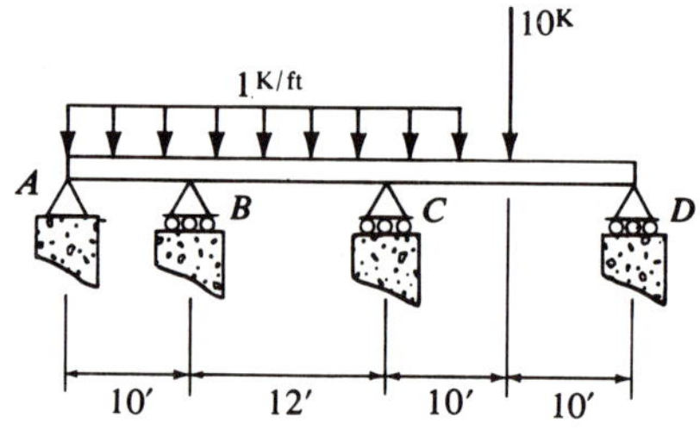

Figure P 8–18

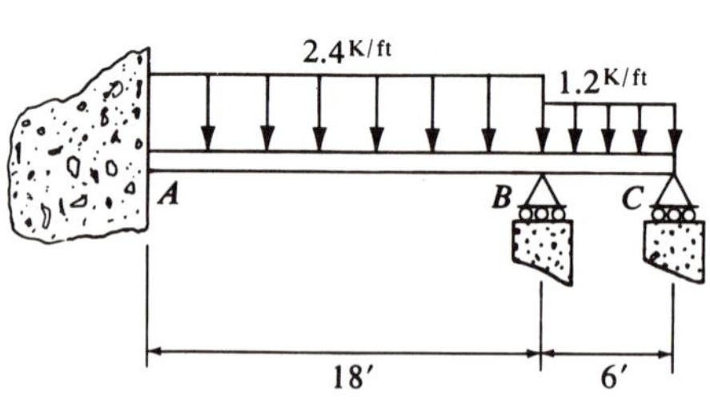

Figure P 8–19

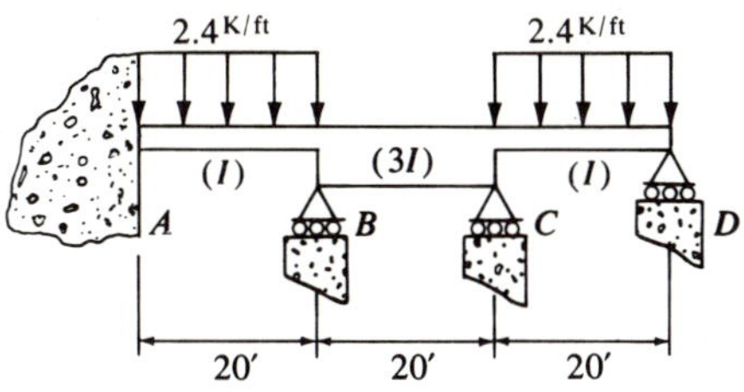

Figure P 8–20

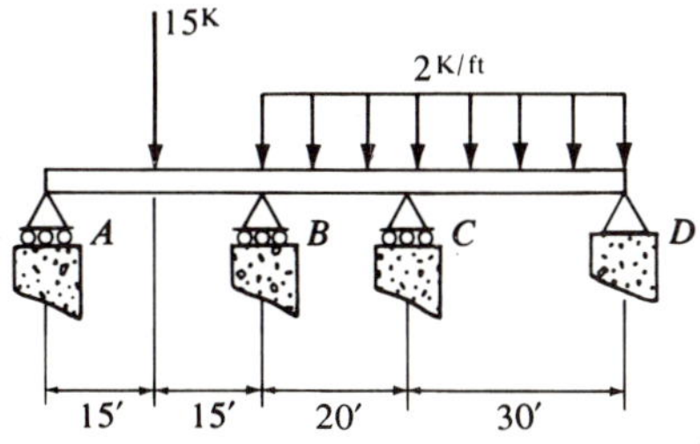

Figure P 8–21

8–22. A tank is to be supported on a low platform as shown. The weight of the tank and its contents produce a load of 500 lb per foot on the supporting beam on each side. Determine the load on each supporting post and draw the shear and moment diagram for the supporting beam.

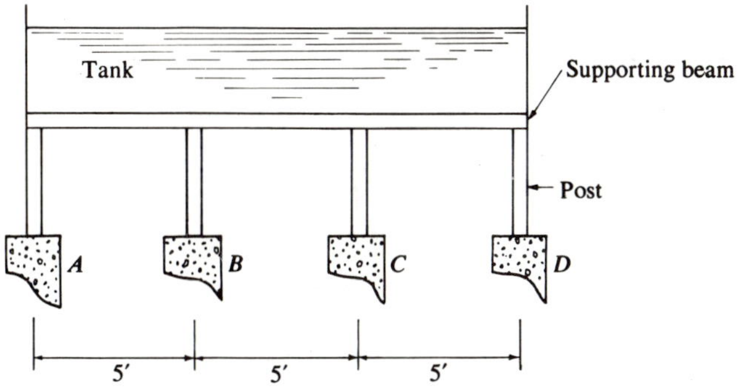

Figure P 8–22

8–23 to 8–30. Determine the reactions for the beams of Problems 8–1 to 8–8 using the conjugate beam method.

8–31 to 8–32. Determine the reactions for the beams shown using the conjugate beam method.

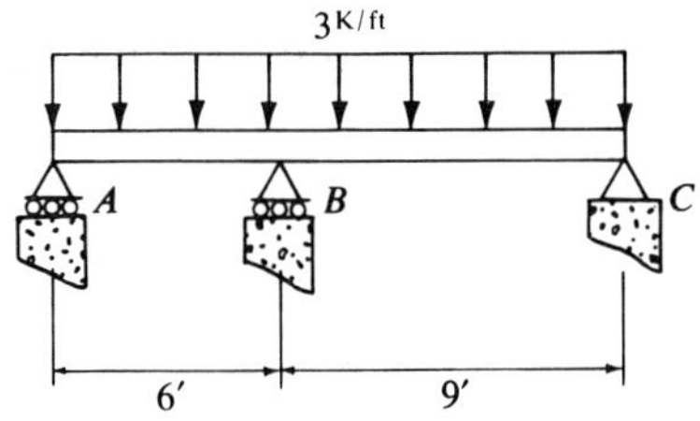

Figure P 8–31

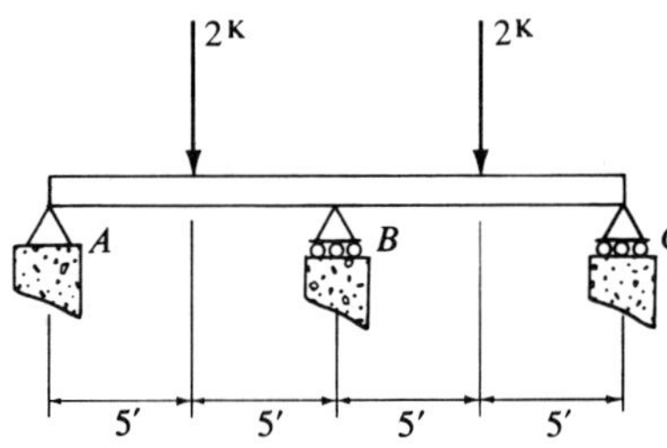

Figure P 8–32

8–33 to 8–36. Determine the reactions for the beams shown using the slope deflection method.

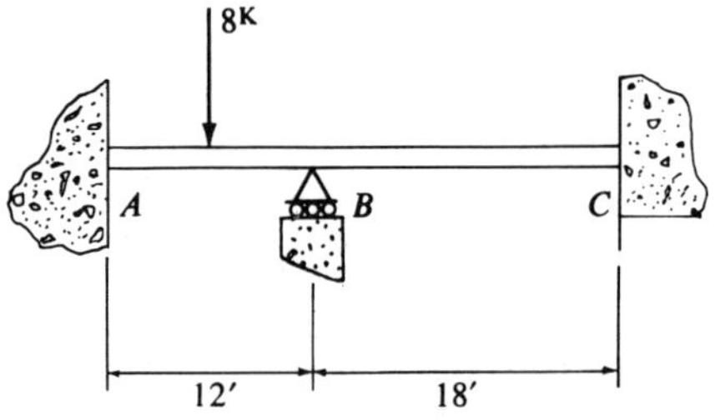

Figure P 8–33

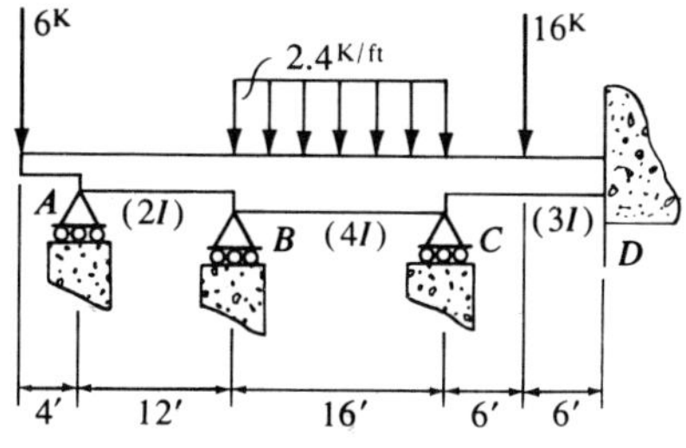

Figure P 8–34

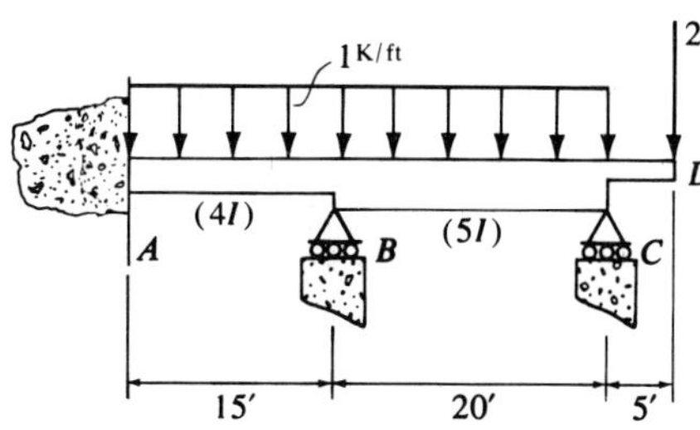

Figure P 8–35

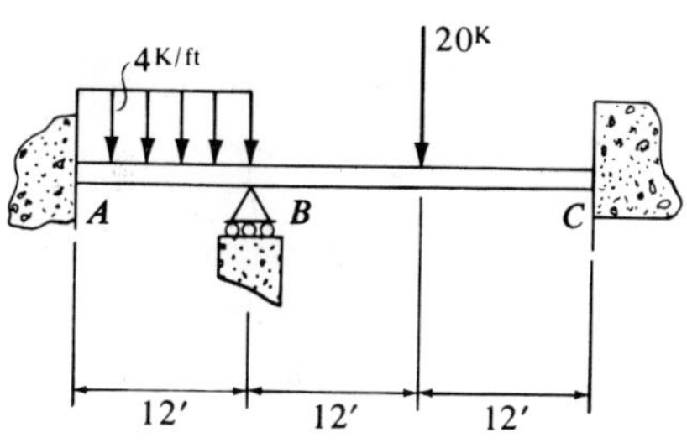

Figure P 8–36

8–37 to 8–39. For the structures shown, find all the end moments using the slope deflection method.

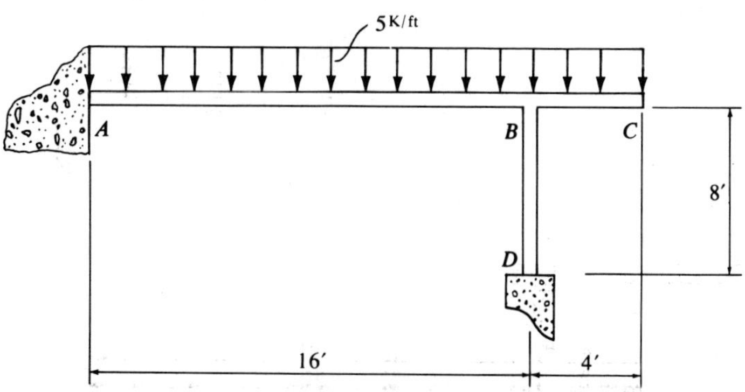

Figure P 8–37

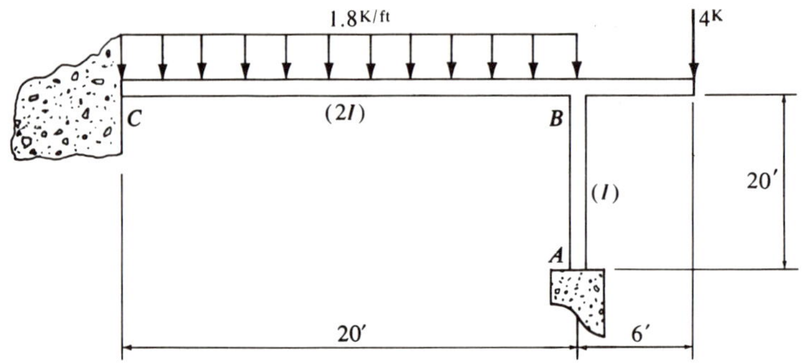

Figure P 8–38

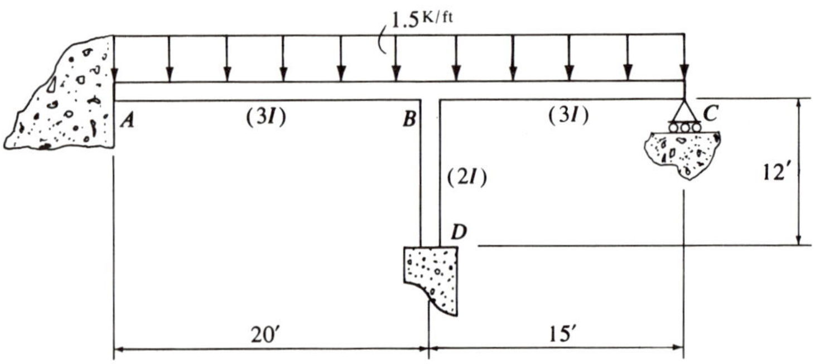

Figure P 8–39

8–40. Determine the reactions for the frame shown using the slope deflection method ($\gamma_w = 62.4$ lb per cubic foot).

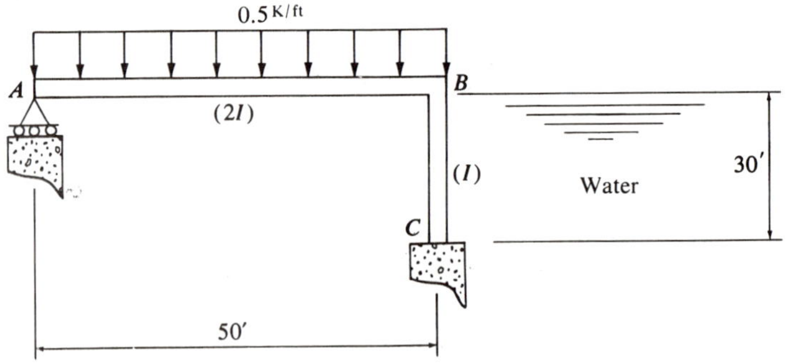

Figure P 8–40

8–41 to 8–44. Find the end moments for each member of the rigid frames shown using the slope deflection method.

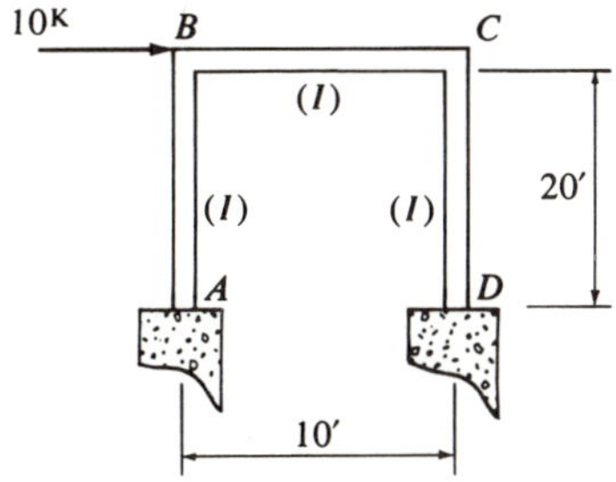

Figure P 8–41

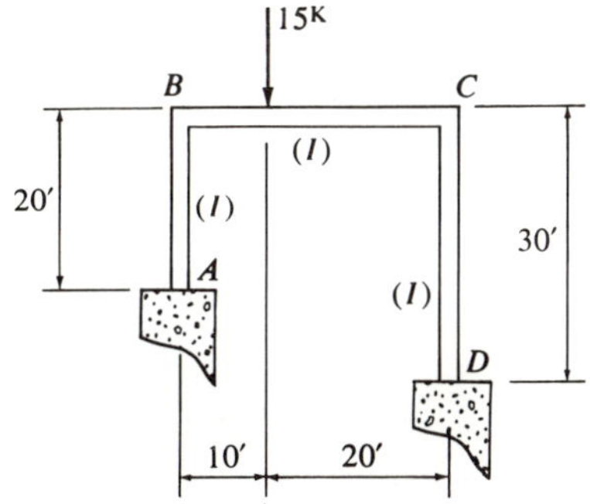

Figure P 8–42

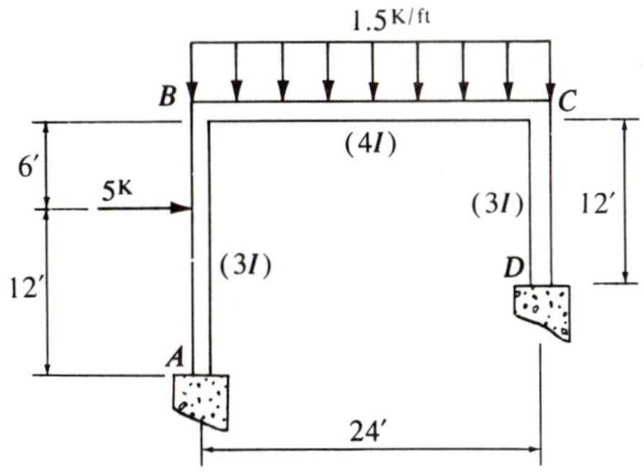

Figure P 8–43

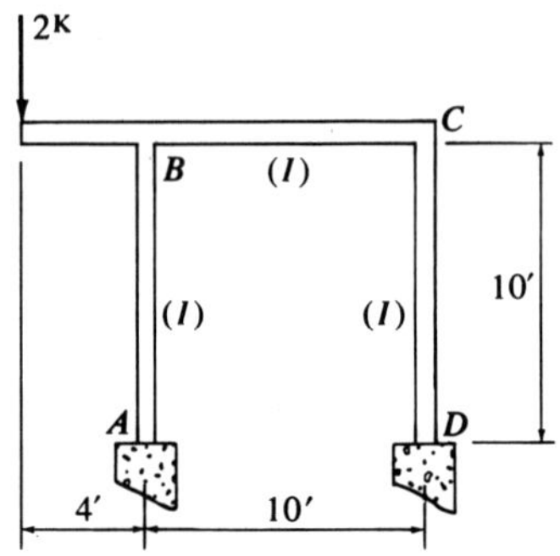

Figure P 8–44

Selected References

Au, Tung. *Elementary structural mechanics.* Englewood Cliffs, N.J.: Prentice-Hall, Inc., 1963.

Beer, F. P., and E. R. Johnston, Jr. *Vector mechanics for engineers: Statics.* Vol. 1. New York: McGraw-Hill Book Company, 1962.

Best, C. L., and W. G. McLean. *Analytical mechanics for engineers: Statics.* Vol. 1. Scranton, Pa.: INTEXT, 1968.

Borg, S. F., and J. J. Gennaro. *Advanced structural analysis.* Princeton: Van Nostrand Reinhold Company, 1959.

Eckhardt, O. W. *Strength of materials.* San Francisco: Holt, Rinehart & Winston, Inc., 1969.

Harris, E. C. *Elements of structural engineering.* New York: The Ronald Press Company, 1954.

Higdon, A., E. H. Ohlsen, and W. B. Stiles. *Mechanics of materials.* New York: John Wiley & Sons, Inc., 1960.

Higdon, A., and W. B. Stiles. *Engineering mechanics: Statics.* Vol. 1. Englewood Cliffs, N.J.: Prentice-Hall, Inc., 1962.

Hsieh, Y. Y. *Elementary theory of structures.* Englewood Cliffs, N.J.: Prentice-Hall, Inc., 1970.

Kinney, J. S. *Indeterminate structural analysis.* Reading, Mass.: Addison-Wesley Publishing Co., Inc., 1957.

McCormac, J. C. *Structural analysis* (2nd ed.). Scranton, Pa.: INTEXT, 1970.

Norris, C. H., and J. B. Wilbur. *Elementary structural analysis.* 2d ed. New York: McGraw-Hill Book Company, 1960.

Parcel, J. I., and R. B. Moorman. *Analysis of statically indeterminate structures.* New York: John Wiley & Sons, Inc., 1955.

Peterson, A. C. *Applied mechanics: strength of materials.* Boston, Mass.: Allyn & Bacon, Inc., 1969.

Shames, I. H. *Engineering mechanics: Statics*. Vol. 1. Englewood Cliffs, N.J.: Prentice-Hall, Inc., 1959.

Timoshenko, S. P. *Strength of materials—Part I*. 3d ed. Princeton: Van Nostrand Reinhold Company, 1956.

Timoshenko, S. P., and D. H. Young. *Elements of strength of materials*. 5th ed. Princeton: Van Nostrand Reinhold Company, 1968.

Wang, C. K. *Statically indeterminate structures*. New York: McGraw-Hill Book Company, 1953.

Wang, C. K., and C. L. Eckel. *Elementary theory of structures*. New York: McGraw-Hill Book Company, 1957.

Answers to Selected Problems

Chapter Two

2–10. $R = 11^K$ @ 3.64′ from A
2–11. Resultant couple of 1200 lb-ft ccw
2–12. 200 lb down @ A, 200 lb left @ A, 200 lb right @ B
2–13. 28.3 lb @ 87°–58′ cw from X-axis
2–14. $R = 360$ lb @ $x = 0.55a$, $z = 0.61a$
2–15. $F_A = 22^K$ down, $M_A = 76$ kip-ft cw
2–16. $R = 26^K$ down @ $x = 1.85'$, $z = 0.77'$
2–17. $F_A = 7^K$ up, $M_A = 145$ kip-ft ccw
2–18. $R = 90^K$ down, $M = 1440$ kip-in. cw
2–19. $R = 100^K$ down, $M_X = 200$ kip-in., $M_Y = 2200$ kip-in.
2–20. $R = P$ down, $MR = P_C + M$ cw
2–21. $R = 4500$ lb down @ 5′ from B
2–22. $R = 8400$ lb down @ 10.3′ from B
2–23. $R_A = 3300$ lb down, $M_A = 53{,}400$ lb-ft cw
2–24. $T_{AB} = 91.5$ lb, $T_{BC} = 131.0$ lb
2–25. $W = 788$ lb
2–26. (a) st. det., (b) unstable, (c) st. det., (d) st. det., (e) unstable, (f) unstable, (g) st. det., (h) unstable, (i) st. indet.
2–27. $B_Y = 6.1^K$ up, $A_Y = 11.9^K$ up, $B_X = 0$
2–28. $B_Y = 60$ lb up, $A_Y = 40$ lb up, $B_X = 0$
2–29. $B_Y = 13^K$ up, $A_Y = 11^K$ down, $A_X = 0$
2–30. $B_Y = 6.7^K$ up, $A_Y = 0.7^K$ down, $B_X = 0$
2–31. $D_Y = 700$ lb up, $D_X = 0$, $A_Y = 120$ lb up, $B_Y = 180$ lb up
2–32. $B_Y = 1150$ lb up, $A_Y = 350$ lb up, $A_X = 0$
2–33. $T_{BC} = 630$ lb t, $A_Y = 105$ lb up
2–34. $T_{CD} = 27.5^K$ t, $A_Y = 7.5^K$ up, $A_X = 22^K$ right
2–35. $D_Y = 4.36^K$ up, $A_Y = 6.4^K$ up, $C_Y = 7.24^K$ up, $C_X = 0$
2–36. $A_Y = (P + W_0L/4)$ up, $A_X = 0$, $M_A = \frac{1}{4}[PL - 5W_0L^2/6]$cw
2–37. $B_Y = \left[\dfrac{M_1 - M_0 + 3(2P_1 + P_3)}{20}\right]$ up

$A_Y = \left[\dfrac{M_1 - M_0 - 14P_1 + 3P_3}{20}\right]$ down, $B_X = P_3$ left

2–38. $A_Y = 22.6^K$ up, $B_Y = 8.4^K$ up, $B_X = 12^K$ right
2–39. $A_Y = 2.58^K$ up, $B_Y = 11.42^K$ up, $B_X = 0$
2–40. $A_Y = W_0L/2$ up, $A_X = 0$, $M_A = W_0L^2/4$ ccw
2–41. $A_Y = 41^K$ up, $B_Y = 5^K$ up, $A_X = 0$, $M_A = 302$ kip-ft ccw
2–42. $A_Y = 39.29^K$ up, $A_X = 2.72^K$ left, $T_{BD} = 3.14^K$ t

Chapter Three

3–1. $V_{LC} = 600$ lb down on AC, $M_{LC} = 2400$ lb-ft ccw on AC
$V_{RC} = 400$ lb up on AC, $M_{RC} = 2400$ lb-ft ccw on AC
3–2. $V_B = 17.6^K$ up on AB, $M_B = 70.4$ kip-ft cw on AB
3–3. $V_{BL} = P/6$ up on AB, $M_{BL} = 3P/2$ cw on AB
$V_{BR} = P/6$ up on AB, $M_{BR} = P/2$ ccw on AB
3–4. (a) $V_D = 0$, $M_D = 0$
(b) $V_{BL} = 12^K$ up on AB, $M_{BL} = 36$ kip-ft cw on AB
(c) $V_{BR} = 12^K$ down on AB, $M_{BR} = 36$ kip-ft cw on AB
3–5. (a) $V_{CL} = 0$, $M_{CL} = 0$
(b) $V_{CR} = 10^K$ up on AB, $M_{BR} = 0$
3–6. (a) $V_C = 0$, $M_C = 100$ lb-in. ccw on AC
(b) $V_D = 0$, $M_D = 0$
3–7. $V_X = wL/8$, $M_X = wL/8(L - x)$
3–8. $V_X = 4x + 10$, $M_X = 2x^2 - 19.8x - 30$
3–9. $V_X = \dfrac{wx^2}{2L} - \dfrac{wL}{6}$, $M_X = \dfrac{wLx}{6} - \dfrac{wx^3}{6L}$
3–10. $V_X = 18 + \frac{3}{2}x^2$, $M_X = 18x + x^3/2 - 400$
3–11. $V_Y = 3.12(2yh - y^2 - h^2)$
$M_Y = 31.2h^2y - 10.4h^3 - 31.2(h - y)y^2 - 20.8y^3$
3–12. $V_X = \dfrac{M}{2L}$, $M_X = \dfrac{Mx}{2L}$, $0 \leq x \leq L$
$V_X = \dfrac{M}{2L}$, $M_X = M(1 - x/2L)$, $x > L$
3–13. Partial; $V_A = +20$ lb, $M_B = +60, -140$ lb-in.
3–14. Partial; $V_C = -42^K$, $M_C = -248$ kip-ft
3–15. Partial; $V_A = 720$ lb, $M_A = -7960$ lb-ft
3–16. Partial; $V_A = -18^K$, $M_C = -192$ kip-ft
3–17. Partial; $V_C = +85, -250$ lb, $M_C = 620$ lb-ft
3–18. Partial; $V_B = -6^K$, $M_B = 0$
3–19. Partial; $V_B = -500$ lb, $M_B = -2500$ lb-ft
3–20. Partial; $V_B = 600, 400$ lb, $M_C = 5800$ lb-ft
3–21. Partial; $V_C = 4^K$, $M_B = -16$ kip-ft
3–22. Partial; $V_B = -14^K$, $M_A = -18$ kip-ft
3–23. Partial; $V_{CL} = 0$, $M_{CL} = -95.5$ kip-ft
3–24. Partial; $V_A = +10.8^K$, $M_{CL} = 129.6$ kip-ft
3–25. Partial; $V_B = +3^K$, $M_B = +24, +33$ kip-ft
3–26. Partial; $V_B = 28^K$, $M_C = +630$ kip-ft
3–27. Partial; $V_B = +11.2^K$, $M = +44$ kip-ft
3–28. Partial; $V_B = -16^K$, $M_{max} = 36.9$ kip-ft
3–29. Partial; $V_A = +25, -30^K$, $M_A = 62.5$ kip-ft
3–30. Partial; $V_A = +18^K$, $M_B = +180, +240$ kip-ft
3–31. Partial; $V_{max} = 4500$ lb, $M_{max} = 54{,}000$ lb-ft
3–32. Partial; $V_A = -1250$ lb, $M_{max} = -7500$ lb-ft
3–33. Partial; $V = +28.6, -21.4^K$ @ left column
$M = +57.12$ kip-ft @ left column
3–34. Partial; $V_A = +5.5^K$, $M_A = -80.5$ kip-ft
3–35. Partial; $V_{max} = -2000$ lb, $M_{max} = -10{,}666$ lb-ft

Chapter Four

4–1. (a) stable, (b) stable, (c) unstable, (d) stable, (e) stable, (f) st. indet., (g) st. indet.
4–2. (a) compound, (b) simple, (c) simple, (d) compound, (e) simple, (f) simple, (g) complex, (h) complex
4–3. $AB = 7.5^K$ c, $BC = 5.02^K$ c, $CD = 4.5^K$ t, $DE = 4.5^K$ t, $AE = 1.2^K$ c, $BD = 2.25^K$ c, $BE = 2.48^K$ t
4–4. $AB = 36^K$ t, $BC = 17.4^K$ t, $CD = 17.4^K$ t, $DE = 0$, $EF = 0$, $FG = 10.2^K$ c, $GH = 10.2^K$ c, $HI = 27.6^K$ c, $JI = 27.6^K$ c, $AJ = 5.8^K$ t, $BI = 0$, $CH = 0$, $DG = 0$, $DF = 10.5^K$ t
$DH = 10.5^K$ c, $BH = 10.5^K$ t, $BJ = 10.5^K$ c
4–5. $AB = 2.25^K$ t, $BC = 3.74^K$ t, $CD = 3.74^K$ t, $DE = 2.25^K$ t, $EF = 1.87^K$ c, $GF = 1.87^K$ c, $AG = 1.87^K$ c, $BG = 2.25^K$ c, $GC = 2.25^K$ t, $CF = 2.25^K$ t, $DF = 2.25^K$ c

4–6. $AB = 41.2^K$ t, $BC = 31^K$ t, $CD = 20.6^K$ t, $DE = 20^K$ c, $FE = 20^K$ c, $GF = 30^K$ c, $BG = 11.2^K$ c, $BF = 2.5^K$ t, $FC = 10.3^K$ c, $CE = 0$
4–7. Maximum bar force in $AB = 1.2W$ t; $W = 5$ kips
4–8. $AB = 125^K$ t, $BC = 13^K$ t, $CD = 62^K$ c, $DH = 50^K$ t, $GH = 0$, $EF = 113^K$ c, $AE = 150^K$ t, $EB = 187.5^K$ c, $BF = 150^K$ t, $FC = 125^K$ c, $CG = 100^K$ t, $GD = 62.5^K$ c, $FG = 38^K$ c
4–9. $AB = 260^K$ t, $BC = 260^K$ t, $CD = 260^K$ t, $ED = 300^K$ c, $FE = 300^K$ c, $GF = 325^K$ c, $AG = 162.5^K$ t, $AF = 31^K$ t, $BF = 0$, $BE = 0$, $CE = 0$
4–10. $AB = 520$ lb c, $BC = 0$, $CD = 0$, $DE = 220$ lb t, $EF = 740$ lb t, $AF = 520$ lb t, $AE = 736$ lb c, $BD = 736$ lb c, $BE = 520$ lb t
4–11. $BC = 48^K$ t, $CL = 26.8^K$ c, $HL = 26.8^K$ t
4–12. $AB = 43.2^K$ t, $BH = 3^K$ c, $DF = 9^K$ c,
4–13. $HD = 1.0^K$ t
4–14. $DE = 38^K$ t, $EJ = 0$, $KP = 39.6^K$ t, $OP = 28^K$ t
4–15. $MI = 0.86^K$ t
4–16. $CD = 47.3^K$ c, $JI = 47.3^K$ t, $LD = 33.5^K$ c, $LI = 33.5^K$ t
4–17. $CD = 178^K$ c, $ND = 55.6^K$ t, $NJ = 80^K$ c, $KJ = 178^K$ t
4–18. $BN = 5^K$ c, $CN = 3.11^K$ t, $ML = 8.45^K$ t, $OC = 4.69^K$ t, $CM = 11.5^K$ c
4–19. See 4–16.
4–20. $BC = 75^K$ t, $HG = 56.3^K$ c, $BG = 31.3^K$ t, $CG = 5^K$ t
4–21. $AB = 8.84^K$ c, $BC = 5.30^K$ c, $DC = 3.28^K$ c, $DF = 0$, $FE = 2.30^K$ t, $GF = 2.30^K$ t, $CG = 4.53^K$ t, $AH = 2.71^K$ t, $HG = 2.71^K$ t, $BH = 0$, $DE = 3.26^K$ c
4–22. $DE = 15^K$ t, $AE = 18.05^K$ t, $AL = 30^K$ c
4–23. $DI = 0$, $CD = 125^K$ c, $CI = 127.5^K$ t
4–24. $CD = 37.5^K$ c, $DV = 46.9^K$ c, $UQ = 15.6^K$ c, $SU = 31.3^K$ c, $UR = 50^K$ t

Chapter Five

5–1. (a) $R_{DV} = 41^K$ up, $R_{AV} = 7^K$ up, $R_{AH} = 8^K$ left, $M_B = 40$ kip-ft, $M_C = -32$ kip-ft
(b) $R_{DV} = 18^K$ up, $R_{AV} = 6^K$ down, $R_{AH} = 24^K$ left, $M_B = 144$ kip-ft
(c) $R_{AV} = 13^K$ up, $R_{AH} = 2^K$ right, $R_{DV} = 23^K$ up, $R_{DH} = 7^K$ left, $M_B = -24$ kip-ft, $M_C = -84$ kip-ft
(d) $R_{AV} = 2.17^K$ down, $R_{AH} = 18^K$ left, $R_{DV} = 10.17^K$ up, $M_B = 72$ kip-ft, $M_C = -18$ kip-ft
5–2. (a) $R_{AV} = 4.43^K$ down, $R_{AH} = 10^K$ left, $R_{DV} = 10.43$ up, $M_B = 80$ kip-ft, $M_C = 44.8$ kip-ft
(b) $R_{AV} = 9^K$ down, $R_{AH} = 12^K$ left, $R_{EV} = 9^K$ up, $M_B = 144$ kip-ft, $M_C = 72$ kip-ft
(c) $R_{AV} = 17.5^K$ up, $R_{EV} = 30.5^K$ up, $R_{EH} = 8^K$ left, $M_C = 26$ kip-ft, $M_D = -160$ kip-ft
(d) $R_{AV} = 11.25^K$ up, $R_{AH} = 5^K$ right, $R_{EV} = 1.25^K$ down, $M_B = -60$ kip-ft, $M_C = -40$ kip-ft
5–3. $R_{AV} = 40.8^K$ up, $R_{CV} = .8^K$ down, $R_{CH} = 62.4^K$ right, $M_B = 416$ kip-ft
5–4. $R_{AV} = 26.3^K$ up, $R_{BV} = 3.7^K$ up, $R_{BH} = 37.5^K$ left, $M_C = -187.5$ kip-ft
5–5. $R_{AV} = 60^K$ up, $R_{AH} = 90^K$ right, $R_{BV} = 180^K$ up, $R_{BH} = 90^K$ left, $M_B = -900$ kip-ft, $N_D = 108.1^K$ c, $V_D = 0^K$
5–6. $R_{AV} = 3.75^K$ down, $R_{AH} = 22.5^K$ left, $R_{BV} = 3.75^K$ up, $R_{BH} = 7.5^K$ left, $M_D = -28.3$ kip-ft, $N_D = 8.38^K$ c, $V_D = 0^K$
5–7. $R_{AV} = 5.06^K$ up, $R_{AH} = 17.8^K$ right, $R_{BV} = 6.9^K$ up, $R_{BH} = 27.8^K$ left, $M_D = -20$ kip-ft, $N_D = 18.44^K$ c, 9.08^K c, $V_D = .51^K$, 3.5^K
5–8. $R_{AV} = 50^K$ up, $R_{BV} = 250^K$ up, $R_{AH} = 0$, $R_{BH} = 100^K$ left, $M_D = 625$ kip-ft, $N_D = 0^K$, $V_D = 141.1^K$
5–9. $R_{AV} = 20^K$ up, $R_{AH} = 6.67^K$ left, $R_{BV} = 20^K$ up, $R_{BH} = 6.67^K$ right, $M_D = 47$ kip-ft, $N_D = 4.7^K$ c, $V_D = 4.7^K$
5–10. $R_{AV} = 40^K$ up, $R_{AH} = 20^K$ right, $R_{BV} = 40^K$ up, $R_{BH} = 20^K$ left, $M_D = -98.7$ kip-ft, $N_D = 34.64^K$ c, $V_D = 39.9^K$
5–11. $R_A = 8.1^K$ up, $R_{AH} = 5.6^K$ right, $R_{BV} = 3.1^K$ down, $R_{BH} = 4.4^K$ right, $M_D = 46$ kip-ft, $N_D = 1.28^K$ t, $V_D = 4.9^K$
5–12. Member BC, 15.6^K t
5–13. Member BE, 30^K t
5–14. Member BD, 2^K t
5–15. Member CB, 2.41^K c
Member AB, 1.33^K t

Chapter Six

6–4. $R_A = 58.0^K$ up, $M_B = -204$ kip-ft, $V_B = 34.0^K$
6–7. $R_A = 38.8^K$ up, $V_{1/3} = 24.8^K$, $M_{MS} = +1068$ kip-ft
6–8. Absolute max, moment $= +988$ kip-ft

6–9. Absolute max. moment = + 547 kip-ft
6–11. $R_A = 37.5^K$ up, $M_A = -90$ kip-ft
6–13. $CD = 47.1^K$ c, $DI = 52.5^K$ c, $FH = 48^K$ c
6–15. $R_A = 8.86^K$ up, $CJ = 4.28^K$ t, $KJ = 24.9^K$ t
6–16. $CD = 7200$ lb c, $DH = 2466$ lb t
6–17. $BL = 1414$ lb t, $CD = 4500$ lb c

Chapter Seven

7–5. $\Delta_{MS} = 189.4/EI$ lb-ft³ down
7–6. $\Delta_{MS} = 61.7/EI$ kip-ft³ down
7–7. $\Delta_{MS} = 3666/EI$ kip-ft³ down
7–8. $\Delta_{MS} = 99/EI$ kip-ft³ down
7–9. $\Delta_{MS} = 3.67$ in. down
7–10. $\Delta_C = 0.123$ in. down
7–11. $\Delta_C = 13{,}333/EI$ kip-ft³ down
7–12. $\Delta_C = 616/EI$ kip-ft³ upward
7–13. $\Delta_C = 0.282$ in. upward; $\Delta_D = 0.136$ in. down
7–14. $\Delta_C = \Delta_D = 1042/EI$ kip-ft³ down; $\Delta_{MS} = 156/EI$ kip-ft³ upward
7–15. $\Delta_{max} = 1917/EI$ kip-ft³ down
7–16. $\Delta_{max} = 256/EI$ kip-ft³ upward
7–17. $\Delta_{max} = 4038/EI$ kip-ft³ down
7–18. $\Delta_{max} = 4725/EI$ kip-ft³ down
7–19. $\Delta_{max} = 0.075$ in. down
7–20. $\Delta_{max} = 1.85$ in. down
7–21. $\Delta_{max} = 413/EI$ kip-ft³ down
7–22. See Problem 7–5
7–23. See Problem 7–8
7–24. See Problem 7–11
7–25. See Problem 7–17
7–26. $\Delta_C = 512/EI$ kip-ft³ down
7–27. $\Delta_A = 1250/EI$ kip-ft³ down
7–28. $\Delta_{MS} = 360/EI$ kip-ft³ upward

Chapter Eight

8–1. $R_{AV} = 1066$ lb up, $R_{BV} = 2934$ lb up, $M_A = -4800$ lb-ft, $M_B = -7200$ lb-ft
8–2. $R_{AV} = .85^K$ up, $R_{BV} = 2.15^K$ up, $M_C = +16$ kip-ft
8–3. $R_{AV} = 9.4^K$ up, $R_{BV} = 19.4^K$ up, $M_A = 32.8$ kip-ft
8–4. $R_{BV} = 119^K$ up, $R_{CV} = 53.8^K$ up, $M_C = -240$ kip-ft
8–5. $R_{AV} = 670$ lb up, $R_{BV} = 330$ lb up, $M_A = -2$ kip-ft, $M_B = -2$ kip-ft
8–6. $R_{AV} = 13.8^K$ up, $R_{BV} = 6.2^K$ up, $M_A = -76$ kip-ft
8–7. $R_{AV} = 63^K$, $R_{BV} = 37^K$, $M_A = -490$ kip-ft
8–8. $R_{AV} = 1.5^K$ up, $R_{BV} = 1.5^K$ up, $M_A = -20$ kip-ft
8–9. $R_{AV} = 223$ lb down, $R_{BV} = 2698$ lb up, $R_{CV} = 1125$ lb up, $M_B = -2320$ lb-ft
8–10. $R_{AV} = 1.775^K$ up, $R_{BV} = 2.45^K$ up, $R_{CV} = 0.775^K$ up, $M_B = -1.35$ kip-ft
8–11. $R_{AV} = 3125$ lb up, $R_{BV} = 1375$ lb up, $R_{CV} = 3125$ lb up, $M_B = -1.875$ kip-ft
8–12. $R_{AV} = 3M/2L$ down, $R_{BV} = 3M/2L$ up, $M_A = M$, $M_B = -M/2$
8–13. $R_{AV} = 2.25^K$ up, $R_{AH} = 6^K$ right, $R_{BV} = 2.25^K$ down, $M_A = -3$ kip-ft
8–14. $R_{AV} = 19.5^K$ up, $R_{BV} = 28.5^K$ up, $M_A = -63$ kip-ft
8–15. $R_{AV} = 8.70^K$ up, $R_{BV} = 43.5^K$ up, $R_{CV} = 13.8^K$ up, $M_B = -50.25$ kip-ft
8–16. $R_{AV} = 2.88^K$ up, $R_{BV} = 3.6^K$ up, $R_{CV} = 1.12^K$ up, $M_B = -1.97$ kip-ft
8–17. $R_{AV} = 6.15^K$ up, $M_B = -14.8$ kip-ft, $V_{BA} = -9.85^K$
8–18. $M_B = -7.65$ kip-ft, $M_C = -28.65$ kip-ft
8–19. $M_A = -74.4$ kip-ft, $M_B = -46.5$ kip-ft
8–20. $M_A = -95$ kip-ft, $M_B = -50$ kip-ft, $M_C = -45.6$ kip-ft
8–21. $M_B = -58$ kip-ft, $M_C = -163$ kip-ft
8–22. $R_{AV} = .5^K$ up, $R_{BV} = 1.35^K$ up, $R_{CV} = 1.44^K$ up, $R_{DV} = .47^K$ up, $M_B = -.587$ kip-ft, $M_C = -.78$ kip-ft
8–23 to 8–30. Same as 8–1 to 8–8
8–31. $R_{AV} = 5.05^K$ up, $R_{BV} = 29.2^K$ up, $R_{CV} = 10.9^K$ up, $M_B = -23.7$ kip-ft

8–32. $R_{AV} = .62^K$ up, $R_{BV} = 2.76^K$ up, $R_{CV} = .62^K$ up, $M_B = -3.8$ kip-ft
8–33. $R_{AV} = 4.9^K$ up, $R_{BV} = 3.5^K$ up, $R_{CV} = .405^K$ up, $M_A = -15.6$ kip-ft, $M_C = -2.4$ kip-ft
8–34. $R_{AV} = 6.0^K$ up, $R_{BV} = 17.4^K$ up, $R_{CV} = 32.4^K$ up, $R_{DV} = 4.6^K$ up, $M_A = -24$ kip-ft, $M_D = -10.8$ kip-ft
8–35. $R_{AV} = 5.96^K$ up, $R_{BV} = 20.2^K$ up, $R_{CV} = 9.84^K$ up, $M_A = -11.2$ kip-ft
8–36. $R_{AV} = 23.6^K$ up, $R_{BV} = 33.9^K$ up, $R_{CV} = 10.5^K$ up, $M_A = -44$ kip-ft, $M_C = -62$ kip-ft
8–37. $M_{AB} = -129$ kip-ft, $M_{BA} = +85.1$ kip-ft, $M_{BC} = -40$ kip-ft, $M_{BD} = -45.2$ kip-ft, $M_{DB} = -22.6$ kip-ft
8–38. $M_{CB} = --72$ kip-ft, $M_{BC} = +36$ kip-ft, $M_{BD} = -24$ kip-ft, $M_{BA} = -12$ kip-ft, $M_{AB} = -6$ kip-ft
8–39. $M_{AB} = -51.4$ kip-ft, $M_{BA} = 47.4$ kip-ft, $M_{BC} = -45.8$ kip-ft, $M_{DB} = -1.38$ kip-ft
8–40. $R_{AV} = 64.8^K$ up, $R_{CV} = 39.8^K$ down, $M_C = +895$ kip-ft, $R_{CH} = 351^K$ right
8–41. $M_{AB} = -54$ kip-ft, $M_{BA} = -46$ kip-ft, $M_{CD} = -46$ kip-ft, $M_{DC} = -54$ kip-ft
8–42. $M_{AB} = -63$ kip-ft, $M_{BA} = 63$ kip-ft, $M_{CD} = -70.2$ kip-ft, $M_{DC} = -52$ kip-ft
8–43. $M_{AB} = 5.7$ kip-ft, $M_{BA} = -14.5$ kip-ft, $M_{CD} = -163$ kip-ft, $M_{DC} = -110$ kip-ft
8–44. $M_{AB} = 0$ kip-ft, $M_{BA} = -2.66$ kip-ft, $M_{BC} = -5.32$ kip-ft, $M_{DC} = 2.66$ kip-ft

INDEX

INDEX